Imaging Radar Polarimetric Rotation Domain Interpretation

Polarimetric rotation domain interpretation is an innovation in radar image processing and understanding. Orientation rotation is a basic operator well known in the classic polarimetry theory, and significant advancement has been made in recent years. This book presents new and advanced concepts, theories, and methodologies in radar polarimetry and bridges the gaps between target scattering diversity, polarimetric radar data, and their practical applications. It provides a comprehensive summarization and investigation of polarimetric rotation domain features and demonstrates novel applications of polarimetric radar target detection, classification, target structure recognition, and urban damage mapping.

FEATURES

- Focuses on basic concepts, key techniques, and various applications of the polarimetric rotation domain interpretation paradigm for the first time in book form
- Explains, represents, and utilizes the radar target scattering diversity effect
- Identifies new methods for target polarimetric scattering mechanism understanding
- Provides a comprehensive investigation of polarimetric roll-invariant features
- Includes novel application developments for imaging radar target detection, structure recognition, and damage mapping

This book is written for researchers and professionals in radar polarimetry, radar imaging, microwave remote sensing, environmental studies, and other related fields. Senior undergraduate and postgraduate students, as well as teachers in the same fields, will benefit from the advancements highlighted in this book.

SAR Remote Sensing

Series Editor: Jong-Sen Lee, Naval Research Laboratory, Washington DC

Synthetic Aperture Radar (SAR) is an indispensable and highly capable Earth remote sensing instrument. It is a day-night and all-weather sensor for all aspects of environmental monitoring, disaster (earthquake, tsunami, forest fire, etc.) assessment, and military applications. Since the 1980s, a plethora of space-borne SAR systems have been launched and are in operation, and new satellites are launched every month. Many SAR imaging modes have been developed and tailored to specific applications. SAR is becoming a vast and continuously evolving field of remote sensing technology and applications. This book series provides timely information on advancements and encompasses innovative SAR scattering theory, processing techniques, and applications in all environments. Books in this series serve and benefit professionals, academics, and students in the remote sensing community.

Imaging from Spaceborne and Airborne SARs, Calibration, and Applications
Masanobu Shimada

Airborne Circularly Polarized SAR: Theory, System Design, Hardware Implementation, and Applications
Josaphat Tetuko Sri Sumantyo, Ming Yam Chua, Cahya Edi Santosa, and Yuta Izumi

Spatial Analysis for Radar Remote Sensing of Tropical Forests
Gianfranco D. De Grandi and Elsa Carla De Grandi

Radar Scattering and Imaging of Rough Surfaces: Modeling and Applications with MATLAB®
Kun-Shan Chen

Polarimetric SAR Imaging: Theory and Applications
Yoshio Yamaguchi

Moon-Based Synthetic Aperture Radar: A Signal Processing Prospect
Zhen Xu and Kun-Shan Chen

Imaging Radar Polarimetric Rotation Domain Interpretation: Theory and Applications
Si-Wei Chen

For more information about this series, please visit: www.routledge.com/SAR-Remote-Sensing/book-series/CRCSRS

Imaging Radar Polarimetric Rotation Domain Interpretation

Theory and Applications

Si-Wei Chen

CRC Press is an imprint of the
Taylor & Francis Group, an **informa** business

Designed cover image: © Pending

First edition published 2025
by CRC Press
2385 NW Executive Center Drive, Suite 320, Boca Raton FL 33431

and by CRC Press
4 Park Square, Milton Park, Abingdon, Oxon, OX14 4RN

CRC Press is an imprint of Taylor & Francis Group, LLC

© 2025 Si-Wei Chen

Reasonable efforts have been made to publish reliable data and information, but the author and publisher cannot assume responsibility for the validity of all materials or the consequences of their use. The authors and publishers have attempted to trace the copyright holders of all material reproduced in this publication and apologize to copyright holders if permission to publish in this form has not been obtained. If any copyright material has not been acknowledged please write and let us know so we may rectify in any future reprint.

Except as permitted under U.S. Copyright Law, no part of this book may be reprinted, reproduced, transmitted, or utilized in any form by any electronic, mechanical, or other means, now known or hereafter invented, including photocopying, microfilming, and recording, or in any information storage or retrieval system, without written permission from the publishers.

For permission to photocopy or use material electronically from this work, access www.copyright.com or contact the Copyright Clearance Center, Inc. (CCC), 222 Rosewood Drive, Danvers, MA 01923, 978–750–8400. For works that are not available on CCC please contact mpkbookspermissions@tandf.co.uk

Trademark notice: Product or corporate names may be trademarks or registered trademarks and are used only for identification and explanation without intent to infringe.

ISBN: 978-1-032-60958-4 (hbk)
ISBN: 978-1-032-60959-1 (pbk)
ISBN: 978-1-003-46129-6 (ebk)

DOI: 10.1201/9781003461296

Typeset in Times
by Apex CoVantage, LLC

Contents

Author

Si-Wei Chen earned a PhD (Hons.) in environmental studies at Tohoku University, Sendai, Japan, in 2012. He is a Professor at the National University of Defense Technology. He has published more than 130 journal and conference articles and coauthored three monographs and holds 28 patents. His research interests include radar polarimetry, polarimetric radar imaging, environmental study, and machine learning.

Dr. Chen was a recipient of the Excellent Young Scholar from the National Natural Science Foundation of China, a recipient of the IEEE Geoscience and Remote Sensing Society (GRSS) Early Career Award, the Natural Science Foundation of Hunan Province for Distinguished Young Scholars, the Young Talent of Hunan Province, China. He won the First Prize of Invention and Innovation Award from China Association of Inventions in 2023, the Second Prize of the National Teaching Achievement Award in 2023, the Second Prize of the National Award for Progress in Science and Technology in 2021, and the First Prize of the Natural Science Award from Hunan Province of China in 2018. He is an Associate Editor of the *IEEE Geoscience and Remote Sensing Letters* and an Editorial Board Member of the *Journal of Remote Sensing, Journal of Electronics and Information Technology, Journal of Radars, Journal of Signal Processing,* and *Journal of National University of Defense Technology.*

Preface

In recent years, there have been rapid developments in imaging radar due to advancements in hardware systems, imaging technology, and signal and information processing theory. Polarization, as a crucial dimension of electromagnetic waves, carries valuable physical information of targets, such as the geometry, structure, material, and so on, during radar measurements. Currently, equipped with full-polarization measurement capability, polarimetric radar imaging systems have been widely adopted in fields of earth observation, disaster reduction and prevention, reconnaissance and surveillance, space exploration, precision guidance, and so on and have become one of the predominant sensors in microwave remote sensing.

The main development trends of polarimetric radar imaging systems are from single-polarization to full-polarization, from narrowband to wideband and ultra-wideband, from single-frequency-band to multi-frequency-band, from single-baseline to multi-baseline, from single-looking-direction to multi-looking-direction, and so on. With the huge acquisition and accumulation of spaceborne/airborne polarimetric radar imaging datasets in terms of various polarizations, frequencies, and imaging modes, along with the rapid advancements of electromagnetic scattering interpretation theory and tools, the understandings and utilizations of radar polarimetric information have transitioned from a qualitative stage to a quantitative stage. Meanwhile, the major scientific problems and core technical challenges that will persist in the current and future periods are how to accurately, robustly, and efficiently extract valuable information and knowledge from the electromagnetic scattering responses of radar targets and how to employ this information and knowledge to guide practical applications.

The electromagnetic scattering responses of a target are closely related to its attitude and orientation, which are referred to as the target scattering diversity. The target scattering diversity effects encompass two main aspects. First, the same target may exhibit quite different scattering characteristics from different illuminations. Second, different targets may exhibit similar scattering characteristics from a specific measurement geometry. Traditional polarimetric interpretation methods often fail to account for target scattering diversity, leading to interpretation ambiguity. Over time, eliminating or mitigating target scattering diversity has been a major research motivation in the radar polarimetry community. However, the underlying physical mechanisms of the target scattering diversity effect can make target electromagnetic scattering characteristics exhibit significant variations in the polarimetric rotation domain or enhance their scattering differences in the polarimetric rotation domain. Intrinsically, these differences within target scattering diversity effects are kinds of valuable and physical information in the polarimetric rotation domain. Therefore, while target scattering diversity presents challenges in electromagnetic scattering modeling, interpretation, and application, it also provides rich information. The explorations and utilizations of target scattering diversity will provide solid foundations and profound insights for disclosing target scattering mechanism differences and extracting newly physical features.

In 2012, the author proposed the concept of polarimetric rotation domain interpretation. The core idea is to extend the initial polarimetric information acquired by a polarimetric radar under certain measurement geometry to the polarimetric rotation domain along the radar line of sight and to develop a series of interpretation tools for target scattering diversity characterization, exploration, and utilization. Over a span of more than 10 years, the author and his team have focused on this topic and established the polarimetric rotation domain interpretation framework. This framework mainly contains the uniform polarimetric matrix rotation theory, two-dimension polarimetric coherence pattern interpretation tool, and two-/three-dimension polarimetric correlation pattern interpretation tool. These tools have been successfully adopted for land cover classification, target detection, structure recognition, and damage mapping. This book serves as a phased summary of these research achievements and consists of seven chapters. Chapter 1 provides an overview of the developments and fundamental knowledge of imaging radar and radar polarimetry. Chapter 2 introduces the polarimetric rotation domain interpretation theory. Chapter 3 summarizes the polarimetric rotation domain roll-invariant features. Chapters 4 to 7 present application studies of the polarimetric rotation domain interpretation theory. In detail, Chapter 4 introduces polarimetric rotation domain land cover classification. Chapter 5 presents polarimetric rotation domain target detection. Chapter 6 addresses polarimetric rotation domain structure recognition. Finally, Chapter 7 introduces polarimetric rotation domain urban damage mapping.

This book aims at the introduction of polarimetric rotation domain interpretation theory and its applications. It is expected to be a possible reference for scientists, researchers, engineers, and graduate students in related fields and to foster new developments in imaging radar polarimetric information processing and applications. Since it focuses on the author group's research achievements, it is important to note that many other excellent achievements in this field have not been included in this book. Finally, limited by the author's knowledge, expertise, and energy, this book may have some mistakes, for which readers' corrections and feedback are sincerely welcome.

Acknowledgments

During the past almost 20 years of study and research, the author received support and help from a number of scientists, researchers, and colleagues worldwide. The author would like to express his deep thanks to Chinese Academician Bi-Tao Jiang, Chinese Academician Zhi Li, Professor Qun Zhang, Professor Jian Yang, Professor Shun-Ping Xiao, Professor Yi Su, and Professor Gang-Yao Kuang, for their long-term guidance and support. Special and deep gratitude is extended to my supervisors, Professor Xue-Song Wang from the National University of Defense Technology and Professor Motoyuki Sato from Tohoku University, for their dedicated guidance, support, and encouragement. The author also acknowledges the help and support from Professor Wolfgang-Martin Boerner from the University of Illinois, Professor Jong-Sen Lee and Professor Thomas L. Ainsworth from the United States Naval Research Laboratory, Professor Yoshio Yamaguchi from Niigata University, and Professor Ridha Touzi from the Natural Resources Centre of Canada. The author sincerely appreciates their contributions and support. Additionally, the author sincerely acknowledges the support from the research team, particularly my graduate students Chen-Song Tao, Xing-Chao Cui, Hao-Liang Li, Ming-Dian Li, and so on, who deeply participated in the research conduction, book compiling, and proofreading.

Plenty of spaceborne and airborne polarimetric radar datasets have been utilized during these studies. The author would like to express his appreciation to the China Land Satellite Remote Sensing Application Center for providing the GaoFen-3 data, the Japan Aerospace Exploration Agency for providing the ALOS data, the National Institute of Information and Communications Technology for providing Pi-SAR data, the European Space Agency and German Aerospace Center for providing E-SAR and F-SAR data, the Jet Propulsion Laboratory for providing the AIRSAR and UAVSAR data, the Canadian Space Agency for providing the Radarsat-2 data, and so on. These valuable and high-quality datasets greatly support this research and make scientific ideas come to fruition.

The contents of this book are mainly from the author's team journal and conference publications in the past years. The author would like to sincerely thank these publishers and conferences for their contributions and support, especially the Institute of Electrical and Electronics Engineers (IEEE) for permission to reuse materials that have appeared in IEEE publications; Taylor & Francis for permission to reuse our materials that have appeared in its publications; the Institute of Electronics, Information and Communication Engineers (IEICE) for permission to reuse our materials that have appeared in an IEICE publication; and so on. In addition, part of the polarimetric rotation domain interpretation contents have been included in the book titled *Target Scattering Mechanism in Polarimetric Synthetic Aperture Radar—Interpretation and Application* from Springer (2018). To be a complete summary of polarimetric rotation domain interpretation theory and applications, these materials have been revised, updated, and included in this book. The author sincerely appreciates the effort and support from Springer.

This book will be published in Chinese and English by the Science Press and CRC Press, respectively, and the author is extremely grateful for their help and support.

The research for this book was supported in part by the National Natural Science Foundation of China (62122091, 61771480, and 41301490), the Natural Science Foundation of Hunan Province (2020JJ2034), the High-Level Scientific and Technological Innovation Talents Project, and so on. The author sincerely appreciates this support.

The publication of this book would not have been possible without the continuing contributions and support from my family. Indeed, this book is dedicated to my family.

Si-Wei Chen
November 18, 2023
Changsha, China

1 Introduction

With the progress of hardware systems, imaging technology, and signal and information processing theory, imaging radar has developed rapidly in recent years. Imaging radar is widely used and plays an important role in fields such as earth observation, disaster mitigation, and space exploration. Humans also benefit widely from these successful applications based on imaging radar technology. At present, polarimetric radar with imaging ability has become one of the mainstream sensors in the microwave remote sensing field. This chapter briefly introduces the development and basic knowledge of radar polarimetry and imaging radar and the main contents of this book.

1.1 BRIEF HISTORY OF RADAR POLARIMETRY

Radar is a system that transmits and receives electromagnetic waves for the detection and location of reflecting objects. The term radar originally comes from the words radio detection and ranging [1]. With the progress and advances of hardware technologies and signal processing theories, the definition and function of a radar have been greatly expanded. Modern radar can generate a radar image to "see" a target directly.

Polarization is an intrinsic property of the electromagnetic wave [2]. Research on polarization has a long history since the seventeenth century, while the discovery of the polarization phenomenon was even earlier, about A.D. 1000 [3]. However, the earliest work on radar polarimetry only dates back to the 1940s [4, 5]. Many pioneers were dedicated to this field and stimulated the development of radar polarimetry. In 1950, G. W. Sinclair introduced the polarimetric scattering matrix to link the transmitted and received waves in Jones vectors and to represent the fully polarimetric information scattered by a coherent scatterer [4]. It is also known as the Sinclair matrix. In 1952, E. M. Kennaugh proposed the optimal polarization theory by demonstrating that there are characteristic polarization states leading to the maximum or minimum receiving power [6]. Meanwhile, the Kennaugh matrix and Mueller matrix, which link the associated Stokes vectors, were also presented [6, 7]. Also, the concept of the polarimetric power scattering matrix, which is suitable to optimize the density of the scattering field, was proposed by C. D. Graves [8]. Until the work of J. R. Huynen [9], radar polarimetry experienced a depression, with only a few notable achievements. His dissertation [9] opened a new stage for radar polarimetry in 1970 and reattracted more interest and attention. However, the deep understanding of the importance of radar polarimetry was limited by the lack of advanced polarimetric radar systems. In the 1970s and early 1980s, the major contributions were from W. M. Boerner, who rediscovered and pointed out the importance of polarization in electromagnetic scattering. He enhanced the work from E. M. Kennaugh and J. R. Huynen and extended the optimal polarization theory [5, 10–17].

DOI: 10.1201/9781003461296-1

In addition, Chinese scholars' research on radar polarimetry dates back to the 1960s, focusing on the polarimetric scattering matrix and polarimetric matched receiving [18]. Around the 1990s, some early representative doctoral dissertations on radar polarimetry included radar target recognition in the polarimetric domain and frequency domains [19], target recognition theory and application of wideband polarimetric radar [20], study of wideband polarization information processing [21], and theoretical problems in radar polarimetry [22]. In addition, [23] is an early monograph on radar polarimetric information processing and application in China.

More detailed reviews of the historical developments of radar polarimetry can be found in [5, 24, 25]. With the great contributions of radar polarimetry pioneers, the basic theory of radar polarimetry is formed and reviewed in [5, 26, 27]. At present, with the continuous emergence of advanced imaging radar systems with full-polarization measurement capabilities and the acquisition of a large number of measured polarimetric radar data with different modes, the basic theories and key technologies such as polarimetric radar imaging and target mechanism interpretation have been rapidly developed and successfully applied in many important fields.

1.2 POLARIMETRIC RADAR IMAGING

1.2.1 Radar Imaging Overview

Synthetic aperture radar (SAR) is a typical imaging radar, and its concept was proposed by Carl Wiley of Goodyear Aerospace in 1951 [28]. It was discovered that the Doppler information can be utilized to obtain high resolution in the azimuth direction perpendicular to the beam illumination. Thereby, two-dimension (2D) radar images are focused for the imaging scene according to range direction pulse compression and azimuth synthetic aperture processing. SAR, as a microwave imaging radar, shows its superiority in remote sensing, since it can work all day and in almost all weather conditions. The first satellite SAR, SEASAT, was launched in 1978, which established the imaging radar as a practical remote sensing system from then on.

With the developed radar polarimetry theory and the progress of SAR imaging techniques, the combination of them as polarimetric SAR (PolSAR) became a natural extension. This combination opened a new door for both radar polarimetry and microwave remote sensing. In 1985, the Jet Propulsion Laboratory successfully implemented the practical fully polarimetric AIRSAR system [3]. It had the capability to simultaneously acquire fully polarimetric data at three frequency bands (P-, L-, C-bands). These datasets were available to the radar community and promoted a large number of PolSAR data analysis and interpretation techniques. Later on, many airborne PolSAR systems came out. One of the representative systems is the Experimental SAR (E-SAR) developed by the Microwaves and Radar Institute of the German Aerospace Research Centre. Several advanced spaceborne PolSAR systems have also been successfully launched recently. More detailed introductions and descriptions of advanced PolSAR systems can be found in [3].

PolSAR extends the SAR system within the polarization dimension, while interferometric SAR (InSAR) extends it within the baseline dimension, encompassing both spatial and temporal dimensions of the SAR transceiver antenna.

Depending on the baseline configuration, InSAR can be utilized in a variety of ways and applications. When the interferometric baseline follows a vertical track, Graham confirmed InSAR's efficiency in terrain elevation inversion [29]. The ground moving target indication (GMTI) approach becomes feasible when the interferometric baseline aligns with the flight track. Additionally, by employing multiple co-track observations to form a differential interferometry mode with various temporal baselines, it becomes possible to monitor and estimate minor deformations. Gabriel et al. successfully validated this technique in 1989 using SEASAT spaceborne SAR data [30]. Comprehensive reviews of InSAR technology advancements are available in [31–33].

Another milestone was achieved by the Shuttle SAR system SIR-C/X-SAR, which acquired repeat-pass and multi-frequency (X-, C-, and L-band) datasets in April and October 1994. Fully polarimetric datasets at the C- and L-bands were also acquired during the second mission [34]. Based on these datasets, S. Cloude and K. Papathanassiou proposed the innovative polarimetric SAR interferometry (PolInSAR) technique and demonstrated forest height inversion via the investigation of polarization effects in SAR interferometry [34–37].

SAR obtains a 2D focused image by projecting three-dimension (3D) targets onto the 2D imaging plane. InSAR has the ability to retrieve the target height by adding one more receiving antenna across the flight track. PolInSAR, as a combination of PolSAR and InSAR, can partially separate target scattering centers from different scattering mechanisms in the elevation direction. In order to reconstruct a high-resolution 3D image, similar to the synthetic aperture in the azimuth direction, an aperture should be synthesized in the elevation direction as well, which leads to the concept of SAR tomography. The first demonstration of polarimetric SAR tomography with an airborne system was reported by A. Reigber and A. Moreira in 2000 [38, 39], which was based on the Fourier transformation technique. With the development of the compressed sensing technique, both SAR and PolSAR tomographics can achieve better imaging quality.

In order to get more information on the imaging scene, other imaging modes have been successfully demonstrated, such as the multi-looking-direction SAR, circular SAR, bistatic SAR, and so on [40–43]. In addition to airborne and spaceborne SAR systems, a ground-based SAR system has also been developed and showed its flexibility for regional-scale monitoring [44].

The overall development trend of radar imaging is from single-polarization to full-polarization, from single-band to multi-band, from narrow band to wide band, from single-baseline to multi-baseline. In addition, from the application perspective, advancements from monostatic to bistatic and from single-looking-direction to multi-looking-direction are also very important. The development of imaging radar has benefited from the progresses of hardware technology and imaging processing theory. The data obtained by these advanced imaging radars, especially those with polarization measurement capabilities, have also greatly promoted relevant theoretical research. In the past 20 years, many excellent monographs on radar polarimetry and imaging radar have been published worldwide [3, 23, 45–52]. These monographs have promoted the accumulation and dissemination of knowledge and the further development of this field.

Concurrently, the advancements in theoretical tools and imaging datasets have promoted application researches of SAR imaging. Numerous advanced methods and techniques have given rise to a plethora of successful applications, including tree height inversion, biomass inversion, topographic subsidence monitoring, crop management, topographic mapping, digital elevation modeling, and natural disaster assessment [3, 33, 35, 47–50, 52–61].

1.2.2 SAR Imaging Principles

SAR is a microwave remote sensor usually deployed on a spaceborne or airborne platform. Compared with optical sensors, SAR can work day and night, since it is nearly unaffected by weather and atmospheric conditions. Besides, with a low-frequency band (e.g. P- and L-band), the transmitted waves can penetrate into forest canopy and dry sand areas, which have potential for biomass estimation and subsurface imaging. These advantages make SAR particularly important in remote sensing.

In order to obtain radar images, during platform movements, SAR repeatedly transmits phase-encoded pulses (e.g. linear frequency modulated signal) and records the radar echoes reflected or backscattered from an imaging scene. These collected signals are the raw data with spreading energy from each scatterer. In order to form an image and focus the spreading energy at the target position, basically two signal processing procedures are needed in the range and azimuth directions. Pulse compression techniques, such as a matched filter, are implemented to obtain a fine resolution in the range direction, which is perpendicular to the flight pass of the sensor. The signal processing in the azimuth direction is more complicated. For real aperture radar, the azimuth resolution is determined by the beam width of the transmitted electromagnetic wave. Usually, a larger antenna aperture size produces a narrower beam width. However, the antenna size is limited by the system design and implementation. In order to achieve fine resolution in the azimuth direction, the "synthetic aperture" concept has been utilized, which is the key feature of SAR system and signal processing. In the azimuth direction, each scatterer in the radar beam reradiates energy with a different Doppler shift, which is exploited to separate the received energy into fine azimuth resolution cells. By utilizing this Doppler shift, the synthesized aperture in terms of several hundred meters or several kilometers can be formed during platform movement. The Doppler shift information is modulated with the signal, especially with the signal phase. After examining the characteristic of this modulation, the matched filter can also be applied to focus the spreading energy in the azimuth direction and obtain fine resolution. The basic SAR imaging algorithm is the range-Doppler method. A demonstration of SAR imaging is illustrated in Figure 1.2.1. The left figure shows the collected raw data from the GaoFen-3 satellite on-board SAR sensor. The right figure is the focused radar image. The spread energy in the raw data is well focused, and the local imaging scene becomes clear. More detailed descriptions of SAR signal and imaging processing algorithms are well summarized in [45].

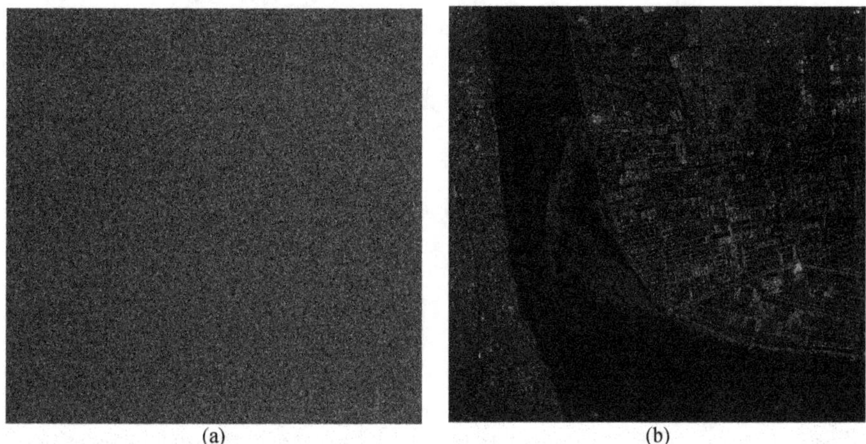

FIGURE 1.2.1 SAR imaging demonstration with GaoFen-3 satellite SAR. (a) Raw data, (b) SAR image.

1.2.3 PolSAR Principle

PolSAR is the further development of single-polarization SAR systems. PolSAR can obtain fully polarimetric information represented by a polarimetric scattering matrix through transmitting and receiving electromagnetic waves with orthogonal polarization states. Using horizontal and vertical polarization basis (H,V) as an example, an illustration of PolSAR imaging is shown in Figure 1.2.2. Meanwhile, there are three main operation modes to acquire a fully polarimetric scattering matrix. The first mode alternatively transmits horizontal and vertical polarization signals while alternatively using horizontal-polarized and vertical-polarized antennas to receive backscattered echoes. In this vein, with four pulse repetition times (PRT), a polarimetric scattering matrix can be obtained. This mode was adopted in very early PolSAR systems since it can reduce the hardware system complexity. The second mode takes turns transmitting horizontal and vertical polarization signals while simultaneously receiving backscattered signals using horizontal-polarized and vertical-polarized antennas. This operation mode can measure a polarimetric scattering matrix with two PRTs. The majority of PolSAR systems and almost all current on-board PolSAR systems adopt this operation mode. The schematic illustration of this operation mode is also displayed in Figure 1.2.2 (b). The third mode is to obtain a polarimetric scattering matrix by simultaneously transmitting horizontal and vertical polarization microwaves while simultaneously collecting echoes using horizontal-polarized and vertical-polarized antennas. This mode has the capability to measure a polarimetric scattering matrix within a PRT. However, it requires more complicated waveform designing and signal processing techniques to separate the coupled echoes from two polarization states. In principle, it is more suitable for ultra-high-speed or ultra-high-fluctuation targets. This fully polarimetric information acquisition mode

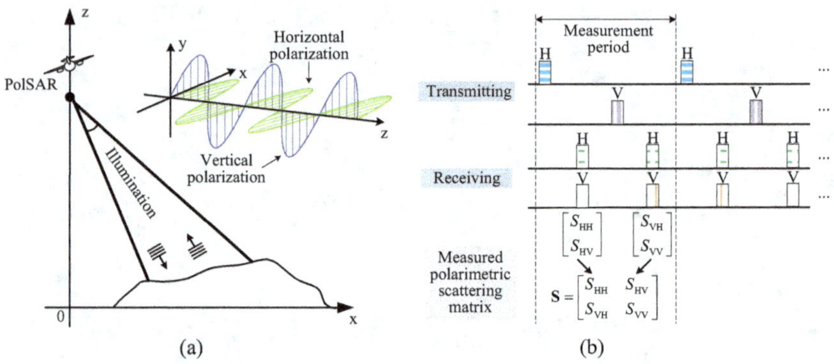

FIGURE 1.2.2 Illustration of PolSAR imaging (flight pass is perpendicular to the plane). (a) Imaging geometry, (b) schematic illustration of the alternative transmitting and receiving orthogonal polarization microwaves for polarimetric scattering matrix measurement.

has been adopted for polarimetric weather radar, while there is no report about its implementation with PolSAR systems.

In order to reduce on-board system complexity while doubling the imaging swath, compact-polarization imaging mode has also been established. Generally, it has three basic implementation methods. The first is to transmit a 45° linear polarization microwave while simultaneously receiving horizontal and vertical polarization backscattered echoes [62]. The second is to transmit circular polarization signals while collecting horizontal and vertical polarization backscattered signals at the same time [63]. The third is to transmit a left circular or right circular polarization microwave while simultaneously receiving left circular and right circular polarimetric backscattering signals [64]. Then, using the projection of 45° linear polarization or circular polarization onto the horizontal and vertical polarization axes, the approximated polarimetric scattering matrix can be reconstructed. It should be emphasized that this approximated polarimetric scattering matrix obtained by compact polarization mode is not fully equivalent to the polarimetric scattering matrix acquired by the corresponding full-polarization mode. One of the main weaknesses of compact-polarization mode is the lack of ability to measure high quality cross-polarization terms. In principle, compact-polarization mode can acquire partially polarimetric information other than fully polarimetric information.

1.3 RADAR POLARIMETRY BASICS

Throughout the development history of radar polarimetry, numerous characterization features and tools have been proposed by pioneering researchers to describe the polarimetric information of electromagnetic waves and targets. Sinclair introduced a polarimetric scattering matrix to link the Jones vectors of transmitted and received electromagnetic waves [4]. Graves proposed the concept of the polarimetric power scattering matrix [8]. Additionally, Kennaugh and Mueller matrices are utilized to connect the Stokes vector of electromagnetic waves [6, 7, 10]. From the polarimetric scattering matrix, Pauli and Lexicographic scattering vectors can be derived, and secondary statistics of the polarimetric coherency matrix and polarimetric covariance matrix can be formed using the vector product and sample average processing methods. Furthermore, researchers have developed geometric description tools, including the polarization ellipse [3, 11], the Poincaré sphere [3, 11], and the polarimetric characteristic curve [47]. Most of these representation and characterization tools and their interrelationships have been summarized and reviewed in [3, 11]. This section mainly introduces the basic and widely used representations and their physical principles within the scope of polarimetric radar imaging.

1.3.1 Polarization of Electromagnetic Waves

The fundamental theory of electromagnetic fields is based on Maxwell's equations [2]

$$\begin{aligned} \nabla\times\mathbf{E} &= -\frac{\partial\mathbf{B}}{\partial t} \quad & \nabla\times\mathbf{H} &= \mathbf{J}+\frac{\partial\mathbf{D}}{\partial t} \\ \nabla\cdot\mathbf{B} &= 0 & \nabla\cdot\mathbf{D} &= \rho_v \end{aligned} \tag{1.3.1}$$

where $\mathbf{E}$, $\mathbf{H}$, $\mathbf{B}$, and $\mathbf{D}$ are the electric field, magnetic field, electric induction, and magnetic induction, respectively. $\mathbf{J}$ and ρ_v are electric current density and electric charge density, respectively.

For time-harmonic fields at a single frequency$\omega = 2\pi f$, $\mathbf{E}$ and $\mathbf{H}$ at an arbitrary position r can be expressed as

$$\mathbf{E} = \mathbf{E}(r)\exp(j\omega t) \quad \mathbf{H} = \mathbf{H}(r)\exp(j\omega t) \tag{1.3.2}$$

Assuming that electromagnetic fields are generated in free space by sources $\mathbf{J}$ and ρ_v in a local region where $\mathbf{J} = 0$ and $\rho_v = 0$ for fields out of this region, then Maxwell's equations become

$$\begin{aligned} \nabla\times\mathbf{E} &= -j\omega\mathbf{B} \quad & \nabla\times\mathbf{H} &= j\omega\mathbf{D} \\ \nabla\cdot\mathbf{B} &= 0 & \nabla\cdot\mathbf{D} &= 0 \end{aligned} \tag{1.3.3}$$

The free-space constitutive relations are

$$\mathbf{D} = \varepsilon_0\mathbf{E} \quad \mathbf{B} = \mu_0\mathbf{H} \tag{1.3.4}$$

where ε_0 and μ_0 are the dielectric permittivity and magnetic permeability for free space.

Substituting (1.3.3) and (1.3.4) into $\nabla\times(\nabla\times\mathbf{E}) = \nabla(\nabla\cdot\mathbf{E}) - \nabla^2\mathbf{E}$, the wave equations can be obtained

$$\nabla^2\mathbf{E} + k^2\mathbf{E} = 0 \quad \nabla^2\mathbf{H} + k^2\mathbf{H} = 0 \tag{1.3.5}$$

where k is the wavenumber and $k^2 = \omega^2\mu_0\varepsilon_0$.

Generally, the vector **E** of a time-harmonic electromagnetic wave varies sinusoidally with time at a fixed point in space. The polarization of the wave is described by the locus of the tip of the vector **E** as time progresses. For plane waves, the electric and magnetic fields are perpendicular to each other, and both are perpendicular to the wave propagation direction. If this plane wave propagates along the positive z direction, the electric field **E** in (1.3.5) becomes

$$\frac{\partial^2 \mathbf{E}(z,t)}{\partial z^2} + k^2 \mathbf{E}(z,t) = 0 \tag{1.3.6}$$

The real components of the instantaneous electric field of the solution for (1.3.6) along the x and y axes are

$$\mathbf{E}(z,t) = \begin{bmatrix} E_x(z,t) \\ E_y(z,t) \end{bmatrix} = \begin{bmatrix} a_x \cos(\omega t - kz + \phi_x) \\ a_y \cos(\omega t - kz + \phi_y) \end{bmatrix} \tag{1.3.7}$$

where a_x and a_y denote the magnitudes, while ϕ_x and ϕ_y denote the phases for the x and y components, respectively.

At a specific position $z = z_0$, the temporal wave trajectory is determined

$$\left[\frac{E_x(z_0,t)}{a_x}\right]^2 - 2\frac{E_x(z_0,t)E_y(z_0,t)}{a_x a_y}\cos\phi + \left[\frac{E_y(z_0,t)}{a_y}\right]^2 = \sin^2\phi \tag{1.3.8}$$

where $\phi = \phi_y - \phi_x$ is the phase difference.

For most cases, the expression (1.3.8) is the equation of an ellipse and describes elliptical polarization. When $\phi = 0$ or $\phi = \pm\pi$, the ellipse shrinks to a line, which indicates linear polarization. When $\phi = \pm\frac{\pi}{2}$ and $\frac{a_x}{a_y} = 1$, the ellipse becomes a circle, which represents circular polarization.

Generally, the locus of (1.3.8) is named the polarization ellipse to describe the wave polarization. The shape of the polarization ellipse can be characterized by three parameters, shown in Figure 1.3.1.

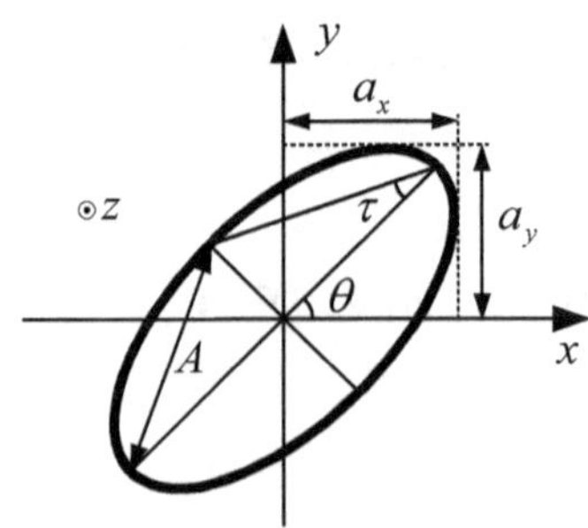

FIGURE 1.3.1 Polarization ellipse.

1) A is the ellipse amplitude and determined as

$$A = \sqrt{a_x^2 + a_y^2} \tag{1.3.9}$$

2) $\theta \in \left[-\frac{\pi}{2}, \frac{\pi}{2}\right]$ is the ellipse orientation and is defined as the angle between the major axis of the ellipse and the x-axis, as

$$\tan(2\theta) = 2\frac{a_x a_y}{a_x^2 - a_y^2}\cos\phi \tag{1.3.10}$$

3) $|\tau| \in \left[0, \frac{\pi}{4}\right]$ is the ellipticity and defined as

$$|\sin(2\tau)| = 2\frac{a_x a_y}{a_x^2 + a_y^2}|\sin\phi| \tag{1.3.11}$$

1.3.2 Polarimetric Scattering Matrix

For active radar, the transmitted electromagnetic wave will interact with a potential target when the wave reaches it. During the interaction, part of the energy carried by the incident wave is absorbed by the target, while the rest is reradiated as a new electromagnetic wave and modulated with the properties of the target itself. In order to characterize the target property from the viewpoint of power exchange, for a point target smaller than the footprint of the radar illumination beam, the radar cross section (RCS) σ is introduced [1]

$$\sigma = 4\pi r^2 \frac{|\mathbf{E}_\mathrm{S}|^2}{|\mathbf{E}_\mathrm{I}|^2} \tag{1.3.12}$$

where $\mathbf{E}_\mathrm{I}$ is the incident electromagnetic wave reaching the target, $\mathbf{E}_\mathrm{S}$ is the scattered wave reradiated by the same target, r is the distance between the radar and the target.

For an extended or distributed target, which is larger than the footprint of the radar illumination beam, the scattering coefficient σ^0 is defined [1]

$$\sigma^0 = \frac{\langle\sigma\rangle}{A_0} = \frac{4\pi r^2}{A_0}\frac{\langle|\mathbf{E}_\mathrm{S}|^2\rangle}{|\mathbf{E}_\mathrm{I}|^2} \tag{1.3.13}$$

where $\langle\rangle$ denotes the sample average, and the scattering coefficient σ^0 is the averaged RCS per unit area A_0 and represents the ratio of the statistically averaged scattered power density to the average incident power density over the surface of the sphere with radius r.

Generally, the RCS σ and scattering coefficient σ^0 carry the intrinsic information of the scatterers and targets, such as the dielectric properties, geometric structures, and so on. Fully polarimetric radar can obtain additional polarimetric information of targets by transmitting and receiving orthogonal electromagnetic waves. The polarization of a plane and monochromatic electromagnetic wave can be represented by the Jones vector. Also, two orthogonal Jones vectors form a polarization basis where any polarization state of a given electromagnetic wave can be expressed. If the Jones vectors of the incident and scattered waves are denoted by $\mathbf{E}_\mathrm{I}$ and $\mathbf{E}_\mathrm{S}$, respectively, with the far field assumption, the scattering process at a specific target can be represented as [4]

$$\mathbf{E}_\mathrm{S} = \frac{e^{-jkr}}{r}\mathbf{S}\mathbf{E}_\mathrm{I} = \frac{e^{-jkr}}{r}\begin{bmatrix} S_\mathrm{XX} & S_\mathrm{XY} \\ S_\mathrm{YX} & S_\mathrm{YY} \end{bmatrix}\mathbf{E}_\mathrm{I} \tag{1.3.14}$$

where $\dfrac{e^{-jkr}}{r}$ accounts for the propagation effects both in amplitude and phase. $\mathbf{S} = \begin{bmatrix} S_\mathrm{XX} & S_\mathrm{XY} \\ S_\mathrm{YX} & S_\mathrm{YY} \end{bmatrix}$ is the polarimetric scattering matrix and also called the Sinclair matrix.

For the horizontal and vertical polarization basis $(\mathrm{H,V})$, the polarimetric scattering matrix becomes

$$\mathbf{S} = \begin{bmatrix} S_\mathrm{HH} & S_\mathrm{HV} \\ S_\mathrm{VH} & S_\mathrm{VV} \end{bmatrix} \tag{1.3.15}$$

where S_VH is the backscattered return from horizontal transmitting and vertical receiving polarization. Other terms are similarly defined.

The total backscattering power $SPAN$ is

$$SPAN = |S_\mathrm{HH}|^2 + |S_\mathrm{HV}|^2 + |S_\mathrm{VH}|^2 + |S_\mathrm{VV}|^2 \tag{1.3.16}$$

1.3.3 Polarization Basis Transformation

In order to understand and extract desired information from the polarimetric scattering matrix, the transformation of the polarization basis is needed to achieve a specific polarization combination. The $(\mathrm{X,Y})$ polarization basis can be obtained from the $(\mathrm{H,V})$ polarization basis [3]

$$\begin{bmatrix} S_\mathrm{XX} & S_\mathrm{XY} \\ S_\mathrm{YX} & S_\mathrm{YY} \end{bmatrix} = \frac{1}{1+|\rho|^2}\begin{bmatrix} e^{j\alpha} & 0 \\ 0 & e^{-j\alpha} \end{bmatrix}\begin{bmatrix} 1 & \rho \\ -\rho^* & 1 \end{bmatrix} \begin{bmatrix} S_\mathrm{HH} & S_\mathrm{HV} \\ S_\mathrm{VH} & S_\mathrm{VV} \end{bmatrix}\begin{bmatrix} 1 & -\rho^* \\ \rho & 1 \end{bmatrix}\begin{bmatrix} e^{-j\alpha} & 0 \\ 0 & e^{j\alpha} \end{bmatrix} \tag{1.3.17}$$

where the polarization ratio is $\rho = \dfrac{\tan\phi + \mathrm{j}\tan\tau}{1 - \mathrm{j}\tan\phi\tan\tau}$, and ρ^* is the conjugate of ρ. The phase parameter is $\alpha = \tan^{-1}(\tan\phi\tan\tau)$. Also, ϕ and τ are the geometric parameters of the polarization ellipse.

For the left and right circular polarization basis $(\mathrm{L,R})$, $\rho = \mathrm{j}$ and $\alpha = 0$ are valid. Then, the corresponding polarimetric scattering matrix is

$$\begin{aligned}\begin{bmatrix} S_{\mathrm{LL}} & S_{\mathrm{LR}} \\ S_{\mathrm{RL}} & S_{\mathrm{RR}} \end{bmatrix} &= \frac{1}{2}\begin{bmatrix} 1 & \mathrm{j} \\ \mathrm{j} & 1 \end{bmatrix}\begin{bmatrix} S_{\mathrm{HH}} & S_{\mathrm{HV}} \\ S_{\mathrm{VH}} & S_{\mathrm{VV}} \end{bmatrix}\begin{bmatrix} 1 & \mathrm{j} \\ \mathrm{j} & 1 \end{bmatrix} \\ &= \frac{1}{2}\begin{bmatrix} S_{\mathrm{HH}} + \mathrm{j}(S_{\mathrm{HV}} + S_{\mathrm{VH}}) - S_{\mathrm{VV}} & \mathrm{j}(S_{\mathrm{HH}} + S_{\mathrm{VV}}) + S_{\mathrm{HV}} - S_{\mathrm{VH}} \\ \mathrm{j}(S_{\mathrm{HH}} + S_{\mathrm{VV}}) - S_{\mathrm{HV}} + S_{\mathrm{VH}} & -S_{\mathrm{HH}} + \mathrm{j}(S_{\mathrm{HV}} + S_{\mathrm{VH}}) + S_{\mathrm{VV}} \end{bmatrix}\end{aligned} \tag{1.3.18}$$

The elements of a polarimetric scattering matrix with the circular polarization basis $(\mathrm{L,R})$ are

$$\begin{aligned} S_{\mathrm{LL}} &= \frac{1}{2}\left[S_{\mathrm{HH}} - S_{\mathrm{VV}} + \mathrm{j}(S_{\mathrm{HV}} + S_{\mathrm{VH}})\right] \\ S_{\mathrm{LR}} &= \frac{1}{2}\left[\mathrm{j}(S_{\mathrm{HH}} + S_{\mathrm{VV}}) + S_{\mathrm{HV}} - S_{\mathrm{VH}}\right] \\ S_{\mathrm{RL}} &= \frac{1}{2}\left[\mathrm{j}(S_{\mathrm{HH}} + S_{\mathrm{VV}}) - S_{\mathrm{HV}} + S_{\mathrm{VH}}\right] \\ S_{\mathrm{RR}} &= \frac{1}{2}\left[-(S_{\mathrm{HH}} - S_{\mathrm{VV}}) + \mathrm{j}(S_{\mathrm{HV}} + S_{\mathrm{VH}})\right] \end{aligned} \tag{1.3.19}$$

1.3.4 Polarimetric Coherency Matrix

In order to interpret target scattering mechanism, a polarimetric scattering matrix can be projected onto the Pauli spin matrices. For bistatic scattering case, the Pauli spin matrices are

$$\mathbf{P}_1 = \frac{1}{\sqrt{2}}\begin{bmatrix} 1 & 0 \\ 0 & 1 \end{bmatrix}, \mathbf{P}_2 = \frac{1}{\sqrt{2}}\begin{bmatrix} 1 & 0 \\ 0 & -1 \end{bmatrix}, \mathbf{P}_3 = \frac{1}{\sqrt{2}}\begin{bmatrix} 0 & 1 \\ 1 & 0 \end{bmatrix}, \mathbf{P}_4 = \frac{1}{\sqrt{2}}\begin{bmatrix} 0 & -\mathrm{j} \\ \mathrm{j} & 0 \end{bmatrix} \tag{1.3.20}$$

Then, the projection of polarimetric scattering matrix becomes

$$\mathbf{S} = k_1\mathbf{P}_1 + k_2\mathbf{P}_2 + k_3\mathbf{P}_3 + k_4\mathbf{P}_4 \tag{1.3.21}$$

The coefficients $k_1 \sim k_4$ are elements that can form the Pauli scattering vector $\mathbf{k}_{\mathrm{P}_{(\mathrm{H,V})}}$, as

$$\mathbf{k}_{\mathrm{P}_{(\mathrm{H,V})}} = \frac{1}{\sqrt{2}}\left[S_{\mathrm{HH}} + S_{\mathrm{VV}} \quad S_{\mathrm{HH}} - S_{\mathrm{VV}} \quad S_{\mathrm{HV}} + S_{\mathrm{VH}} \quad \mathrm{j}(S_{\mathrm{HV}} - S_{\mathrm{VH}})\right]^{\mathrm{T}} \tag{1.3.22}$$

For the monostatic scattering case, which satisfies the reciprocity condition($S_{\mathrm{HV}} \approx S_{\mathrm{VH}}$), the Pauli spin matrices contain

$$\mathbf{P}_1 = \frac{1}{\sqrt{2}}\begin{bmatrix} 1 & 0 \\ 0 & 1 \end{bmatrix}, \mathbf{P}_2 = \frac{1}{\sqrt{2}}\begin{bmatrix} 1 & 0 \\ 0 & -1 \end{bmatrix}, \mathbf{P}_3 = \frac{1}{\sqrt{2}}\begin{bmatrix} 0 & 1 \\ 1 & 0 \end{bmatrix} \tag{1.3.23}$$

The Pauli scattering vector becomes

$$\mathbf{k}_{\mathrm{P}_{(\mathrm{H,V})}} = \frac{1}{\sqrt{2}}\left[S_{\mathrm{HH}} + S_{\mathrm{VV}} \quad S_{\mathrm{HH}} - S_{\mathrm{VV}} \quad 2S_{\mathrm{HV}}\right]^{\mathrm{T}} \tag{1.3.24}$$

The superiority of the Pauli scattering vector representation lies in its capability to directly indicate a canonical scattering mechanism. For example, Pauli spin matrices $\mathbf{P}_1$, $\mathbf{P}_2$, and $\mathbf{P}_3$ represent canonical odd-bounce scattering, double-bounce scattering, and volume scattering, respectively. In this vein, the elements in $\mathbf{k}_{\mathrm{P}_{(\mathrm{H,V})}}$ indicate the scattering coefficients of these canonical scattering mechanisms accordingly.

With the Pauli scattering vector, the second-order statistics of the polarimetric coherency matrix $\mathbf{T}$ can be formed. The polarimetric coherency matrix inherits the representation advantage from the Pauli scattering vector, and it is defined as

$$\begin{aligned}\mathbf{T}_{(\mathrm{H,V})} &= \left\langle \mathbf{k}_{\mathrm{P}_{(\mathrm{H,V})}} \mathbf{k}_{\mathrm{P}_{(\mathrm{H,V})}}^{\mathrm{H}} \right\rangle = \begin{bmatrix} T_{11} & T_{12} & T_{13} \\ T_{21} & T_{22} & T_{23} \\ T_{31} & T_{32} & T_{33} \end{bmatrix} \\ &= \frac{1}{2}\begin{bmatrix} \left\langle \left|S_{\mathrm{HH}} + S_{\mathrm{VV}}\right|^2 \right\rangle & \left\langle (S_{\mathrm{HH}} + S_{\mathrm{VV}})(S_{\mathrm{HH}} - S_{\mathrm{VV}})^* \right\rangle & \left\langle 2(S_{\mathrm{HH}} + S_{\mathrm{VV}})S_{\mathrm{HV}}^* \right\rangle \\ \left\langle (S_{\mathrm{HH}} + S_{\mathrm{VV}})^*(S_{\mathrm{HH}} - S_{\mathrm{VV}}) \right\rangle & \left\langle \left|S_{\mathrm{HH}} - S_{\mathrm{VV}}\right|^2 \right\rangle & \left\langle 2(S_{\mathrm{HH}} - S_{\mathrm{VV}})S_{\mathrm{HV}}^* \right\rangle \\ \left\langle 2(S_{\mathrm{HH}} + S_{\mathrm{VV}})^* S_{\mathrm{HV}} \right\rangle & \left\langle 2(S_{\mathrm{HH}} - S_{\mathrm{VV}})^* S_{\mathrm{HV}} \right\rangle & \left\langle 4\left|S_{\mathrm{HV}}\right|^2 \right\rangle \end{bmatrix}\end{aligned} \tag{1.3.25}$$

where $\mathbf{k}_{\mathrm{P}_{(\mathrm{H,V})}}^{\mathrm{H}}$ is the conjugate transpose of $\mathbf{k}_{\mathrm{P}_{(\mathrm{H,V})}}$. T_{ij} is the (i, j) entry of $\mathbf{T}$.

1.3.5 Polarimetric Covariance Matrices

A polarimetric scattering matrix can be alternatively represented by another scattering vector named the Lexicographic scattering vector, with each element of the polarimetric scattering matrix as its entity. With the linear polarization basis $(\mathrm{H,V})$ and the reciprocity condition, the Lexicographic scattering vector is

$$\mathbf{k}_{\mathrm{L}_{(\mathrm{H,V})}} = \left[S_{\mathrm{HH}} \quad \sqrt{2}S_{\mathrm{HV}} \quad S_{\mathrm{VV}}\right]^{\mathrm{T}} \tag{1.3.26}$$

The corresponding polarimetric covariance matrix $\mathbf{C}_{(\mathrm{H,V})}$ is defined as

$$\begin{aligned}\mathbf{C}_{(\mathrm{H,V})} &= \left\langle \mathbf{k}_{\mathrm{L}_{(\mathrm{H,V})}} \mathbf{k}_{\mathrm{L}_{(\mathrm{H,V})}}^{\mathrm{H}} \right\rangle = \begin{bmatrix} C_{11} & C_{12} & C_{13} \\ C_{21} & C_{22} & C_{23} \\ C_{31} & C_{32} & C_{33} \end{bmatrix} \\ &= \begin{bmatrix} \left\langle \left|S_{\mathrm{HH}}\right|^2 \right\rangle & \sqrt{2}\left\langle S_{\mathrm{HH}}S_{\mathrm{HV}}^* \right\rangle & \left\langle S_{\mathrm{HH}}S_{\mathrm{VV}}^* \right\rangle \\ \sqrt{2}\left\langle S_{\mathrm{HH}}^*S_{\mathrm{HV}} \right\rangle & 2\left\langle \left|S_{\mathrm{HV}}\right|^2 \right\rangle & \sqrt{2}\left\langle S_{\mathrm{HV}}S_{\mathrm{VV}}^* \right\rangle \\ \left\langle S_{\mathrm{HH}}^*S_{\mathrm{VV}} \right\rangle & \sqrt{2}\left\langle S_{\mathrm{HV}}^*S_{\mathrm{VV}} \right\rangle & \left\langle \left|S_{\mathrm{VV}}\right|^2 \right\rangle \end{bmatrix}\end{aligned} \tag{1.3.27}$$

The relationship between the Pauli and Lexicographic scattering vectors is a unitary transform

$$\mathbf{k}_{\mathrm{L}_{(\mathrm{H,V})}} = \mathbf{U}_{\mathrm{P}_{(\mathrm{H,V})}2\mathrm{L}_{(\mathrm{H,V})}} \mathbf{k}_{\mathrm{P}_{(\mathrm{H,V})}} \tag{1.3.28}$$

where $\mathbf{U}_{\mathrm{P}_{(\mathrm{H,V})}2\mathrm{L}_{(\mathrm{H,V})}}$ is a unitary matrix, as

$$\mathbf{U}_{\mathrm{P}_{(\mathrm{H,V})}2\mathrm{L}_{(\mathrm{H,V})}} = \frac{1}{\sqrt{2}}\begin{bmatrix} 1 & 1 & 0 \\ 0 & 0 & \sqrt{2} \\ 1 & -1 & 0 \end{bmatrix} \tag{1.3.29}$$

Therefore, the link between the polarimetric covariance matrix $\mathbf{C}_{(\mathrm{H,V})}$ and polarimetric coherency matrix $\mathbf{T}_{(\mathrm{H,V})}$ is a similar transform

$$\begin{aligned} \mathbf{C}_{(\mathrm{H,V})} &= \mathbf{U}_{\mathrm{P}_{(\mathrm{H,V})}2\mathrm{L}_{(\mathrm{H,V})}} \mathbf{T}_{(\mathrm{H,V})} \mathbf{U}^{\mathrm{H}}_{\mathrm{P}_{(\mathrm{H,V})}2\mathrm{L}_{(\mathrm{H,V})}} = \mathbf{U}_{\mathrm{P}_{(\mathrm{H,V})}2\mathrm{L}_{(\mathrm{H,V})}} \mathbf{T}_{(\mathrm{H,V})} \mathbf{U}^{-1}_{\mathrm{P}_{(\mathrm{H,V})}2\mathrm{L}_{(\mathrm{H,V})}} \\ &= \frac{1}{2}\begin{bmatrix} T_{11}+T_{22}+2\,\mathrm{Re}[T_{12}] & \sqrt{2}(T_{13}+T_{23}) & T_{11}-T_{22}-\mathrm{j}2\,\mathrm{Im}[T_{12}] \\ \sqrt{2}(T_{13}+T_{23})^* & 2T_{33} & \sqrt{2}(T_{13}-T_{23})^* \\ T_{11}-T_{22}+\mathrm{j}2\,\mathrm{Im}[T_{12}] & \sqrt{2}(T_{13}-T_{23}) & T_{11}+T_{22}-2\,\mathrm{Re}[T_{12}] \end{bmatrix} \end{aligned} \tag{1.3.30}$$

where $\mathbf{U}^{-1}_{\mathrm{P}_{(\mathrm{H,V})}2\mathrm{L}_{(\mathrm{H,V})}}$ is the inverse of $\mathbf{U}_{\mathrm{P}_{(\mathrm{H,V})}2\mathrm{L}_{(\mathrm{H,V})}}$, and $\mathbf{U}^{-1}_{\mathrm{P}_{(\mathrm{H,V})}2\mathrm{L}_{(\mathrm{H,V})}} = \mathbf{U}^{\mathrm{H}}_{\mathrm{P}_{(\mathrm{H,V})}2\mathrm{L}_{(\mathrm{H,V})}}$ for the unitary matrix.

The polarimetric covariance matrix with a circular polarization basis is also commonly used for radar polarimetric data interpretation. For a left and right $(\mathrm{L,R})$ circular polarization basis, with the reciprocity condition $(S_{\mathrm{LR}} \approx S_{\mathrm{RL}})$, from (1.3.19), the corresponding scattering vector $\mathbf{k}_{\mathrm{L}_{(\mathrm{L,R})}}$ is

$$\mathbf{k}_{\mathrm{L}_{(\mathrm{L,R})}} = \begin{bmatrix} S_{\mathrm{LL}} \\ \sqrt{2}S_{\mathrm{LR}} \\ S_{\mathrm{RR}} \end{bmatrix} = \frac{1}{2}\begin{bmatrix} S_{\mathrm{HH}} - S_{\mathrm{VV}} + \mathrm{j}2S_{\mathrm{HV}} \\ \mathrm{j}\sqrt{2}(S_{\mathrm{HH}} + S_{\mathrm{VV}}) \\ -(S_{\mathrm{HH}} - S_{\mathrm{VV}}) + \mathrm{j}2S_{\mathrm{HV}} \end{bmatrix} \tag{1.3.31}$$

The relationship between $\mathbf{k}_{\mathrm{L}_{(\mathrm{L,R})}}$ and the Pauli scattering vectors is also a unitary transform

$$\mathbf{k}_{\mathrm{L}_{(\mathrm{L,R})}} = \mathbf{U}_{\mathrm{P}_{(\mathrm{H,V})}2\mathrm{L}_{(\mathrm{L,R})}} \mathbf{k}_{\mathrm{P}_{(\mathrm{H,V})}} \tag{1.3.32}$$

where $\mathbf{U}_{\mathrm{P}_{(\mathrm{H,V})}2\mathrm{L}_{(\mathrm{L,R})}}$ is a unitary matrix, as

$$\mathbf{U}_{\mathrm{P}_{(\mathrm{H,V})}2\mathrm{L}_{(\mathrm{L,R})}} = \frac{\sqrt{2}}{2}\begin{bmatrix} 0 & 1 & \mathrm{j} \\ \mathrm{j}\sqrt{2} & 0 & 0 \\ 0 & -1 & \mathrm{j} \end{bmatrix} \tag{1.3.33}$$

The polarimetric covariance matrix with circular polarization basis (L,R) is

$$\mathbf{C}_{(\text{L,R})} = \left\langle \mathbf{k}_{\text{L}_{(\text{L,R})}} \mathbf{k}^{\text{H}}_{\text{L}_{(\text{L,R})}} \right\rangle = \begin{bmatrix} \left\langle |S_{\text{LL}}|^2 \right\rangle & \sqrt{2}\left\langle S_{\text{LL}} S^*_{\text{LR}} \right\rangle & \left\langle S_{\text{LL}} S^*_{\text{RR}} \right\rangle \\ \sqrt{2}\left\langle S^*_{\text{LL}} S_{\text{LR}} \right\rangle & 2\left\langle |S_{\text{LR}}|^2 \right\rangle & \sqrt{2}\left\langle S_{\text{LR}} S^*_{\text{RR}} \right\rangle \\ \left\langle S^*_{\text{LL}} S_{\text{RR}} \right\rangle & \sqrt{2}\left\langle S^*_{\text{LR}} S_{\text{RR}} \right\rangle & \left\langle |S_{\text{RR}}|^2 \right\rangle \end{bmatrix} \tag{1.3.34}$$

Therefore, the link between the polarimetric covariance matrix $\mathbf{C}_{(\text{L,R})}$ and polarimetric coherency matrix $\mathbf{T}_{(\text{H,V})}$ is also a similar transform

$$\mathbf{C}_{(\text{L,R})} = \mathbf{U}_{\text{P}_{(\text{H,V})}2\text{L}_{(\text{L,R})}} \mathbf{T}_{(\text{H,V})} \mathbf{U}^{\text{H}}_{\text{P}_{(\text{H,V})}2\text{L}_{(\text{L,R})}} = \mathbf{U}_{\text{P}_{(\text{H,V})}2\text{L}_{(\text{L,R})}} \mathbf{T}_{(\text{H,V})} \mathbf{U}^{-1}_{\text{P}_{(\text{H,V})}2\text{L}_{(\text{L,R})}}$$
$$= \frac{1}{2}\begin{bmatrix} T_{22}+T_{33}+2\,\text{Im}[T_{23}] & \sqrt{2}(T_{13}+\text{j}T_{12})^* & T_{33}-T_{22}-\text{j}2\,\text{Re}[T_{23}] \\ \sqrt{2}(T_{13}+\text{j}T_{12}) & 2T_{11} & \sqrt{2}(T_{13}-\text{j}T_{12}) \\ T_{33}-T_{22}+\text{j}2\,\text{Re}[T_{23}] & \sqrt{2}(T_{13}-\text{j}T_{12})^* & T_{22}+T_{33}-2\,\text{Im}[T_{23}] \end{bmatrix} \tag{1.3.35}$$

where $\mathbf{U}^{-1}_{\text{P}_{(\text{H,V})}2\text{L}_{(\text{L,R})}}$ is the inverse of $\mathbf{U}_{\text{P}_{(\text{H,V})}2\text{L}_{(\text{L,R})}}$, and $\mathbf{U}^{-1}_{\text{P}_{(\text{H,V})}2\text{L}_{(\text{L,R})}} = \mathbf{U}^{\text{H}}_{\text{P}_{(\text{H,V})}2\text{L}_{(\text{L,R})}}$ for the unitary matrix.

1.3.6 Polarimetric Power Scattering Matrix

Considering polarimetric radar optimal power reception, the polarimetric power scattering matrix $\mathbf{G}_{(\text{H,V})}$, which is also known as the Graves matrix, was proposed [8] as

$$\mathbf{G}_{(\text{H,V})} = \mathbf{S}^{\text{H}}_{(\text{H,V})}\mathbf{S}_{(\text{H,V})} = \begin{bmatrix} |S_{\text{HH}}|^2 + |S_{\text{VH}}|^2 & S^*_{\text{HH}}S_{\text{HV}} + S^*_{\text{VH}}S_{\text{VV}} \\ S^*_{\text{HV}}S_{\text{HH}} + S^*_{\text{VV}}S_{\text{VH}} & |S_{\text{VV}}|^2 + |S_{\text{HV}}|^2 \end{bmatrix} \tag{1.3.36}$$

where $\mathbf{S}^{\text{H}}_{(\text{H,V})}$ is the conjugate transpose of $\mathbf{S}_{(\text{H,V})}$.

In addition, the target backscattering can also be represented in another power matrix form. The Kennaugh matrix $\mathbf{K}_{(\text{H,V})}$ is defined [7] as

$$\mathbf{K}_{(\text{H,V})} = \mathbf{A}^*\left(\mathbf{S}_{(\text{H,V})} \otimes \mathbf{S}^*_{(\text{H,V})}\right)\mathbf{A}^{-1} \tag{1.3.37}$$

where the superscript * stands for the conjugate and the operator ⊗ indicates the Kronecher product. The transformation matrix **A** is

$$\mathbf{A}=\begin{bmatrix}1&0&0&1\\1&0&0&-1\\0&1&1&0\\0&\mathrm{j}&-\mathrm{j}&0\end{bmatrix}\tag{1.3.38}$$

In the forward-scattering system, the Kennaugh matrix can also be transformed as the Mueller matrix $\mathbf{M}_{(\mathrm{H,V})}$ [7], as

$$\mathbf{M}_{(\mathrm{H,V})}=\mathbf{A}\left(\mathbf{S}_{(\mathrm{H,V})}\otimes\mathbf{S}^{*}_{(\mathrm{H,V})}\right)\mathbf{A}^{-1}\tag{1.3.39}$$

The elements of the Kennaugh and Mueller matrices in terms of the Sinclair matrix elements are as follows

$$\mathbf{K}_{(\mathrm{H,V})}=\begin{bmatrix}\frac{1}{2}\left(|S_{\mathrm{HH}}|^2+|S_{\mathrm{HV}}|^2+|S_{\mathrm{VH}}|^2+|S_{\mathrm{VV}}|^2\right) & \frac{1}{2}\left(|S_{\mathrm{HH}}|^2-|S_{\mathrm{HV}}|^2+|S_{\mathrm{VH}}|^2-|S_{\mathrm{VV}}|^2\right) & \mathrm{Re}\left[S_{\mathrm{HH}}S^{*}_{\mathrm{HV}}+S_{\mathrm{VH}}S^{*}_{\mathrm{VV}}\right] & \mathrm{Im}\left[S_{\mathrm{HH}}S^{*}_{\mathrm{HV}}+S_{\mathrm{VH}}S^{*}_{\mathrm{VV}}\right]\\ \frac{1}{2}\left(|S_{\mathrm{HH}}|^2-|S_{\mathrm{HV}}|^2+|S_{\mathrm{VH}}|^2-|S_{\mathrm{VV}}|^2\right) & \frac{1}{2}\left(|S_{\mathrm{HH}}|^2-|S_{\mathrm{HV}}|^2-|S_{\mathrm{VH}}|^2+|S_{\mathrm{VV}}|^2\right) & \mathrm{Re}\left[S_{\mathrm{HH}}S^{*}_{\mathrm{HV}}-S_{\mathrm{VH}}S^{*}_{\mathrm{VV}}\right] & \mathrm{Im}\left[S_{\mathrm{HH}}S^{*}_{\mathrm{HV}}-S_{\mathrm{VH}}S^{*}_{\mathrm{VV}}\right]\\ \mathrm{Re}\left[S_{\mathrm{HH}}S^{*}_{\mathrm{HV}}+S_{\mathrm{VH}}S^{*}_{\mathrm{VV}}\right] & \mathrm{Re}\left[S_{\mathrm{HH}}S^{*}_{\mathrm{HV}}-S_{\mathrm{VH}}S^{*}_{\mathrm{VV}}\right] & \mathrm{Re}\left[S_{\mathrm{HH}}S^{*}_{\mathrm{VV}}+S_{\mathrm{HV}}S^{*}_{\mathrm{VH}}\right] & \mathrm{Re}\left[S_{\mathrm{HH}}S^{*}_{\mathrm{VV}}+S_{\mathrm{VH}}S^{*}_{\mathrm{HV}}\right]\\ \mathrm{Im}\left[S_{\mathrm{HH}}S^{*}_{\mathrm{HV}}+S_{\mathrm{VH}}S^{*}_{\mathrm{VV}}\right] & \mathrm{Im}\left[S_{\mathrm{HH}}S^{*}_{\mathrm{HV}}-S_{\mathrm{VH}}S^{*}_{\mathrm{VV}}\right] & \mathrm{Re}\left[S_{\mathrm{HH}}S^{*}_{\mathrm{VV}}+S_{\mathrm{VH}}S^{*}_{\mathrm{HV}}\right] & -\mathrm{Re}\left[S_{\mathrm{HH}}S^{*}_{\mathrm{VV}}-S_{\mathrm{HV}}S^{*}_{\mathrm{VH}}\right]\end{bmatrix}\tag{1.3.40}$$

$$\mathbf{M}_{(\mathrm{H,V})}=\begin{bmatrix}\frac{1}{2}\left(|S_{\mathrm{HH}}|^2+|S_{\mathrm{HV}}|^2+|S_{\mathrm{VH}}|^2+|S_{\mathrm{VV}}|^2\right) & \frac{1}{2}\left(|S_{\mathrm{HH}}|^2-|S_{\mathrm{HV}}|^2+|S_{\mathrm{VH}}|^2-|S_{\mathrm{VV}}|^2\right) & \mathrm{Re}\left[S_{\mathrm{HH}}S^{*}_{\mathrm{HV}}+S_{\mathrm{VH}}S^{*}_{\mathrm{VV}}\right] & \mathrm{Im}\left[S_{\mathrm{HH}}S^{*}_{\mathrm{HV}}+S_{\mathrm{VH}}S^{*}_{\mathrm{VV}}\right]\\ \frac{1}{2}\left(|S_{\mathrm{HH}}|^2+|S_{\mathrm{HV}}|^2-|S_{\mathrm{VH}}|^2-|S_{\mathrm{VV}}|^2\right) & \frac{1}{2}\left(|S_{\mathrm{HH}}|^2-|S_{\mathrm{HV}}|^2-|S_{\mathrm{VH}}|^2+|S_{\mathrm{VV}}|^2\right) & \mathrm{Re}\left[S_{\mathrm{HH}}S^{*}_{\mathrm{HV}}-S_{\mathrm{VH}}S^{*}_{\mathrm{VV}}\right] & \mathrm{Im}\left[S_{\mathrm{HH}}S^{*}_{\mathrm{HV}}-S_{\mathrm{VH}}S^{*}_{\mathrm{VV}}\right]\\ \mathrm{Re}\left[S_{\mathrm{HH}}S^{*}_{\mathrm{VH}}+S_{\mathrm{HV}}S^{*}_{\mathrm{VV}}\right] & \mathrm{Re}\left[S_{\mathrm{HH}}S^{*}_{\mathrm{VH}}-S_{\mathrm{HV}}S^{*}_{\mathrm{VV}}\right] & \mathrm{Re}\left[S_{\mathrm{HH}}S^{*}_{\mathrm{VV}}+S_{\mathrm{HV}}S^{*}_{\mathrm{VH}}\right] & \mathrm{Re}\left[S_{\mathrm{HH}}S^{*}_{\mathrm{VV}}+S_{\mathrm{VH}}S^{*}_{\mathrm{HV}}\right]\\ -\mathrm{Im}\left[S_{\mathrm{HH}}S^{*}_{\mathrm{VH}}+S_{\mathrm{HV}}S^{*}_{\mathrm{VV}}\right] & -\mathrm{Im}\left[S_{\mathrm{HH}}S^{*}_{\mathrm{VH}}-S_{\mathrm{HV}}S^{*}_{\mathrm{VV}}\right] & -\mathrm{Im}\left[S_{\mathrm{HH}}S^{*}_{\mathrm{VV}}-S_{\mathrm{VH}}S^{*}_{\mathrm{HV}}\right] & \mathrm{Re}\left[S_{\mathrm{HH}}S^{*}_{\mathrm{VV}}-S_{\mathrm{HV}}S^{*}_{\mathrm{VH}}\right]\end{bmatrix}\tag{1.3.41}$$

where $\mathrm{Re}[\cdot]$ and $\mathrm{Im}[\cdot]$ are to obtain the real and imaginary part of an entity.

1.4 SUMMARY

With the massive acquisitions of airborne and spaceborne PolSAR data in terms of different polarizations, different frequencies, and different imaging modes, and with the rapid developments of basic theories and tools for electromagnetic scattering interpretation, we are now in the golden age of imaging radar and microwave remote sensing. The understanding and utilization of microwave remote sensing information is also moving from the qualitative stage to the quantitative stage. However, how to accurately, robustly, and efficiently extract target information and how to use it for applications are still the main scientific and technical challenges. At the same time, these challenges also promote in-depth and innovative developments in the fields of electromagnetic wave propagation and scattering, target scattering characterization, modeling and interpretation, and so on.

From the previous introduction, the developments of SAR in imaging radar are mostly rapid, and its applications are mostly extensive. Therefore, this book mainly focuses on full-polarization SAR to introduce the new advances in the field of radar polarimetry: the recently established polarimetric rotation domain interpretation theory and its advancements for imaging radar target scattering understanding and applications. Certainly, in addition to PolSAR, polarimetric rotation domain interpretation theory can also be extended for other typical polarimetric radar imaging systems.

In general, this book focuses on the introduction of the developments of polarimetric rotation domain interpretation theory and its applications. It is intended to serve as a resource for scientists, engineers, and graduate students in related disciplines and to collectively foster further advancements in imaging radar polarimetric information processing and applications. This book includes seven chapters. Chapter 1 is an overview, which mainly introduces the development and basic knowledge of radar polarimetry and imaging radar. Also, it briefly introduces the contents of this book. Chapter 2 introduces polarimetric rotation domain interpretation theory. Chapter 3 introduces the polarimetric rotation domain roll-invariant features. Chapters 4–7 introduce the applications of polarimetric rotation domain theory. In detail, Chapter 4 introduces polarimetric rotation domain land cover classification. Chapter 5 introduces polarimetric rotation domain target detection. Chapter 6 introduces polarimetric rotation domain target structure recognition. Chapter 7 introduces polarimetric rotation domain urban damage mapping.

REFERENCES

1. M. L. Skolnik, *Introduction to Radar Systems (Third Edition)*. New York: McGraw-Hill, 2001.
2. L. C. Shen and J. A. Kong, *Applied Electromagnetism (Third Edition)*. Monterey, CA: CL-Engineering, 1995.
3. J. S. Lee and E. Pottier, *Polarimetric Radar Imaging: From Basics to Applications*. Boca Raton: CRC Press, 2009.
4. G. Sinclair, "The transmission and reception of elliptically polarized waves," *Proceedings of the Institute of Radio Engineers,* vol. 38, no. 2, pp. 148–151, 1950.
5. W. M. Boerner, *Historical Development of Radar Polarimetry, Incentives for This Workshop, and Overview of Contributions to These Proceedings*. Amsterdam, The Netherlands: Kluwer Academic Publisher, 1992.

6. E. M. Kennaugh, "Polarization properties of radar reflection," Ph.D. dissertation, The Ohio State University, USA, 1952.
7. A. Guissard, "Mueller and Kennaugh matrices in radar polarimetry," *IEEE Transactions on Geoscience and Remote Sensing,* vol. 32, no. 3, pp. 590–597, 1994.
8. C. D. Graves, "Radar polarization power scattering matrix," *Proceedings of the IRE,* vol. 44, no. 2, pp. 248–252, 1956.
9. J. R. Huynen, "Phenomenological theory of radar targets," Ph.D. dissertation, Technical University of Delft, Netherlands, 1970.
10. A. B. Kostinski and W. M. Boerner, "On foundations of radar polarimetry," *IEEE Transactions on Antennas and Propagation,* vol. 34, no. 12, pp. 1395–1404, 1986.
11. W. M. Boerner, W. L. Yan, A. Q. Xi, and Y. Yamaguchi, "On the basic principles of radar polarimetry-the target characteristic polarization state theory of Kennaugh, Huynen's polarization fork concept, and its extension to the partially polarized case," *Proceedings of the IEEE,* vol. 79, no. 10, pp. 1538–1550, 1991.
12. M. Boerner, M. B. Elarini, and S. S. Saatchi, "Utilization of the optimal polarization concept in radar meteorology," *Bulletin of the American Meteorological Society,* vol. 62, no. 6, pp. 941–941, 1981.
13. W. M. Boerner, "Use of polarization in electromagnetic inverse scattering," *Radio Science,* vol. 16, no. 6, pp. 1037–1045, 1981.
14. W. M. Boerner, M. B. Elarini, C. Y. Chan, and P. M. Mastoris, "Polarization dependence in electromagnetic inverse problems," *IEEE Transactions on Antennas and Propagation,* vol. 29, no. 2, pp. 262–271, 1981.
15. W. M. Boerner and A. Q. Xi, "The characteristic radar target polarization state theory for the coherent monostatic and reciprocal case using the generalized polarization transformation ratio formulation," *International Journal of Electronics and Communications,* vol. 44, no. 4, pp. 273–281, 1990.
16. A. P. Agrawal and W. M. Boerner, "Redevelopment of Kennaugh target characteristic polarization state theory using the polarization transformation ratio formalism for the coherent case," *IEEE Transactions on Geoscience and Remote Sensing,* vol. 27, no. 1, pp. 3–14, 1989.
17. M. Davidovitz and W. M. Boerner, "Extension of Kennaugh optimal polarization concept to the asymmetric scattering matrix case," *IEEE Transactions on Antennas and Propagation,* vol. 34, no. 4, pp. 569–574, 1986.
18. Y. A. Ke, "Radar scattering matrix and polarization matching reception," *Acta Electronica Sinica*, no. 3, pp. 1–11, 1963.
19. Z. W. Zhuang, "Radar target recognition in the polarimetric domain and frequency domains," Ph.D. dissertation, Beijing Institute of Technology, Beijing, 1989.
20. S. P. Xiao, "Target recognition theory and application of wideband polarimetric radar," Ph.D. dissertation, National University of Defense Technology, Changsha, 1995.
21. X. S. Wang, "Study on wide band polarization information processing," Ph.D. dissertation, National University of Defense Technology, Changsha, 1999.
22. J. Yang, "On theoretical problems in radar polarimetry," Ph.D. dissertation, Niigata University, Japan, 1999.
23. Z. W. Zhuang, S. P. Xiao, and X. S. Wang, *Radar Polarization Information Processing and Application*. Beijing: National Defense Industry Press, 1999.
24. E. Pottier, "On Dr J. R. Huynen's main contributions in the development of polarimetric radar technique," *Proceedings of the SPIE,* vol. 1748, pp. 72–85, 1992.
25. W. M. Boerner, "Recent advances in extra-wide-band polarimetry, interferometry and polarimetric interferometry in synthetic aperture remote sensing and its applications," *IEE Proceedings-Radar Sonar and Navigation,* vol. 150, no. 3, pp. 113–124, 2003.
26. D. Giuli, "Polarization diversity in radars," *Proceedings of the IEEE,* vol. 74, no. 2, pp. 245–269, 1986.
27. H. A. Zebker and J. J. Van Zyl, "Imaging radar polarimetry-a review," *Proceedings of the IEEE*, vol. 79, no. 11, pp. 1583–1606, 1991.

28. C. Wiley, "Pulse Doppler radar method and means," USA Patent, 1954.
29. L. C. Graham, "Synthetic interferometer radar for topographic mapping," *Proceedings of the IEEE,* vol. 62, no. 6, pp. 763–768, 1974.
30. A. K. Gabriel, R. M. Goldstein, and H. A. Zebker, "Mapping small elevation changes over large areas-differential radar interferometry," *Journal of Geophysical Research-Solid Earth and Planets,* vol. 94, no. B7, pp. 9183–9191, 1989.
31. R. Bamler and P. Hartl, "Synthetic aperture radar interferometry," *Inverse Problems,* vol. 14, no. 4, pp. R1–R54, 1998.
32. P. A. Rosen, S. Hensley, I. R. Joughin, F. K. Li, S. N. Madsen, E. Rodriguez, et al., "Synthetic aperture radar interferometry-Invited paper," *Proceedings of the IEEE,* vol. 88, no. 3, pp. 333–382, 2000.
33. G. Krieger, I. Hajnsek, K. P. Papathanassiou, M. Younis, and A. Moreira, "Interferometric synthetic aperture radar (SAR) missions employing formation flying," *Proceedings of the IEEE,* vol. 98, no. 5, pp. 816–843, 2010.
34. K. P. Papathanassiou, "Polarimetric SAR interferometry," Ph.D. dissertation, Technique University of Graz, Austria, 1999.
35. S. R. Cloude and K. P. Papathanassiou, "Polarimetric SAR interferometry," *IEEE Transactions on Geoscience and Remote Sensing,* vol. 36, no. 5, pp. 1551–1565, 1998.
36. K. P. Papathanassiou and S. R. Cloude, "Single-baseline polarimetric SAR interferometry," *IEEE Transactions on Geoscience and Remote Sensing,* vol. 39, no. 11, pp. 2352–2363, 2001.
37. S. R. Cloude and K. P. Papathanassiou, "Three-stage inversion process for polarimetric SAR interferometry," *IEE Proceedings-Radar Sonar and Navigation,* vol. 150, no. 3, pp. 125–134, 2003.
38. A. Reigber and A. Moreira, "First demonstration of airborne SAR tomography using multibaseline L-band data," *IEEE Transactions on Geoscience and Remote Sensing,* vol. 38, no. 5, pp. 2142–2152, 2000.
39. A. Reigber, "Airborne polarimetric SAR tomography," Ph.D. dissertation, University of Stuttgart, Germany, 2001.
40. F. Xu and Y. Q. Jin, "Automatic reconstruction of building objects from multiaspect meter-resolution SAR images," *IEEE Transactions on Geoscience and Remote Sensing,* vol. 45, no. 7, pp. 2336–2353, 2007.
41. M. Rodriguez Cassola, P. Prats, D. Schulze, N. Tous-Ramon, U. Steinbrecher, L. Marotti, et al., "First bistatic spaceborne SAR experiments with TanDEM-X," *IEEE Geoscience and Remote Sensing Letters,* vol. 9, no. 1, pp. 33–37, 2012.
42. O. Ponce, P. Prats, M. Rodriguez-Cassola, R. Scheiber, and A. Reigber, "Processing of circular SAR trajectories with fast factorized back-projection," in *IEEE International Geoscience and Remote Sensing Symposium*, Vancouver, 2011, 24–29 July, pp. 3692–3695.
43. Y. R. Wu, "Concept on multidimensional space joint-observation SAR," *Journal of Radars,* vol. 2, no. 2, pp. 135–142, 2013.
44. Z. S. Zhou, W. M. Boerner, and M. Sato, "Development of a ground-based polarimetric broadband SAR system for noninvasive ground-truth validation in vegetation monitoring," *IEEE Transactions on Geoscience and Remote Sensing,* vol. 42, no. 9, pp. 1803–1810, 2004.
45. I. G. Cumming and F. H. Wong, *Digital Processing of Synthetic Aperture Radar Data: Algorithms and Implementation*. Boston, London: Artech House, 2005.
46. S. W. Chen, X. S. Wang, S. P. Xiao, and M. Sato, *Target Scattering Mechanism in Polarimetric Synthetic Aperture Radar-Interpretation and Application*. Singapore: Springer, 2018.
47. J. J. van Zyl and Y. Kim, *Synthetic Aperture Radar Polarimetry*. Hoboken, NJ: Wiley, 2011.

48. Y. Q. Jin and F. Xu, *Polarization Scattering and SAR Remote Sensing Information Theory and Method*. Beijing: Science Press, 2008.
49. H. Zhang, C. Wang, M. Liu, and H. Z. Li, *Polarimetric SAR Theory, Method and Application*. Beijing: Science Press, 2015.
50. J. Yang and J. J. Ying, *Polarization Radar Theory and Remote Sensing Application*. Beijing: Science Press, 2020.
51. Y. Yamaguchi, *Polarimetric SAR Imaging: Theory and Applications*. Boca Raton, FL: CRC Press, 2020.
52. S. R. Cloude, *Polarisation Application in Remote Sensing*. New York: Oxford University Press, 2009.
53. D. L. Schuler, T. L. Ainsworth, J. S. Lee, and G. De Grandi, "Topographic mapping using polarimetric SAR data," *International Journal of Remote Sensing,* vol. 19, no. 1, pp. 141–160, 1998.
54. F. Garestier, P. Dubois Fernandez, X. Dupuis, P. Paillou, and I. Hajnsek, "PolInSAR analysis of X-band data over vegetated and urban areas," *IEEE Transactions on Geoscience and Remote Sensing,* vol. 44, no. 2, pp. 356–364, 2006.
55. C. L. Martínez and K. P. Papathanassiou, "Cancellation of scattering mechanisms in PolInSAR: Application to underlying topography estimation," *IEEE Transactions on Geoscience and Remote Sensing,* vol. 51, no. 2, pp. 953–965, 2013.
56. I. Hajnsek, E. Pottier, and S. R. Cloude, "Inversion of surface parameters from polarimetric SAR," *IEEE Transactions on Geoscience and Remote Sensing,* vol. 41, no. 4, pp. 727–744, 2003.
57. V. Akbari and C. Brekke, "Iceberg detection in open and ice-infested waters using C-Band polarimetric synthetic aperture radar," *IEEE Transactions on Geoscience and Remote Sensing,* vol. 56, no. 1, pp. 407–421, 2018.
58. M. Sato, S. W. Chen, and M. Satake, "Polarimetric SAR analysis of tsunami damage following the March 11, 2011 East Japan earthquake," *Proceedings of the IEEE,* vol. 100, no. 10, pp. 2861–2875, 2012.
59. S. W. Chen and M. Sato, "Tsunami damage investigation of built-up areas using multitemporal spaceborne full polarimetric SAR images," *IEEE Transactions on Geoscience and Remote Sensing,* vol. 51, no. 4, pp. 1985–1997, 2013.
60. S. W. Chen, X. S. Wang, and M. Sato, "Urban damage level mapping based on scattering mechanism investigation using fully polarimetric SAR data for the 3.11 East Japan earthquake," *IEEE Transactions on Geoscience and Remote Sensing,* vol. 54, no. 12, pp. 6919–6929, 2016.
61. F. Cigna, B. Osmanoglu, E. Cabral-Cano, T. H. Dixon, J. A. Avila-Olivera, V. H. Garduno-Monroy, et al., "Monitoring land subsidence and its induced geological hazard with synthetic aperture radar interferometry: A case study in Morelia, Mexico," *Remote Sensing of Environment,* vol. 117, pp. 146–161, 2012.
62. J. C. Souyris, P. Imbo, R. Fjortoft, S. Mingot, and J. S. Lee, "Compact polarimetry based on symmetry properties of geophysical media: The pi/4 mode," *IEEE Transactions on Geoscience and Remote Sensing,* vol. 43, no. 3, pp. 634–646, 2005.
63. R. K. Raney, "Hybrid-polarity SAR architecture," *IEEE Transactions on Geoscience and Remote Sensing,* vol. 45, no. 11, pp. 3397–3404, 2007.
64. N. Stacy, "Compact polarimetric analysis of X-band SAR data," in *European Conference on Synthetic Aperture Radar*, Dresden, Germany, 2006, 16–18 May, pp. 1–4.

2 Polarimetric Rotation Domain Interpretation Theory

2.1 INTRODUCTION

Polarimetric radar which can acquire full-polarization information has become mainstream in microwave remote sensing [1]. Exploration and utilization of polarization diversity in microwave remote sensing have advantages to better interpret scattering mechanisms and lead to many successful applications, including crop monitoring [2–4], target detection and classification [1, 5–11], natural disaster evaluation [12–16], and so on.

Polarimetric scattering response is closely related to the attitude and orientation of the target [17, 18], which is known as the target scattering diversity. The target scattering diversity effect contains main two aspects. First, the same target may exhibit quite different scattering characteristics under different observation directions. Second, different targets may share similar scattering mechanisms under specific illumination angles. For conventional model-based polarimetric target decompositions, the scattering models usually do not account for target scattering diversity [19]. Meanwhile, these polarimetric target decompositions often suffer from scattering mechanism interpretation ambiguity [19]. In this vein, eigenvalue-eigenvector-based decompositions and derived polarimetric roll-invariant features are popularly preferred for land cover classification application [1, 8, 20]. In addition, in order to mitigate scattering mechanism ambiguity, the concept of orientation compensation, also known as deorientation processing, has been emphasized and incorporated into model-based polarimetric target decomposition schemes, and improved decomposition results can be obtained. The commonly used orientation compensation approach is to minimize the power of the cross-polarization term by rotating the polarimetric coherency matrix at a specific rotation angle. Theoretically, this rotation angle is equivalent to the polarization orientation angle [18, 21, 22]. There are many methods for polarization orientation estimation [17]. Although improved performance is achieved, as first pointed out in [18], scattering mechanism ambiguity is still serious for manmade targets with large orientation angles from polarimetric target decompositions.

On the other hand, although target scattering diversity effect makes polarimetric radar data modeling and interpretation difficult, it also contains rich and valuable information. In this vein, instead of traditional orientation compensation to partially mitigate this target scattering diversity effect, a more suitable solution is to explore

DOI: 10.1201/9781003461296-2

and utilize the embedded hidden information. Meanwhile, proper exploration of such target scattering diversity can provide insights for new physical feature retrieval and powerful tools for revealing different targets' intrinsic properties. This is the main motivation of the developed polarimetric rotation domain interpretation theory.

The core idea of the developed polarimetric rotation domain interpretation theory is to extend the initial polarimetric information obtained by the polarimetric radar at a given observation geometry to the polarimetric rotation domain along the radar line of sight and to propose a series of interpretation tools for target scattering diversity representation, characterization, and utilization. From the mathematical processing viewpoint, polarimetric rotation domain interpretation theory is to extend the acquired polarimetric matrices at a given rotation state $(\theta = 0)$ to the polarimetric rotation domain $(\theta \in [-\pi, \pi))$. Mathematically, the rotation transformation of polarimetric matrices does not bring information gains by itself. However, from the electromagnetic scattering perspective, the target scattering diversity effect and the physical mechanism that different targets normally exhibit various scattering diversity effects can make the target electromagnetic scattering information present significant differences in the polarimetric rotation domain or can enhance the differentiations of various targets' scattering features. Essentially, the differences within the target scattering diversity are a kind of valuable information, which can be mined and utilized with the proper methods. These are physical foundations of the developed polarimetric rotation domain interpretation theory. Since 2012, when the concept of the polarimetric rotation domain interpretation was first proposed [23], with more than 10 years of further developments and accumulations by the author and the author's research team, a relatively complete polarimetric rotation domain interpretation framework has been established. It mainly includes uniform polarimetric matrix rotation theory [24–26], a two-dimension polarimetric coherence pattern interpretation tool [26–29], two-dimension polarimetric correlation pattern interpretation tool [30, 31], and three-dimension polarimetric correlation pattern interpretation tool [32, 33]. It has been successfully applied in the fields of land cover classification [4, 34–38], target detection [9, 30, 31, 39–45], structure recognition [46–55], disaster evaluation [56–58], and so on. This chapter will focus on the introduction of the polarimetric rotation domain interpretation theory and tools.

2.2 POLARIMETRIC MATRIX IN THE POLARIMETRIC ROTATION DOMAIN

In order to investigate target scattering diversity, the author introduced the concept of polarimetric rotation domain interpretation in 2012 [23]. The core idea is to extend a polarimetric matrix acquired at a given geometry to the polarimetric rotation domain along the radar line of sight. Then, with the development of suitable interpretation tools, rich and hidden information embedded in target scattering diversity effects can be explored, which can provide new insights and technique solutions for many polarimetric radar applications.

2.2.1 Polarimetric Matrix Rotation

In the polarimetric rotation domain, the polarimetric scattering matrix becomes

$$\mathbf{S}(\theta)=\mathbf{R}_2(\theta)\mathbf{S}\mathbf{R}_2^{\mathrm{T}}(\theta),\theta\in[-\pi,\pi) \tag{2.2.1}$$

where the superscript ^{T} denotes the transpose, and the rotation matrix is $\mathbf{R}_2(\theta)=\begin{bmatrix}\cos\theta & \sin\theta\\ -\sin\theta & \cos\theta\end{bmatrix}$.

For the (H, V) polarization basis, the elements of $\mathbf{S}(\theta)$ are

$$S_{\mathrm{HH}}(\theta)=S_{\mathrm{HH}}\cos^2\theta+S_{\mathrm{HV}}\cos\theta\sin\theta+S_{\mathrm{VH}}\cos\theta\sin\theta+S_{\mathrm{VV}}\sin^2\theta \tag{2.2.2}$$

$$S_{\mathrm{HV}}(\theta)=-S_{\mathrm{HH}}\cos\theta\sin\theta+S_{\mathrm{HV}}\cos^2\theta-S_{\mathrm{VH}}\sin^2\theta+S_{\mathrm{VV}}\cos\theta\sin\theta \tag{2.2.3}$$

$$S_{\mathrm{VH}}(\theta)=-S_{\mathrm{HH}}\cos\theta\sin\theta-S_{\mathrm{HV}}\sin^2\theta+S_{\mathrm{VH}}\cos^2\theta+S_{\mathrm{VV}}\cos\theta\sin\theta \tag{2.2.4}$$

$$S_{\mathrm{VV}}(\theta)=S_{\mathrm{HH}}\sin^2\theta-S_{\mathrm{HV}}\cos\theta\sin\theta-S_{\mathrm{VH}}\cos\theta\sin\theta+S_{\mathrm{VV}}\cos^2\theta \tag{2.2.5}$$

The polarimetric coherency matrix in the polarimetric rotation domain is

$$\mathbf{T}(\theta)=\mathbf{R}_3(\theta)\mathbf{T}\mathbf{R}_3^{\mathrm{T}}(\theta),\theta\in(-\pi,\pi) \tag{2.2.6}$$

where the rotation matrix is $\mathbf{R}_3(\theta)=\begin{bmatrix}1 & 0 & 0\\ 0 & \cos 2\theta & \sin 2\theta\\ 0 & -\sin 2\theta & \cos 2\theta\end{bmatrix}$.

The elements of $\mathbf{T}(\theta)$ are

$$T_{11}(\theta)=T_{11} \tag{2.2.7}$$

$$T_{12}(\theta)=T_{12}\cos 2\theta+T_{13}\sin 2\theta \tag{2.2.8}$$

$$T_{13}(\theta)=-T_{12}\sin 2\theta+T_{13}\cos 2\theta \tag{2.2.9}$$

$$T_{23}(\theta)=\frac{1}{2}(T_{33}-T_{22})\sin 4\theta+\mathrm{Re}[T_{23}]\cos 4\theta+\mathrm{j}\,\mathrm{Im}[T_{23}] \tag{2.2.10}$$

$$T_{22}(\theta)=T_{22}\cos^2 2\theta+T_{33}\sin^2 2\theta+\mathrm{Re}[T_{23}]\sin 4\theta \tag{2.2.11}$$

$$T_{33}(\theta)=T_{22}\sin^2 2\theta+T_{33}\cos^2 2\theta-\mathrm{Re}[T_{23}]\sin 4\theta \tag{2.2.12}$$

The powers of the off-diagonal terms are

$$|T_{12}(\theta)|^2=|T_{12}|^2\cos^2 2\theta+|T_{13}|^2\sin^2 2\theta+\mathrm{Re}\left[T_{12}T_{13}^*\right]\sin 4\theta \tag{2.2.13}$$

$$|T_{13}(\theta)|^2 = |T_{12}|^2 \sin^2 2\theta + |T_{13}|^2 \cos^2 2\theta - \mathrm{Re}\left[T_{12}T_{13}^*\right]\sin 4\theta \tag{2.2.14}$$

$$\begin{aligned} |T_{23}(\theta)|^2 &= \frac{1}{4}(T_{33} - T_{22})^2 \sin^2 4\theta + \mathrm{Re}^2[T_{23}]\cos^2 4\theta \\ &+ \frac{1}{2}(T_{33} - T_{22})\mathrm{Re}[T_{23}]\sin 8\theta + \mathrm{Im}^2[T_{23}] \end{aligned} \tag{2.2.15}$$

where $\mathrm{Re}\left[T_{ij}\right]$ and $\mathrm{Im}\left[T_{ij}\right]$ are the real and imaginary parts of T_{ij}.

2.2.2 Cascade Rotation Property

For the rotation matrix $\mathbf{R}_3(\theta)$, the cascade rotation property can be derived

$$\mathbf{R}_3(\theta_1 + \theta_2) = \mathbf{R}_3(\theta_1)\mathbf{R}_3(\theta_2) = \mathbf{R}_3(\theta_2)\mathbf{R}_3(\theta_1) \tag{2.2.16}$$

$\mathbf{R}_3(\theta)$ belongs to the special unitary group, and this property can also be found in the group theory and polarization algebra [1, 59]. The proof of (2.2.16) is straightforward.

The cascade rotation property for a polarimetric coherency matrix $\mathbf{T}$ can be described as follows:

$$\begin{aligned} \mathbf{T}(\theta_1 + \theta_2) &= \mathbf{R}_3(\theta_1 + \theta_2)\mathbf{T}\mathbf{R}_3^{\mathrm{T}}(\theta_1 + \theta_2) \\ &= \mathbf{R}_3(\theta_1)\mathbf{T}(\theta_2)\mathbf{R}_3^{\mathrm{T}}(\theta_1) \\ &= \mathbf{R}_3(\theta_2)\mathbf{T}(\theta_1)\mathbf{R}_3^{\mathrm{T}}(\theta_2) \end{aligned} \tag{2.2.17}$$

From (2.2.17), it is clear that rotating $\mathbf{T}$ with angle $\theta_1 + \theta_2$ is equivalent to rotating $\mathbf{T}$ through angle θ_1 (θ_2) at first and then θ_2 (θ_1) afterwards. This property indicates that after a further rotation with angle θ_2 (θ_1), the characters of the matrix $\mathbf{T}(\theta_1)$ $(\mathbf{T}(\theta_2))$ cannot be maintained by the rotated matrix $\mathbf{T}(\theta_1 + \theta_2)$ in a period of the rotation angle θ. For example, if rotating $\mathbf{T}$ with angle θ_1, $\mathbf{T}(\theta_1)$ achieves the minimum cross-polarization term. Then, if further rotating $\mathbf{T}(\theta_1)$ with angle θ_2, the rotated matrix becomes $\mathbf{T}(\theta_1 + \theta_2)$, and the cross-polarization term is no longer minimized. In other words, the properties of $\mathbf{T}(\theta)$ will be relatively independent at each rotation state.

2.3 UNIFORM POLARIMETRIC MATRIX ROTATION THEORY

The target scattering diversity makes scattering mechanism modeling and interpretation more difficult. In order to reduce orientation dependence and smooth scattering mechanism ambiguity, the orientation compensation technique [17, 60, 61], also known as deorientation processing [18, 62, 63], has been incorporated into model-based polarimetric target decompositions, and improved decomposition results can be achieved [18, 19, 22, 64, 65]. The commonly used approach for orientation compensation is to minimize the power of the cross-polarization term by rotating the polarimetric coherency matrix at a specific rotation angle. Therefore, orientation compensation processing and the derivation of the polarization orientation angle can be unified in the polarimetric coherency matrix rotation procedure.

The impacts of orientation compensation processing on the polarimetric coherency matrix and polarimetric target decomposition have been comprehensively studied [18, 21, 22]. The used rotation angle from orientation compensation processing can be made available by polarimetric coherency matrix rotation with a rotation angle that can minimize the cross-polarization term. Thereafter, valuable questions worth thinking about are as follows. What is the variation law of each element of a polarimetric coherency matrix if some other rotation angles are used? How can we reveal, represent, and utilize this variation law? These thoughts stimulate development of the uniform polarimetric matrix rotation theory. The core idea is to extend the polarimetric matrix to the polarimetric rotation domain. Then, the uniform expression of the polarimetric matrix elements in the polarimetric rotation domain can be revealed, and a series of new polarimetric features can be derived. Mainly with the polarimetric coherency matrix representation, this section will introduce uniform polarimetric matrix rotation theory [24–26].

2.3.1 Uniform Representation

With simple mathematic transformations, all the elements of a rotated polarimetric coherency matrix can be generally represented as

$$f(\theta) = A\sin\left[\omega(\theta + \theta_0)\right] + B \tag{2.3.1}$$

where A is oscillation amplitude, B is oscillation center, ω is angular frequency, and θ_0 is initial angle. These features can form a new polarimetric feature set $\{A, B, \omega, \theta_0\}$.

Furthermore, expressions (2.2.7)–(2.2.15) can be divided into two categories mathematically. Among them, expressions (2.2.7)–(2.2.10) belong to the first category, while expressions (2.2.11)–(2.2.15) belong to the second class.

For the first category, expressions (2.2.7)–(2.2.10) share the same general form

$$f(\theta) = a\sin\omega_0\theta + b\cos\omega_0\theta + c \tag{2.3.2}$$

Then, the polarimetric feature set $\{A, B, \omega, \theta_0\}$ can be obtained as

$$\begin{cases} A = \sqrt{a^2 + b^2}, & B = c \\ \omega = \omega_0, & \theta_0 = \dfrac{1}{\omega}\text{Angle}(a + \mathrm{j}b) \end{cases} \tag{2.3.2}$$

where $\text{Angle}\{a\}$ is the operator to obtain the phase of a in the complex axis. The range of $\text{Angle}\{\cdot\}$ is $[-\pi, \pi)$. The advantage of this notation is to avoid phase wrapping when using the inverse trigonometric functions.

For the second category, expressions (2.2.11)–(2.2.15) share another uniform representation

$$f(\theta) = a\sin^2\omega_0\theta + b\cos^2\omega_0\theta + c\sin 2\omega_0\theta + d \tag{2.3.4}$$

Then, the polarimetric feature set $\{A, B, \omega, \theta_0\}$ can be obtained as

$$\begin{cases} A = \sqrt{c^2 + \frac{1}{4}(b-a)^2}, & B = \frac{1}{2}(a+b) + d \\ \omega = 2\omega_0, & \theta_0 = \frac{1}{\omega}\text{Angle}\left\{c + \mathrm{j}\frac{1}{2}(b-a)\right\} \end{cases} \tag{2.3.5}$$

Note that both (2.3.2) and (2.3.4) can be simplified and have a uniform representation as (2.3.1). Therefore, the polarimetric matrix rotation behavior for each element is completely characterized by the new polarimetric feature set $\{A, B, \omega, \theta_0\}$. The uniform representation for the polarimetric matrix rotation is thus developed. The main effect of polarimetric matrix rotation leads to the oscillation of each matrix element. Thereby, the new feature set $\{A, B, \omega, \theta_0\}$ is named the polarimetric rotation domain oscillation feature set. The derived feature set from the polarimetric coherency matrix is summarized in Table 2.3.1.

2.3.2 Polarimetric Rotation Domain Oscillation Features

The polarimetric rotation domain oscillation feature set contains rich information associated with polarimetric coherency matrix rotation effects. Intrinsically, the characteristics of these features directly relate to the scattering phenomena in the polarimetric rotation domain and can potentially reflect the target properties. From Table 2.3.1, the elements of a polarimetric coherency matrix can be classified into five groups:

1) $\text{Re}[T_{12}(\theta)]$ and $\text{Re}[T_{13}(\theta)]$;
2) $\text{Im}[T_{12}(\theta)]$ and $\text{Im}[T_{13}(\theta)]$;
3) $\text{Re}[T_{23}(\theta)]$, $T_{22}(\theta)$ and $T_{33}(\theta)$;
4) $|T_{12}(\theta)|^2$ and $|T_{13}(\theta)|^2$;
5) $|T_{23}(\theta)|^2$.

In each group, the polarimetric rotation domain oscillation features are the same or share equivalent information. Moreover, the following dependency relationships can be derived

TABLE 2.3.1
Polarimetric Rotation Domain Oscillation Feature Set

	Amplitude Features		Angle Features	
$\mathrm{T}(\theta)$ elements	$A=\sqrt{(\cdot)}$	B	ω	$\theta_0=\frac{1}{\omega}\text{Angle}\{\cdot\}$
$\mathrm{Re}\left[T_{12}(\theta)\right]$	$\mathrm{Re}^2[T_{12}]+\mathrm{Re}^2[T_{13}]$	0	2	$\mathrm{Re}[T_{13}]+\mathrm{j}\,\mathrm{Re}[T_{12}]$
$\mathrm{Re}\left[T_{13}(\theta)\right]$	$\mathrm{Re}^2[T_{12}]+\mathrm{Re}^2[T_{13}]$	0	2	$-\mathrm{Re}[T_{12}]+\mathrm{j}\,\mathrm{Re}[T_{13}]$
$\mathrm{Im}\left[T_{12}(\theta)\right]$	$\mathrm{Im}^2[T_{12}]+\mathrm{Im}^2[T_{13}]$	0	2	$\mathrm{Im}[T_{13}]+\mathrm{j}\,\mathrm{Im}[T_{12}]$
$\mathrm{Im}\left[T_{13}(\theta)\right]$	$\mathrm{Im}^2[T_{12}]+\mathrm{Im}^2[T_{13}]$	0	2	$-\mathrm{Im}[T_{12}]+\mathrm{j}\,\mathrm{Im}[T_{13}]$
$\mathrm{Re}\left[T_{23}(\theta)\right]$	$\frac{1}{4}(T_{33}-T_{22})^2+\mathrm{Re}^2[T_{23}]$	0	4	$\frac{1}{2}(T_{33}-T_{22})+\mathrm{j}\,\mathrm{Re}[T_{23}]$
$T_{22}(\theta)$	$\frac{1}{4}(T_{33}-T_{22})^2+\mathrm{Re}^2[T_{23}]$	$\frac{1}{2}(T_{22}+T_{33})$	4	$\mathrm{Re}[T_{23}]+\mathrm{j}\frac{1}{2}(T_{22}-T_{33})$
$T_{33}(\theta)$	$\frac{1}{4}(T_{33}-T_{22})^2+\mathrm{Re}^2[T_{23}]$	$\frac{1}{2}(T_{22}+T_{33})$	4	$-\mathrm{Re}[T_{23}]+\mathrm{j}\frac{1}{2}(T_{33}-T_{22})$
$\lvert T_{12}(\theta)\rvert^2$	$\mathrm{Re}^2\left[T_{12}T_{13}^*\right]+\frac{1}{4}\left(\lvert T_{13}\rvert^2-\lvert T_{12}\rvert^2\right)^2$	$\frac{1}{2}\left(\lvert T_{12}\rvert^2+\lvert T_{13}\rvert^2\right)$	4	$\mathrm{Re}\left[T_{12}T_{13}^*\right]+\mathrm{j}\frac{1}{2}\left(\lvert T_{12}\rvert^2-\lvert T_{13}\rvert^2\right)$
$\lvert T_{13}(\theta)\rvert^2$	$\mathrm{Re}^2\left[T_{12}T_{13}^*\right]+\frac{1}{4}\left(\lvert T_{13}\rvert^2-\lvert T_{12}\rvert^2\right)^2$	$\frac{1}{2}\left(\lvert T_{12}\rvert^2+\lvert T_{13}\rvert^2\right)$	4	$-\mathrm{Re}\left[T_{12}T_{13}^*\right]+\mathrm{j}\frac{1}{2}\left(\lvert T_{13}\rvert^2-\lvert T_{12}\rvert^2\right)$
$\lvert T_{23}(\theta)\rvert^2$	$\frac{1}{4}\left\{\frac{1}{4}(T_{33}-T_{22})^2+\mathrm{Re}^2[T_{23}]\right\}^2$	$\frac{1}{2}\left\{\frac{1}{4}(T_{33}-T_{22})^2+\mathrm{Re}^2[T_{23}]\right\}+\mathrm{Im}^2[T_{23}]$	8	$\frac{1}{2}(T_{33}-T_{22})\mathrm{Re}[T_{23}]+\mathrm{j}\frac{1}{2}\left[\mathrm{Re}^2[T_{23}]-\frac{1}{4}(T_{33}-T_{22})^2\right]$

$$\mathrm{Re}\left[T_{12}(\theta)\right]=\mathrm{Re}\left[T_{13}(\theta+\pi/4)\right] \tag{2.3.6}$$

$$\mathrm{Im}\left[T_{12}(\theta)\right]=\mathrm{Im}\left[T_{13}(\theta+\pi/4)\right] \tag{2.3.7}$$

$$T_{22}(\theta)=T_{33}(\theta+\pi/4)=\mathrm{Re}\left[T_{23}(\theta+\pi/8)\right]+B\_T_{22} \tag{2.3.8}$$

$$\left|T_{12}(\theta)\right|^2=\left|T_{13}(\theta+\pi/4)\right|^2 \tag{2.3.9}$$

where B_T_{ij} indicates the oscillation center B of $T_{ij}(\theta)$. The other terms A_T_{ij}, ω_T_{ij}, and $\theta_0\_T_{ij}$ can be similarly defined.

In the following investigations, only the features derived from $\mathrm{Re}\left[T_{12}(\theta)\right]$, $\mathrm{Im}\left[T_{12}(\theta)\right]$, $T_{22}(\theta)$, $\left|T_{12}(\theta)\right|^2$, and $\left|T_{23}(\theta)\right|^2$ are considered.

1) Oscillation Amplitude A

From Table 2.3.1, the number of independent oscillation amplitude feature is four. The feature $A\_\left|T_{23}\right|^2$ can be derived from A_T_{22} as

$$A\_\left|T_{23}\right|^2=\frac{1}{4}\left(A\_T_{22}\right)^2 \tag{2.3.10}$$

If rewriting the feature A_T_{22} with the elements of the polarimetric scattering matrix, it can be observed that A_T_{22} is sensitive to the reflection symmetry condition that the cross-correlations between co-polarization and cross-polarization terms approach zero $\left(\left\langle S_{\mathrm{HH}}S_{\mathrm{HV}}^*\right\rangle\approx\left\langle S_{\mathrm{VV}}S_{\mathrm{HV}}^*\right\rangle\approx 0\right)$. The expression of feature A_T_{22} is

$$\begin{aligned}A\_T_{22}&=\frac{1}{4}\left(T_{33}-T_{22}\right)^2+\mathrm{Re}^2\left[T_{23}\right]\\&=\frac{1}{4}\left(\left\langle\left|S_{\mathrm{HH}}-S_{\mathrm{VV}}\right|^2-4\left|S_{\mathrm{HV}}\right|^2\right\rangle\right)^2+4\left\{\mathrm{Re}\left[\left\langle\left(S_{\mathrm{HH}}-S_{\mathrm{VV}}\right)S_{\mathrm{HV}}^*\right\rangle\right]\right\}^2\end{aligned} \tag{2.3.11}$$

Basically, for natural distributed targets, such as grass, forests, and agriculture, the reflection symmetry condition is satisfied and A_T_{22} becomes

$$A\_T_{22}\approx\frac{1}{4}\left(\left\langle\left|S_{\mathrm{HH}}-S_{\mathrm{VV}}\right|^2-4\left|S_{\mathrm{HV}}\right|^2\right\rangle\right)^2 \tag{2.3.12}$$

However, for manmade targets, such as ships, airplanes, and buildings, the reflection symmetry condition is violated, and the term $\left\{\mathrm{Re}\left[\left\langle\left(S_{\mathrm{HH}}-S_{\mathrm{VV}}\right)S_{\mathrm{HV}}^*\right\rangle\right]\right\}^2$ can be relatively large. Consequently, the derived feature A_T_{22} has good performance to discriminate manmade and natural targets in theory. Besides, from (2.3.11), A_T_{22} is also polarimetric roll-invariant. Moreover, if investigating the feature A_T_{22} in depth, it is very interesting to find out that it also relates to the co-polarization correlation coefficient on a circular polarization basis.

2) Oscillation Center B

The oscillation center B is zero for $\mathrm{Re}\left[T_{12}(\theta)\right]$ and $\mathrm{Im}\left[T_{12}(\theta)\right]$, since their values vary symmetrically from the negative to the positive, while the power terms ($T_{22}(\theta)$, $\left|T_{12}(\theta)\right|^2$ and $\left|T_{23}(\theta)\right|^2$) all have positive oscillation centers. From Table 2.3.1, another dependency relationship is

$$B\_\left|T_{12}\right|^2=\frac{1}{2}\left(\left|T_{12}\right|^2+\left|T_{13}\right|^2\right)=\frac{1}{2}\left(A\_\mathrm{Re}\left[T_{12}\right]+A\_\mathrm{Im}\left[T_{12}\right]\right) \tag{2.3.13}$$

Therefore, the number of independent and inconstant oscillation center features is two.

3) Angular Frequency ω

The angular frequency features are constant and have three different values of $\omega\_\mathrm{Re}\left[T_{12}\right]=\omega\_\mathrm{Im}\left[T_{12}\right]=2$, $\omega\_T_{22}=\omega\_\left|T_{12}\right|^2=4$, and $\omega\_\left|T_{23}\right|^2=8$. Accordingly, the oscillation periods $2\pi/\omega$ are π, $\pi/2$, and $\pi/4$, respectively. Therefore, they are independent of scattering mechanisms for various scatters.

4) Initial Angle θ_0

From Table 2.3.1, the number of independent initial angle features θ_0 is five. In the context of the fully polarimetric acquisition, the phase information of an inner channel or cross-channel is sensitive to target scattering mechanisms. Several polarimetric angle features will be derived and investigated in the following sections.

2.3.3 Further Derived Polarimetric Rotation Domain Angle Features and Interpretation

In the polarimetric rotation domain, there are several interesting rotation angles which are worth further investigation: the stationary angle θ_{sta}; the minimization and maximization angles θ_{min}, θ_{max}; and the null angle θ_{null}. From the developed uniform representation (2.3.1), all these polarimetric angle features can be easily derived from the initial angle θ_0 and the angular frequency ω. Considering the periodic property of sinusoidal function, the following discussion is limited to the principal value range $\left[-\pi/\omega,\pi/\omega\right)$.

1) Stationary Angle θ_{sta}

The stationary angle θ_{sta} is the non-zero rotation angle which keeps the rotated element the same as that without rotation: $f\left(\theta_{\mathrm{sta}}\right)=f(0)$. The stationary angle θ_{sta} is

$$\theta_{\mathrm{sta}}=\begin{cases}\pi/\omega-\theta_0 & \text{if } 0\le\theta_0<\pi/\omega\\ -\pi/\omega-\theta_0 & \text{if } -\pi/\omega\le\theta_0<0\end{cases} \tag{2.3.14}$$

2) Minimization and Maximization Angles $\theta_{\min}$ and $\theta_{\max}$

The minimization and maximization angles $\theta_{\min}$ and $\theta_{\max}$ are two other interesting and important polarimetric rotation domain angle features, which minimize or maximize the elements in the polarimetric rotation domain: $f(\theta_{\min}) = -A + B$ and $f(\theta_{\max}) = A + B$, respectively. The minimization angle $\theta_{\min}$ is

$$\theta_{\min} = \begin{cases} 3\pi/2\omega - \theta_0 & \text{if } \pi/2\omega \le \theta_0 < \pi/\omega \\ -\pi/2\omega - \theta_0 & \text{if } -\pi/\omega \le \theta_0 < \pi/2\omega, \end{cases} \tag{2.3.15}$$

while the maximization angle $\theta_{\max}$ is

$$\theta_{\max} = \begin{cases} \pi/2\omega - \theta_0 & \text{if } \pi/2\omega \le \theta_0 < \pi/\omega \\ -3\pi/2\omega - \theta_0 & \text{if } -\pi/\omega \le \theta_0 < \pi/2\omega \end{cases} \tag{2.3.16}$$

These polarimetric angle features are available when inputting the corresponding ω and θ_0. The general characteristics of these angles are summarized as follows:

a) Since $T_{22}(\theta) + T_{33}(\theta)$ and $|T_{12}(\theta)|^2 + |T_{13}(\theta)|^2$ are polarimetric roll-invariant terms, the minimization angle $\theta_{\min}$ of $T_{22}(\theta)\left(|T_{12}(\theta)|^2\right)$ is the maximization angle $\theta_{\max}$ for $T_{33}(\theta)\left(|T_{13}(\theta)|^2\right)$, respectively, and vice versa;

b) $\theta_{\min}$ of $T_{33}(\theta)$ can minimize both $T_{33}(\theta)$ and $|T_{23}(\theta)|^2$, since $\mathrm{Im}[T_{23}]$ is polarimetric roll-invariant and $\theta_{\min}$ of $T_{33}(\theta)$ is equal to the null angle of $\mathrm{Re}[T_{23}]$.

In addition, the derived minimization angle $\theta_{\min}\_T_{33}$ is equivalent to the classic polarization orientation angle, which was originally derived from the polarimetric covariance matrix with the circular polarization basis [17, 60, 61]. Using deorientation theory [63], a polarimetric angle feature was obtained during the minimization procedure for the cross-polarization term T_{33}. This angle is theoretically equivalent to the polarization orientation angle. From the polarimetric matrix rotation viewpoint, both the polarization orientation angle [17, 60, 61] and rotation angle obtained from deorientation theory [63] can be unified by the uniform polarimetric matrix rotation theory.

3) Null Angle θ_{null}

The null angle θ_{null} is the rotation angle which rotates polarimetric matrix elements to be zero: $f(\theta_{\text{null}}) = 0$. Since $B > 0$ always holds for all the power terms, the null angle θ_{null} is only available for the real and imaginary parts of the off-diagonal terms. The null angle θ_{null} is

$$\theta_{\text{null}} = -\theta_0 \tag{2.3.17}$$

From (1.3.25), the elements T_{12} and T_{13} are

$$\begin{aligned} T_{12} &= \frac{1}{2}\left\langle (S_{\mathrm{HH}} + S_{\mathrm{VV}})(S_{\mathrm{HH}} - S_{\mathrm{VV}})^* \right\rangle \\ &= \frac{1}{2}\left(\left\langle |S_{\mathrm{HH}}|^2 - |S_{\mathrm{VV}}|^2 \right\rangle\right) + \mathrm{j}\,\mathrm{Im}\left[\left\langle S_{\mathrm{HH}}^* S_{\mathrm{VV}} \right\rangle\right] \end{aligned} \tag{2.3.18}$$

$$T_{13} = \left\langle (S_{\mathrm{HH}} + S_{\mathrm{VV}}) S_{\mathrm{HV}}^* \right\rangle \tag{2.3.19}$$

Thereby, the null angles of $\mathrm{Re}\left[T_{12}(\theta)\right]$ and $\mathrm{Im}\left[T_{12}(\theta)\right]$ are

$$\begin{aligned} \theta_{\mathrm{null}}\_\mathrm{Re}\left[T_{12}\right] &= -\frac{1}{2}\mathrm{Angle}\left\{\mathrm{Re}\left[T_{13}\right] + \mathrm{j}\,\mathrm{Re}\left[T_{12}\right]\right\} \\ &= \frac{1}{2}\mathrm{Angle}\left\{\mathrm{Re}\left[\left\langle (S_{\mathrm{HH}} + S_{\mathrm{VV}}) S_{\mathrm{HV}}^* \right\rangle\right] \right. \\ &\qquad \left. + \mathrm{j}\frac{1}{2}\left(\left\langle |S_{\mathrm{VV}}|^2 - |S_{\mathrm{HH}}|^2 \right\rangle\right)\right\} \end{aligned} \tag{2.3.20}$$

$$\begin{aligned} \theta_{\mathrm{null}}\_\mathrm{Im}\left[T_{12}\right] &= -\frac{1}{2}\mathrm{Angle}\left\{\mathrm{Im}\left[T_{13}\right] + \mathrm{j}\,\mathrm{Im}\left[T_{12}\right]\right\} \\ &= \frac{1}{2}\mathrm{Angle}\left\{\mathrm{Im}\left[\left\langle (S_{\mathrm{HH}} + S_{\mathrm{VV}}) S_{\mathrm{HV}}^* \right\rangle\right] + \mathrm{j}\,\mathrm{Im}\left[\left\langle S_{\mathrm{HH}} S_{\mathrm{VV}}^* \right\rangle\right]\right\} \end{aligned} \tag{2.3.21}$$

where the main range of $\theta_{\mathrm{null}}\_\mathrm{Re}\left[T_{12}\right]$ and $\theta_{\mathrm{null}}\_\mathrm{Im}\left[T_{12}\right]$ is $\left[-\pi/2, \pi/2\right)$.

When a polarimetric coherency matrix is rotated with angle $\theta_{\mathrm{null}}\_\mathrm{Re}\left[T_{12}\right]$, a specific rotation state with $\mathrm{Re}\left[T_{12}\right] = 0$ is achieved, which leads to equal powers of the two co-polarization components, as $|S_{\mathrm{HH}}|^2 = |S_{\mathrm{VV}}|^2$ is achieved. This specific state is very suitable for model-based polarimetric target decomposition. For example, Yamaguchi decomposition [66] introduced a branch condition for vertical or horizontal dipole judgment according to the amplitude balance of $|S_{\mathrm{HH}}|^2$ and $|S_{\mathrm{VV}}|^2$ for adaptive selection of volume scattering models. Thereby, if the decomposition is applied at this rotation state, the branch judgment can be eliminated.

Besides, if rotating a polarimetric coherency matrix at the angle $\theta_{\mathrm{null}}\_\mathrm{Im}\left[T_{12}\right]$, another interesting rotation state is obtained where $\mathrm{Im}\left(\left\langle S_{\mathrm{HH}}^* S_{\mathrm{VV}} \right\rangle\right) = 0$. The phase difference $\mathrm{Angle}\left(\left\langle S_{\mathrm{HH}} S_{\mathrm{VV}}^* \right\rangle\right)$ is a good indicator for the single- or double-bounce scattering mechanism, and it is one of the target Euler features proposed by J. R. Huynen [62, 67, 68]. In model-based polarimetric target decomposition, when the volume scattering contribution is subtracted, the dominance of the single- or double-bounce scattering mechanism is determined by the feature $\mathrm{Angle}\left(\left\langle S_{\mathrm{HH}} S_{\mathrm{VV}}^* \right\rangle\right)$ and indicated by the sign of $\mathrm{Re}\left[\left\langle S_{\mathrm{HH}} S_{\mathrm{VV}}^* \right\rangle\right]$ [66, 69]. If $\mathrm{Re}\left[\left\langle S_{\mathrm{HH}} S_{\mathrm{VV}}^* \right\rangle\right] \geq 0$ ($\mathrm{Angle}\left(\left\langle S_{\mathrm{HH}} S_{\mathrm{VV}}^* \right\rangle\right)$ approaches zero), single-bounce scattering is dominant, while if $\mathrm{Re}\left[\left\langle S_{\mathrm{HH}} S_{\mathrm{VV}}^* \right\rangle\right] < 0$ ($\mathrm{Angle}\left(\left\langle S_{\mathrm{HH}} S_{\mathrm{VV}}^* \right\rangle\right)$ approaches π), then double-bounce scattering is dominant.

However, when $\text{Angle}\left(\left\langle S_{\text{HH}}S_{\text{VV}}^{*}\right\rangle\right)$ approaches $\pm\pi/2$ where $\text{Re}\left[\left\langle S_{\text{HH}}S_{\text{VV}}^{*}\right\rangle\right]\approx 0$ and $\text{Im}\left[\left\langle S_{\text{HH}}S_{\text{VV}}^{*}\right\rangle\right]\neq 0$, the direct application of this judgment strategy with the original polarimetric coherency matrix could be unstable. If the polarimetric coherency matrix is rotated to achieve $\text{Im}\left(\left\langle S_{\text{HH}}^{*}S_{\text{VV}}\right\rangle\right)=0$ at first using the uniform polarimetric matrix rotation theory, the unstable situation of the judgment strategy can be well avoided.

In addition, the characteristics of the derived features $\theta_{\text{null}}\_\text{Re}\left[T_{12}\right]$ and $\theta_{\text{null}}\_\text{Im}\left[T_{12}\right]$ will be further investigated in the following chapters for application developments. Since they are related to the initial angle features, these further derived polarimetric angle features also belong to the polarimetric rotation domain oscillation feature set.

2.3.4 Polarimetric Covariance Matrix in the Polarimetric Rotation Domain

Polarimetric covariance matrix formulation is also commonly adopted for target scattering mechanism modeling and investigation. The characteristics of the polarimetric covariance matrix in the polarimetric rotation domain can be derived from the polarimetric coherency matrix due to their similarity transformation relationship. From Section 1.3.5, the similarity transformation between the polarimetric covariance matrix $\mathbf{C}_{(\text{H,V})}$ and polarimetric coherency matrix $\mathbf{T}$ is

$$\begin{aligned}\mathbf{C}_{(\text{H,V})}&=\mathbf{U}_{\text{P}_{(\text{H,V})}2\text{L}_{(\text{H,V})}}\mathbf{T}\,\mathbf{U}_{\text{P}_{(\text{H,V})}2\text{L}_{(\text{H,V})}}^{\text{H}}=\mathbf{U}_{\text{P}_{(\text{H,V})}2\text{L}_{(\text{H,V})}}\mathbf{T}\,\mathbf{U}_{\text{P}_{(\text{H,V})}2\text{L}_{(\text{H,V})}}^{-1}\\&=\frac{1}{2}\begin{bmatrix}T_{11}+T_{22}+2\text{Re}\left[T_{12}\right] & \sqrt{2}\left(T_{13}+T_{23}\right) & T_{11}-T_{22}-\text{j}2\,\text{Im}\left[T_{12}\right]\\ \sqrt{2}\left(T_{13}+T_{23}\right)^{*} & 2T_{33} & \sqrt{2}\left(T_{13}-T_{23}\right)^{*}\\ T_{11}-T_{22}+\text{j}2\,\text{Im}\left[T_{12}\right] & \sqrt{2}\left(T_{13}-T_{23}\right) & T_{11}+T_{22}-2\text{Re}\left[T_{12}\right]\end{bmatrix}\end{aligned}\tag{2.3.22}$$

where $\mathbf{U}_{\text{P}_{(\text{H,V})}2\text{L}_{(\text{H,V})}}$ is a unitary matrix for Pauli vector to Lexicographic vector transformation, as

$$\mathbf{U}_{\text{P}_{(\text{H,V})}2\text{L}_{(\text{H,V})}}=\frac{1}{\sqrt{2}}\begin{bmatrix}1 & 1 & 0\\ 0 & 0 & \sqrt{2}\\ 1 & -1 & 0\end{bmatrix}\tag{2.3.23}$$

The linear relationship between $\mathbf{C}_{(\text{H,V})}$ and $\mathbf{T}$ verifies that the elements of $\mathbf{C}_{(\text{H,V})}$ can also be represented in terms of sinusoidal functions. Note that all elements of $\mathbf{C}_{(\text{H,V})}$ are polarimetric roll-invariant.

In addition, the polarimetric covariance matrix with a circular polarization basis is also suitable for PolSAR data interpretation. In a left and right (L,R) circular polarization basis, with the reciprocity condition $\left(S_{\text{LR}}=S_{\text{RL}}\right)$, the corresponding polarimetric scattering vector is [1]

$$\mathbf{k}_{\mathrm{L}_{(\mathrm{L,R})}} = \begin{bmatrix} S_{\mathrm{LL}} & \sqrt{2}S_{\mathrm{LR}} & S_{\mathrm{RR}} \end{bmatrix}^{\mathrm{T}}$$
$$= \frac{1}{2}\Big[S_{\mathrm{HH}} - S_{\mathrm{VV}} + \mathrm{j}2S_{\mathrm{HV}} \quad \mathrm{j}\sqrt{2}\left(S_{\mathrm{HH}} + S_{\mathrm{VV}}\right) \tag{2.3.24}$$
$$-\left(S_{\mathrm{HH}} - S_{\mathrm{VV}}\right) + \mathrm{j}2S_{\mathrm{HV}} \Big]^{\mathrm{T}}$$

The polarimetric covariance matrix with a $(\mathrm{L,R})$ polarization basis is

$$\mathbf{C}_{(\mathrm{L,R})} = \left\langle \mathbf{k}_{\mathrm{L}_{(\mathrm{L,R})}} \mathbf{k}_{\mathrm{L}_{(\mathrm{L,R})}}^{\mathrm{H}} \right\rangle \tag{2.3.25}$$

From Section 1.3.5, the link between the polarimetric covariance matrix $\mathbf{C}_{(\mathrm{L,R})}$ and polarimetric coherency matrix $\mathbf{T}$ is also a similarity transformation

$$\mathbf{C}_{(\mathrm{L,R})} = \mathbf{U}_{\mathrm{P}_{(\mathrm{H,V})}2\mathrm{L}_{(\mathrm{L,R})}} \mathbf{T} \mathbf{U}_{\mathrm{P}_{(\mathrm{H,V})}2\mathrm{L}_{(\mathrm{L,R})}}^{\mathrm{H}} = \mathbf{U}_{\mathrm{P}_{(\mathrm{H,V})}2\mathrm{L}_{(\mathrm{L,R})}} \mathbf{T} \mathbf{U}_{\mathrm{P}_{(\mathrm{H,V})}2\mathrm{L}_{(\mathrm{L,R})}}^{-1}$$
$$= \frac{1}{2}\begin{bmatrix} T_{22} + T_{33} + 2\,\mathrm{Im}\left[T_{23}\right] & \sqrt{2}\left(T_{13} + \mathrm{j}T_{12}\right)^{*} & T_{33} - T_{22} - \mathrm{j}2\,\mathrm{Re}\left[T_{23}\right] \\ \sqrt{2}\left(T_{13} + \mathrm{j}T_{12}\right) & 2T_{11} & \sqrt{2}\left(T_{13} - \mathrm{j}T_{12}\right) \\ T_{33} - T_{22} + \mathrm{j}2\,\mathrm{Re}\left[T_{23}\right] & \sqrt{2}\left(T_{13} - \mathrm{j}T_{12}\right)^{*} & T_{22} + T_{33} - 2\,\mathrm{Im}\left[T_{23}\right] \end{bmatrix} \tag{2.3.26}$$

where $\mathbf{U}_{\mathrm{P}_{(\mathrm{H,V})}2\mathrm{L}_{(\mathrm{L,R})}}$ is a unitary matrix for Pauli vector to circular polarization scattering vector transformation, as

$$\mathbf{U}_{\mathrm{P}_{(\mathrm{H,V})}2\mathrm{L}_{(\mathrm{L,R})}} = \begin{bmatrix} 0 & -1 & \mathrm{j}\sqrt{2} \\ \mathrm{j}\sqrt{2} & 0 & 0 \\ 1 & 0 & \mathrm{j}\sqrt{2} \end{bmatrix} \tag{2.3.27}$$

From the polarimetric roll-invariant terms of polarimetric coherency matrix $\mathbf{T}$, it is easily verified that the diagonal elements of $\mathbf{C}_{(\mathrm{L,R})}$ are polarimetric roll-invariant. The amplitudes of the off-diagonal elements are also polarimetric roll-invariant. Therefore, only the real and imaginary parts of the off-diagonal elements are polarimetric roll-invariant.

In another aspect, using (2.3.24), the rotated scattering vector with rotation angle θ can be derived as

$$\mathbf{k}_{\mathrm{L}_{(\mathrm{L,R})}}(\theta) = \begin{bmatrix} S_{\mathrm{LL}}e^{\mathrm{j}2\theta} & \sqrt{2}S_{\mathrm{LR}} & S_{\mathrm{RR}}e^{-\mathrm{j}2\theta} \end{bmatrix}^{\mathrm{T}} \tag{2.3.28}$$

Then, the rotated polarimetric covariance matrix $\mathbf{C}_{(\mathrm{L,R})}(\theta)$ is

$$\mathbf{C}_{(\mathrm{L,R})}(\theta)=\left\langle \mathbf{k}_{\mathrm{L}_{(\mathrm{L,R})}}(\theta)\mathbf{k}^{\mathrm{H}}_{\mathrm{L}_{(\mathrm{L,R})}}(\theta)\right\rangle$$

$$=\begin{bmatrix} \left\langle |S_{\mathrm{LL}}|^2\right\rangle & \sqrt{2}\left\langle\left(S_{\mathrm{LL}}S^*_{\mathrm{LR}}\right)e^{\mathrm{j}2\theta}\right\rangle & \left\langle\left(S_{\mathrm{LL}}S^*_{\mathrm{RR}}\right)e^{\mathrm{j}4\theta}\right\rangle \\ \sqrt{2}\left\langle\left(S^*_{\mathrm{LL}}S_{\mathrm{LR}}\right)e^{-\mathrm{j}2\theta}\right\rangle & 2\left\langle |S_{\mathrm{LR}}|^2\right\rangle & \sqrt{2}\left\langle\left(S_{\mathrm{LR}}S^*_{\mathrm{RR}}\right)e^{\mathrm{j}2\theta}\right\rangle \\ \left\langle\left(S^*_{\mathrm{LL}}S_{\mathrm{RR}}\right)e^{-\mathrm{j}4\theta}\right\rangle & \sqrt{2}\left\langle\left(S^*_{\mathrm{LR}}S_{\mathrm{RR}}\right)e^{-\mathrm{j}2\theta}\right\rangle & \left\langle |S_{\mathrm{RR}}|^2\right\rangle \end{bmatrix} \quad (2.3.29)$$

From the polarization orientation angle theory [17], if assuming the scattering vector $\mathbf{k}_{\mathrm{L}_{(\mathrm{L,R})}}$ in (2.3.24) is without any rotation or its orientation effect has been compensated, then the phase of the co-polarization correlation term is $\mathrm{Angle}\left\langle S_{\mathrm{LL}}S^*_{\mathrm{RR}}\right\rangle=0$. Thereby, the orientation term $e^{j4\theta}$ of $\mathbf{C}_{(\mathrm{L,R})}(\theta)$ is solely induced by the rotation of the polarization state of the backscattering wave, which is caused by the scatters' modulation. This is the basic principle for polarization orientation angle estimation in circular polarization basis. Similar to the conclusions derived from (2.3.26), in (2.3.29), the diagonal elements ($\left\langle |S_{\mathrm{LL}}|^2\right\rangle$, $\left\langle |S_{\mathrm{LR}}|^2\right\rangle$, and $\left\langle |S_{\mathrm{RR}}|^2\right\rangle$) and the amplitudes of the off-diagonal elements ($\left|\left\langle S_{\mathrm{LL}}S^*_{\mathrm{LR}}\right\rangle\right|$, $\left|\left\langle S_{\mathrm{LR}}S^*_{\mathrm{RR}}\right\rangle\right|$, and $\left|\left\langle S_{\mathrm{LL}}S^*_{\mathrm{RR}}\right\rangle\right|$) are polarimetric roll-invariant. Their analytical expressions can be directly obtained from (2.3.29).

Besides, from (2.3.11) and (2.3.26), one interesting relationship is obtained

$$A\_T_{22}=\frac{1}{4}\left(T_{33}-T_{22}\right)^2+\mathrm{Re}^2\left[T_{23}\right]=\left|C_{13(\mathrm{L,R})}\right|^2 \quad (2.3.30)$$

Moreover, $C_{13(\mathrm{L,R})}$ is the numerator of the co-polarization correlation coefficient:

$$\rho_{\mathrm{RR\text{-}LL}}=\frac{\left\langle S_{\mathrm{LL}}S^*_{\mathrm{RR}}\right\rangle}{\sqrt{\left\langle S_{\mathrm{LL}}S^*_{\mathrm{LL}}\right\rangle}\sqrt{\left\langle S_{\mathrm{RR}}S^*_{\mathrm{RR}}\right\rangle}}=\frac{C_{13(\mathrm{L,R})}}{\sqrt{C_{11(\mathrm{L,R})}}\sqrt{C_{33(\mathrm{L,R})}}} \quad (2.3.31)$$

This correlation coefficient $\rho_{\mathrm{RR\text{-}LL}}$ has been demonstrated to be effective for manmade target characterization [70, 71]. Thereby, the efficiency of the feature A_T_{22} for manmade and forest target discrimination is further supported by these studies.

2.3.5 Polarimetric Rotation Domain Amplitude Feature Demonstration

In this section, the derived oscillation amplitude and oscillation center features from polarimetric coherency matrix are used for demonstration and investigation using multi-frequency Pi-SAR PolSAR data.

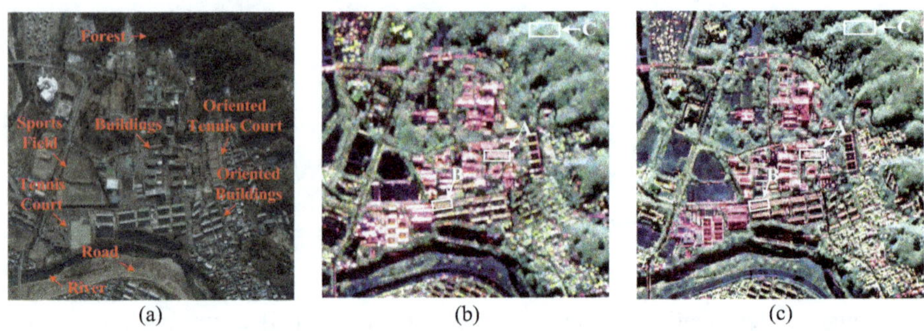

FIGURE 2.3.1 Study area of Kawauchi campus, Tohoku University, Sendai, Japan. (a) Optical image from Google Earth. Pi-SAR L-band (b) and X-band (c) PolSAR images with Pauli scattering components (HH-VV, HV and HH+VV). Building A, oriented building B, and forest C, marked with rectangular boxes, are selected for further investigation.

2.3.5.1 Multi-Frequency Pi-SAR Data Description

Pi-SAR L- and X-band PolSAR data collected over Sendai, Japan, on February 12, 2005, are used to investigate the derived polarimetric rotation domain oscillation features. The study area mainly covers the Kawauchi campus of Tohoku University. It contains various types of land covers, such as buildings, forests, baseball and football fields, tennis courts, roads, a river, and so on. The tennis courts are surrounded by metal fences. The corresponding optical image from Google Earth, acquired on January 22, 2003, is shown in Figure 2.3.1 (a). The PolSAR RGB composite images with a Pauli basis are shown in Figure 2.3.1 (b)–(c). The SimiTest speckle filter [72], with good speckle reduction and good detail preservation performance, has been applied using a 15 × 15 moving window to smooth the speckle effect. The illumination direction is from the top to the bottom. Two buildings and one forest area, labeled A, B, and C, respectively, are selected for further investigation. Building A is aligned parallel to the flight pass, while the orientation of building B is about 8° to the flight direction.

2.3.5.2 Polarimetric Feature Demonstration

As previously discussed, the oscillation amplitudes and oscillation centers of each element of a polarimetric coherency matrix present six independent features: $A\_\mathrm{Re}[T_{12}]$, $A\_\mathrm{Im}[T_{12}]$, A_T_{22}, B_T_{22}, $A\_|T_{12}|^2$, and $B\_|T_{23}|^2$. These derived features are illustrated in Figures 2.3.2 and 2.3.3 for L- and X-band Pi-SAR PolSAR data, respectively. The total backscattering power $SPAN$ and the power of the diagonal terms (T_{11} T_{22} and T_{33}) are also shown in Figures 2.3.2 and 2.3.3 for comparison accordingly. It is clear that buildings and tennis courts (surrounded by metal fences) which are parallel to the flight pass exhibit strong backscattered power, since the dihedral structures formed by the ground-wall and ground-fence reflect most of the transmitted microwave energy back to the SAR receiver. However, for the oriented buildings and tennis courts, the backscattered power values are obviously reduced. In images of $SPAN$, T_{11}, and T_{33}, oriented manmade structures show responses very

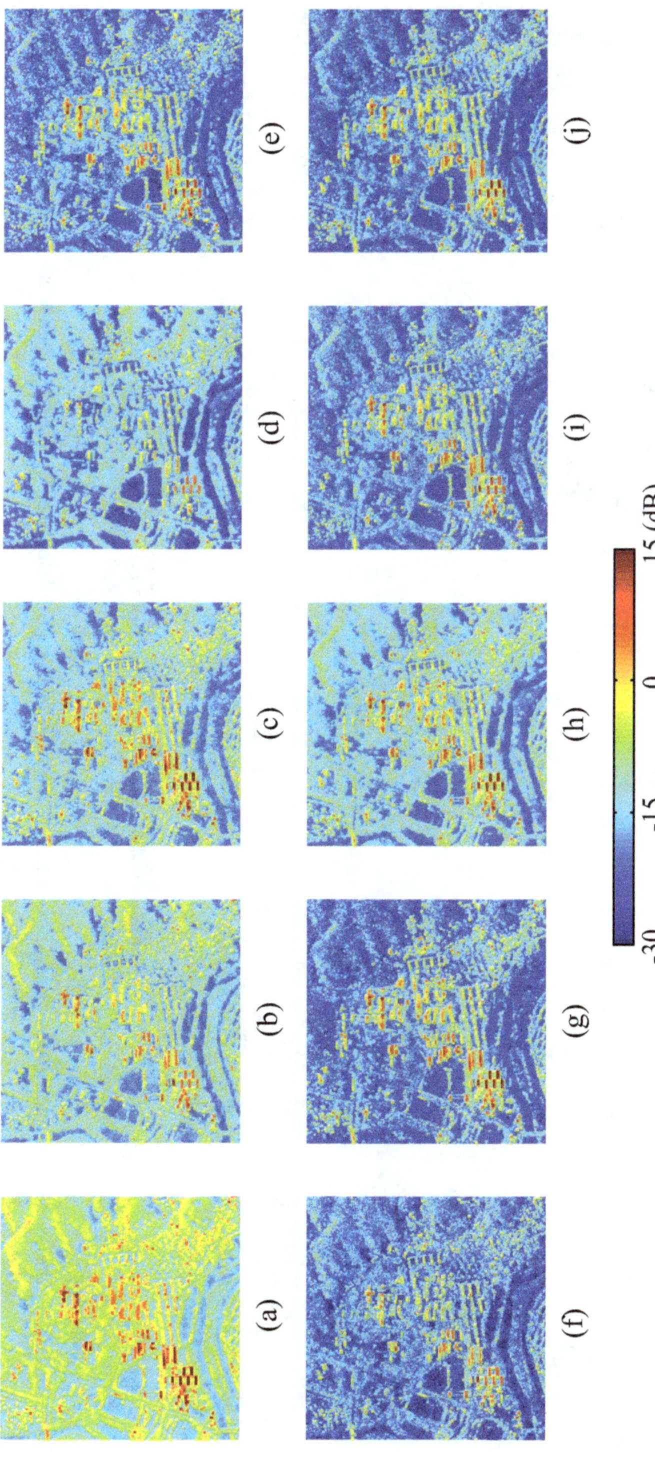

FIGURE 2.3.2 Traditional polarimetric features and polarimetric rotation domain amplitude features for Pi-SAR L-band data. (a) $SPAN$, (b) T_{11}, (c) T_{22}, (d) T_{33}, (e) $A\_\mathrm{Re}[T_{12}]$, (f) $A\_\mathrm{Im}[T_{12}]$, (g) A_T_{22}, (h) B_T_{22}, (i) $A\_|T_{12}|^2$, and (j) $B\_|T_{23}|^2$.

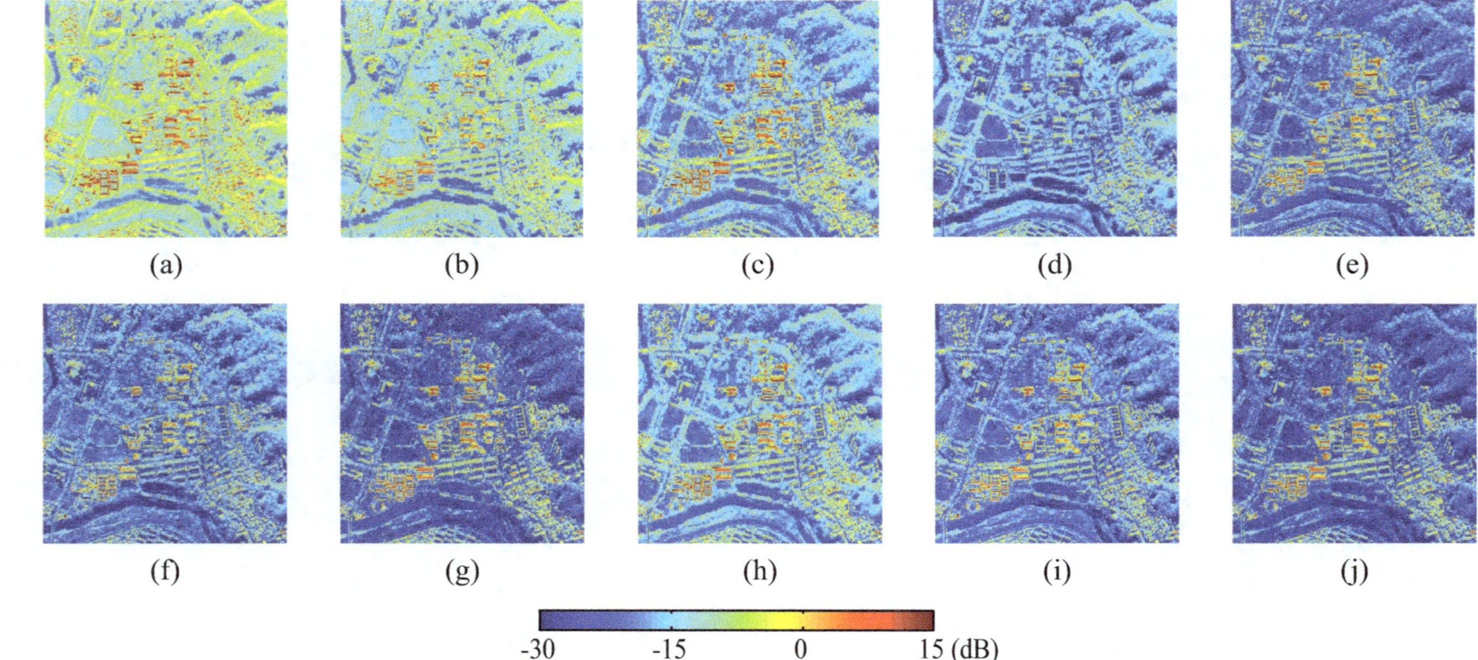

FIGURE 2.3.3 Traditional polarimetric features and polarimetric rotation domain amplitude features for Pi-SAR X-band data. (a) $SPAN$, (b) T_{11}, (c) T_{22}, (d) T_{33}, (e) $A\_\text{Re}[T_{12}]$, (f) $A\_\text{Im}[T_{12}]$, (g) A_T_{22}, (h) B_T_{22}, (i) $A\_|T_{12}|^2$, and (j) $B\_|T_{23}|^2$.

similar to those of forests. This phenomenon could produce scattering mechanism ambiguity in PolSAR data interpretation, such as occurs in model-based polarimetric target decomposition. However, natural targets usually exhibit reflection symmetry, while manmade targets tilting to the radar line of sight violate the reflection symmetry condition. Therefore, in the polarimetric rotation domain, this difference between forests and manmade targets can be sensed by the oscillation amplitudes. From Figures 2.3.2 and 2.3.3, visually the oscillation amplitude A_T_{22} provides the best contrast between manmade structures and forests for both L- and X-band data.

For quantitative comparison, the power ratios between building A and forest C, and building B and forest C are calculated and shown in Figure 2.3.4. Generally, the backscattered power from oriented buildings is weaker than that from buildings parallel to the flight pass. Therefore, the ratio between building A and forest C is much higher than that between building B and forest C. The cross-polarization term T_{33} gives the lowest ratio for both cases among all these features. T_{22}, which relates to the double-bounce scattering, achieves the highest ratio values among T_{11}, T_{22}, T_{33}, and the total backscattering power $SPAN$. Meanwhile, the derived oscillation amplitude A_T_{22} produces the highest ratio value among all these features and leads to the strongest contrast between buildings and forest among these features. In detail, the ratios between building A and forest C from A_T_{22} are 2.9 and 3.2 dB higher than those from T_{22} for L- and X-band data, while the ratios between building B and forest C from A_T_{22} are 3.1 and 2.9 dB higher than those from T_{22} for L- and X-band data, respectively. Moreover, for the used data, it is observed that the backscattered powers from forest C are consistently higher at the L-band than those at the X-band for the ten features shown in Figure 2.3.4. Meanwhile, except for the T_{33} term, the backscattered powers from the two buildings are consistently lower at the L-band than those at the X-band. Therefore, the ratios from the L-band are consistently lower than the corresponding terms from the X-band, which indicates the frequency dependence of target backscattering. For the forests, the reason may

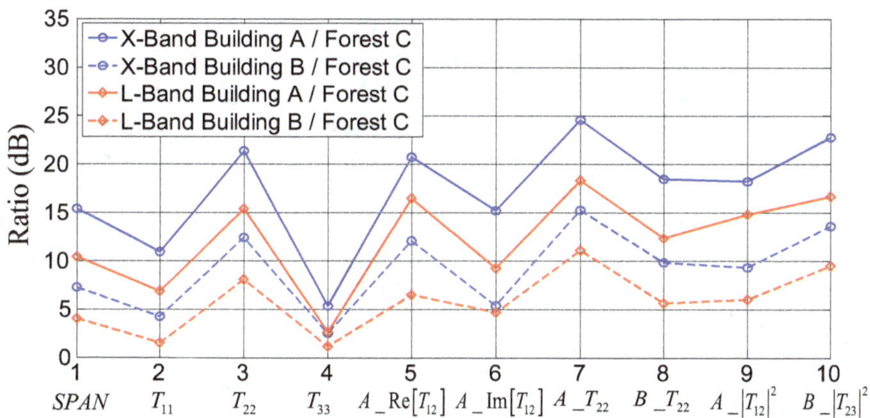

FIGURE 2.3.4 Comparison of power ratios between building A and forest C, and oriented building B and forest C for traditional polarimetric features and polarimetric rotation domain amplitude features, respectively.

be that microwave at the L-band has deeper penetration depth in forests than that at the X-band, and more backscattering powers can be expected directly from the big branches and ground-trunk structures. Thereby, the backscattered powers can be stronger than those from the canopy with volume scattering. For the buildings, further efforts are indeed needed for clarification.

Thereby, theoretical analysis and experimental investigations validate the good performance of A_T_{22} for manmade target enhancement. Thus, for the detection and extraction of manmade targets (especially those oriented to the flight pass), it is better to use the oscillation amplitude A_T_{22}.

2.3.6 Polarimetric Rotation Domain Angle Feature Demonstration

In this section, the derived polarimetric angle features from the polarimetric coherency matrix are used for demonstration and investigation using multi-frequency AIRSAR PolSAR data.

2.3.6.1 Multi-Frequency AIRSAR Data Description

AIRSAR P-, L-, and C-band PolSAR data simultaneously collected on June 15, 1991, over Flevoland, the Netherlands, are used for demonstration of the derived polarimetric rotation domain angle features. The slant range and azimuth pixel resolutions are 6.7 and 12.2 m, respectively. The looking angle ranges from 23.0° to 62.1°. The advanced SimiTest speckle filter [72] has been applied to smooth the speckle effect with a 15 × 15 moving window. The filtered multi-frequency PolSAR RGB composite images with a Pauli basis are shown in Figure 2.3.5. This study area contains various types of land covers, such as agriculture, forests, roads, and so on. The agriculture fields mainly include stembeans, rapeseed, peas, potatoes, lucerne, wheat, and beets. A ground-truth map is generated for these crops, shown in Figure 2.3.5 (d). It is observed that various land covers may have quite different responses in terms of microwave frequency.

2.3.6.2 Polarimetric Features Demonstration

A number of polarimetric rotation domain angle features have been derived in Section 2.3.3. The interesting minimization and null angle features $\theta_{\min}\_T_{33}$, $\theta_{\text{null}}\_\text{Re}[T_{12}]$, and $\theta_{\text{null}}\_\text{Im}[T_{12}]$ are studied. These features are retrieved from the multi-frequency data and shown in Figure 2.3.6. Besides, the mean alpha angle feature $\bar{\alpha}$ from eigenvalue-eigenvector-based decomposition [8] is also extracted for comparison. The majority values of the $\bar{\alpha}$ feature are around $\pi/4$ and indicate the volume scattering mechanism for L-band data, while more $\bar{\alpha}$ values approach $\pi/2$ and indicate double-bounce scattering at P-band due to the deeper penetration. Compared to the P- and L-bands, the $\bar{\alpha}$ feature exhibits more surface scattering at the C-band. The minimization angle $\theta_{\min}\_T_{33}$ is equivalent to the polarization orientation angle. The characteristics of the polarization orientation angle and its application in terrain mapping and building orientation estimation have been studied in detail [1, 18, 21, 22]. The polarization orientation angle can reflect the scatterer orientation relative to the radar line of sight. Since electromagnetic waves of different frequency bands have different penetration characteristics, the backscattered energy

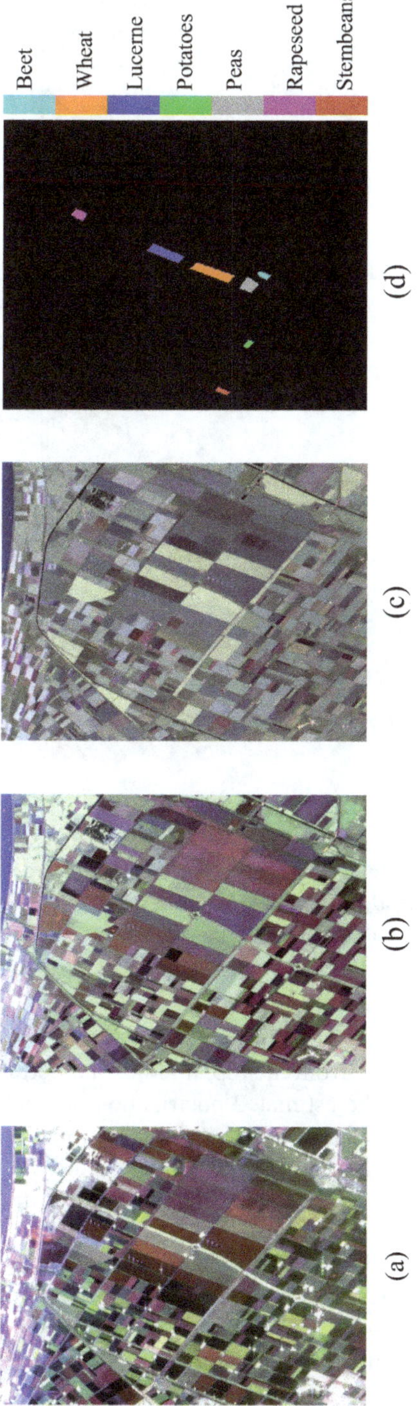

FIGURE 2.3.5 AIRSAR PolSAR images of Flevoland, the Netherlands. (a) P-band, (b) L-band, (c) C-band, (d) ground-truth map.

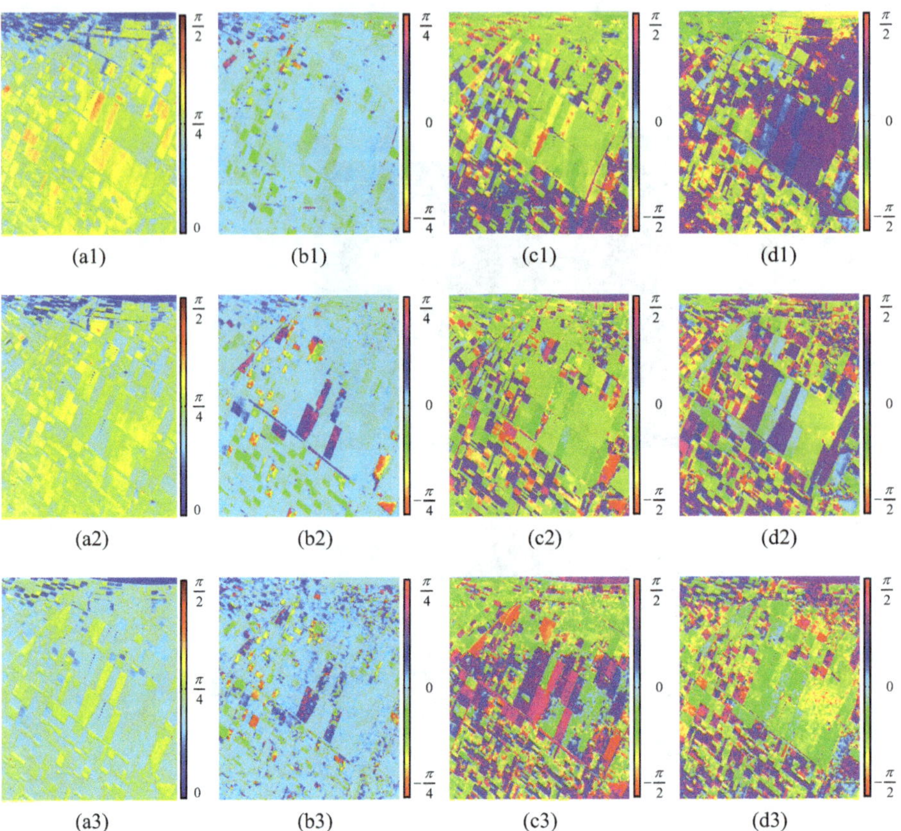

FIGURE 2.3.6 Derived polarimetric angle features from P- (top row), L- (middle row), and C- (bottom row) band AIRSAR PolSAR data, respectively. (a1)–(a3) Mean alpha angle $\bar{\alpha}$, (b1)–(b3) minimization angle $\theta_{\text{min}}\_T_{33}$ (polarization orientation angle), (c1)–(c3) null angle $\theta_{\text{null}}\_\text{Re}[T_{12}]$, and (d1)–(d2) null angle $\theta_{\text{null}}\_\text{Im}[T_{12}]$.

received by PolSAR can come from the top, middle, and bottom of the crops. From Figure 2.3.6, it is clear that the estimated polarization orientation angles vary with the frequency bands, since the interactions of the microwave and scatterers can be at the top, middle, and bottom layers of the crops due to different penetration depths. Besides, for crops where the scatterers have more uniform distributions, the corresponding polarization orientation angles are more homogeneous. Compared to the P- and L-bands, the polarization orientation angles estimated from C-band data fluctuate more. Compared to the mean alpha angle $\bar{\alpha}$ and the polarization orientation angle, the derived null angles $\theta_{\text{null}}\_\text{Re}[T_{12}]$ and $\theta_{\text{null}}\_\text{Im}[T_{12}]$ exhibit more sensitivity to various crops. As interpreted in Section 2.3.3, these two polarimetric angle features can indicate the reflection symmetry condition $\left(\left\langle (S_{\text{HH}}+S_{\text{VV}})S_{\text{HV}}^{*}\right\rangle\right)$, the sign of the co-polarization magnitude $\left(\left\langle |S_{\text{VV}}|^2-|S_{\text{HH}}|^2\right\rangle\right)$, and the phase differences

$\left(\text{Im}\left[\langle S_{\text{HH}}S_{\text{VV}}^*\rangle\right]\right)$. Therefore, their values have clear physical meanings and can be explored for land cover classification.

2.4 TWO-DIMENSION POLARIMETRIC COHERENCE PATTERN INTERPRETATION TOOL

Polarimetric coherence, which has the potential to reveal physical properties of scatterers, is an important source for polarimetric radar data investigation. Target structure and orientation relative to the polarimetric radar illumination direction are key factors affecting the polarimetric coherence degree. The relative orientation between a sensor and a target can be adjusted using rotating processing along the radar line of sight. In 2016, a visualization and characterization tool named the polarimetric coherence pattern for two arbitrary polarization channels was established [27–29]. The main idea is to extend the traditional polarimetric coherence at a given rotation state ($\theta = 0$) to the rotation domain ($\theta \in [-\pi, \pi)$) along the radar line of sight for hidden information exploration. This interpretation tool is able to view the variation of polarimetric coherence in the polarimetric rotation domain containing rich target scattering diversity information, which is seldom considered. On this basis, a series of polarimetric features are derived to characterize the properties of a polarimetric coherence pattern, which provide new insights for the interpretation and application of target scattering mechanisms.

2.4.1 Definition of Polarimetric Coherence Pattern

Theoretically, for two arbitrary polarization channels s_1 and s_2, the polarimetric coherence is defined as

$$|\gamma_{1-2}| = \frac{\left|E\left(s_1 s_2^*\right)\right|}{\sqrt{E\left(|s_1|^2\right)E\left(|s_2|^2\right)}} \tag{2.4.1}$$

where s_2^* is the conjugate of s_2, and $E(s)$ is the expectation of s. The value of $|\gamma_{1-2}|$ is within $[0,1]$.

In practice, sample averaging of sufficient samples with similar properties is used to estimate the polarimetric coherence [72, 73]

$$|\gamma_{1-2}| = \frac{\left|\langle s_1 s_2^*\rangle\right|}{\sqrt{\langle |s_1|^2\rangle\langle |s_2|^2\rangle}} \tag{2.4.2}$$

where $\langle\cdot\rangle$ indicates the sample average.

The core idea aims at extending this original polarimetric coherence at a certain imaging geometry to the polarimetric rotation domain. A visualization and characterization tool named a two-dimension polarimetric coherence pattern (polarimetric coherence pattern for short), which covers all rotation angles in $[-\pi,\pi)$, is developed to investigate the characteristics of polarimetric coherence in the polarimetric rotation domain. The definition of the polarimetric coherence pattern is

$$|\gamma_{1-2}(\theta)| = \frac{|\langle s_1(\theta) s_2^*(\theta)\rangle|}{\sqrt{\langle |s_1(\theta)|^2\rangle \langle |s_2(\theta)|^2\rangle}} \qquad \theta \in [-\pi,\pi) \tag{2.4.3}$$

When the speckle effect is well smoothed during the estimation procedure, the rotation effect of polarimetric coherence is fully determined by polarimetric elements $s_1(\theta)$ and $s_2(\theta)$. The polarimetric coherence pattern $|\gamma_{1-2}(\theta)|$ has the advantage of investigating polarimetric coherence at an arbitrary angle in the polarimetric rotation domain. Also, it provides a visualization solution to view the variation of polarimetric coherence in terms of rotation angles.

For $(\mathrm{H,V})$ polarization basis and subject to the reciprocal condition, six typical polarimetric coherence features of $|\gamma_{\text{HH-VV}}|$, $|\gamma_{\text{HH-HV}}|$, $|\gamma_{\text{VV-HV}}|$, $|\gamma_{\text{(HH+VV)-(HH-VV)}}|$, $|\gamma_{\text{(HH+VV)-(HV)}}|$, and $|\gamma_{\text{(HH-VV)-(HV)}}|$ are available from the Lexicographic scattering vector $\mathbf{k}_{\mathrm{L}_{(\mathrm{H,V})}} = \begin{bmatrix} S_{\mathrm{HH}} & \sqrt{2}S_{\mathrm{HV}} & S_{\mathrm{VV}} \end{bmatrix}^{\mathrm{T}}$ and the Pauli scattering vector $\mathbf{k}_{\mathrm{P}_{(\mathrm{H,V})}} = \frac{1}{\sqrt{2}}\begin{bmatrix} S_{\mathrm{HH}}+S_{\mathrm{VV}} & S_{\mathrm{HH}}-S_{\mathrm{VV}} & 2S_{\mathrm{HV}} \end{bmatrix}^{\mathrm{T}}$. Six polarimetric coherence patterns can be obtained accordingly. Furthermore, the following two equivalence relationships can be verified

$$|\gamma_{\text{HH-HV}}(\theta)| = |\gamma_{\text{VV-HV}}(\theta + \pi/2)| \tag{2.4.4}$$

$$|\gamma_{\text{(HH+VV)-(HH-VV)}}(\theta)| = |\gamma_{\text{(HH+VV)-(HV)}}(\theta + \pi/4)| \tag{2.4.5}$$

Therefore, polarimetric coherence patterns $|\gamma_{\text{HH-HV}}(\theta)|$ and $|\gamma_{\text{VV-HV}}(\theta)|$ and $|\gamma_{\text{(HH+VV)-(HH-VV)}}(\theta)|$ and $|\gamma_{\text{(HH+VV)-(HV)}}(\theta)|$ are respectively equivalent. In the following investigation, four independent polarimetric coherence patterns ($|\gamma_{\text{HH-VV}}(\theta)|$, $|\gamma_{\text{HH-HV}}(\theta)|$, $|\gamma_{\text{(HH+VV)-(HH-VV)}}(\theta)|$, and $|\gamma_{\text{(HH-VV)-(HV)}}(\theta)|$) will be considered. Meanwhile, the main ranges of $|\gamma_{\text{(HH+VV)-(HH-VV)}}(\theta)|$ and $|\gamma_{\text{HH-VV}}(\theta)|$ are both $[-\pi/4,\pi/4)$, while they are $[-\pi/8,\pi/8)$ and $[-\pi/2,\pi/2)$ for $|\gamma_{\text{(HH-VV)-(HV)}}(\theta)|$ and $|\gamma_{\text{HH-HV}}(\theta)|$, respectively.

2.4.2 Visualization and Characterization

A polarimetric coherence pattern provides a visualization tool to view the characteristics of polarimetric coherence in the polarimetric rotation domain. An example of polarimetric coherence pattern $|\gamma_{\text{HH-VV}}(\theta)|$ is shown in Figure 2.4.1. It can be seen that polarimetric coherence may vary significantly in the polarimetric rotation

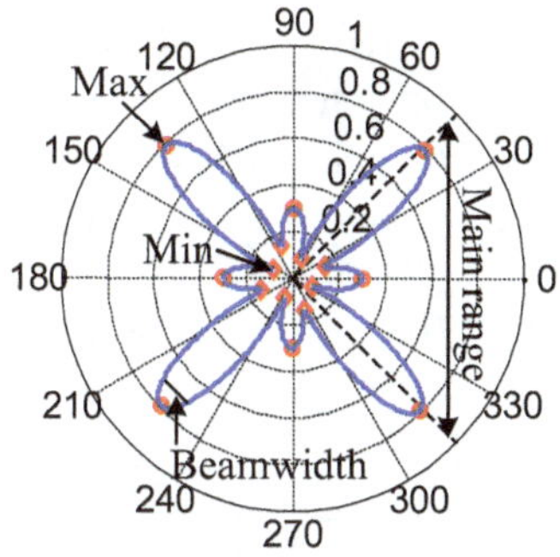

FIGURE 2.4.1 Illustration of a two-dimension polarimetric coherence pattern example.

domain. The variation pattern contains rich hidden information and provides the potential to understand the scattering property, especially the target scattering diversity. In this vein, a set of features has been proposed to quantitatively characterize a polarimetric coherence pattern [27–29]. Exploration of these hidden features in the polarimetric rotation domain will provide valuable insights for polarimetric radar data understanding and investigation. Using $|\gamma_{1-2}(\theta)|$ as an example, the definitions of these features are given as follows

1) Original Coherence $|\gamma_{1-2}(\theta)|_{\text{org}}$: it is the coherence without any rotation processing:

$$|\gamma_{1-2}(\theta)|_{\text{org}} = |\gamma_{1-2}(0)| = |\gamma_{1-2}| \tag{2.4.6}$$

This is the normally used polarimetric coherence feature in the literature and indicates a target decorrelation effect for two polarization channels under the original imaging geometry.

2) Coherence Degree $|\gamma_{1-2}(\theta)|_{\text{mean}}$: it is the mean value of coherence in the polarimetric rotation domain and indicates the general coherence level:

$$|\gamma_{1-2}(\theta)|_{\text{mean}} = \frac{1}{2\pi}\int_{-\pi}^{\pi} |\gamma_{1-2}(\theta)| \mathrm{d}\theta \tag{2.4.7}$$

Coherence degree $|\gamma_{1-2}(\theta)|_{\text{mean}}$ is a measure for targets' average decorrelation effect in the polarimetric rotation domain. The higher the coherence degree, the lower the decorrelation phenomenon.

3) Coherence Fluctuation $|\gamma_{1-2}(\theta)|_{\text{std}}$: it is the standard deviation of coherence in the polarimetric rotation domain and indicates the coherence fluctuation:

$$|\gamma_{1-2}(\theta)|_{\text{std}} = \sqrt{\frac{1}{2\pi}\int_{-\pi}^{\pi} \left(|\gamma_{1-2}(\theta)| - |\gamma_{1-2}(\theta)|_{\text{mean}}\right)^2 \mathrm{d}\theta} \tag{2.4.8}$$

Coherence fluctuation $\left|\gamma_{1-2}(\theta)\right|_{\text{std}}$ can identify target scattering diversity. Generally, the higher the coherence fluctuation, the more obvious the target scattering diversity in the polarimetric rotation domain. For polarimetric roll-invariant scatterers, coherence fluctuation $\left|\gamma_{1-2}(\theta)\right|_{\text{std}}$ will shrink to zero.

4) Maximum Coherence $\left|\gamma_{1-2}(\theta)\right|_{\max}$: it is the maximum coherence in the polarimetric rotation domain:

$$\left|\gamma_{1-2}(\theta)\right|_{\max} = \max_{\theta\in[-\pi,\pi)}\left(\left|\gamma_{1-2}(\theta)\right|\right) \tag{2.4.9}$$

It indicates the upper limit of polarimetric coherence of two polarization channels via orientation adjustment. Maximum coherence $\left|\gamma_{1-2}(\theta)\right|_{\max}$ is especially important for agriculture fields by clearly enhancing their relatively low original coherence $\left|\gamma_{1-2}(\theta)\right|_{\text{org}}$ due to a serious decorrelation effect.

5) Minimum Coherence $\left|\gamma_{1-2}(\theta)\right|_{\min}$: it is the minimum coherence in the polarimetric rotation domain:

$$\left|\gamma_{1-2}(\theta)\right|_{\min} = \min_{\theta\in[-\pi,\pi)}\left(\left|\gamma_{1-2}(\theta)\right|\right) \tag{2.4.10}$$

Minimum coherence $\left|\gamma_{1-2}(\theta)\right|_{\min}$ gives the lower limit of polarimetric coherence through rotation processing in the polarimetric rotation domain.

6) Coherence Contrast $\left|\gamma_{1-2}(\theta)\right|_{\text{contrast}}$: it is defined as

$$\left|\gamma_{1-2}(\theta)\right|_{\text{contrast}} = \left|\gamma_{1-2}(\theta)\right|_{\max} - \left|\gamma_{1-2}(\theta)\right|_{\min} \tag{2.4.11}$$

It reflects the absolute contrast in the polarimetric rotation domain. In addition, coherence contrast $\left|\gamma_{1-2}(\theta)\right|_{\text{contrast}}$ can also partially reflect target scattering diversity in the polarimetric rotation domain. For polarimetric roll-invariant targets without orientation dependency, coherence contrast $\left|\gamma_{1-2}(\theta)\right|_{\text{contrast}}$ will become zero.

7) Coherence Anisotropy $\left|\gamma_{1-2}(\theta)\right|_{\text{A}}$: it is defined as

$$\left|\gamma_{1-2}(\theta)\right|_{\text{A}} = \frac{\left|\gamma_{1-2}(\theta)\right|_{\max} - \left|\gamma_{1-2}(\theta)\right|_{\min}}{\left|\gamma_{1-2}(\theta)\right|_{\max} + \left|\gamma_{1-2}(\theta)\right|_{\min}} \tag{2.4.12}$$

It reflects the relative contrast in the polarimetric rotation domain. Coherence anisotropy $\left|\gamma_{1-2}(\theta)\right|_{\text{A}}$ is a complementary feature to coherence contrast $\left|\gamma_{1-2}(\theta)\right|_{\text{contrast}}$. For polarimetric roll-variant targets with relatively low coherence degree $\left|\gamma_{1-2}(\theta)\right|_{\text{mean}}$, coherence anisotropy $\left|\gamma_{1-2}(\theta)\right|_{\text{A}}$ can further enhance the contrast and show better discriminant performance compared with coherence contrast $\left|\gamma_{1-2}(\theta)\right|_{\text{contrast}}$.

8) Coherence Beamwidth $|\gamma_{1-2}(\theta)|_{\text{bw0.95}}$: it is defined as the rotation angle range within which the coherence values are no less than $0.95\times|\gamma_{1-2}(\theta)|_{\text{max}}$, which is

$$|\gamma_{1-2}(\theta)|_{\text{bw0.95}}=\left\{\theta\,\middle|\,|\gamma_{1-2}(\theta)|_{\text{max}}\geq|\gamma_{1-2}(\theta)|\geq 0.95\times|\gamma_{1-2}(\theta)|_{\text{max}}\right\} \tag{2.4.13}$$

Coherence beamwidth is an index reflecting the sensitivity of the orientation dependency effect. The smaller $|\gamma_{1-2}(\theta)|_{\text{bw0.95}}$ is, the greater the decorrelation effect and orientation dependency.

9) Maximum Rotation Angle $\theta_{|\gamma_{1-2}(\theta)|_{\text{max}}}$: it is defined as the rotation angle within the main range which produces the maximum coherence $|\gamma_{1-2}(\theta)|_{\text{max}}$, which is

$$\theta_{|\gamma_{1-2}(\theta)|_{\text{max}}}=\left\{\theta\,\middle|\,|\gamma_{1-2}(\theta)|=|\gamma_{1-2}(\theta)|_{\text{max}}\right\} \tag{2.4.14}$$

Maximum rotation angle indicates the most suitable state for a target in the polarimetric rotation domain, which leads to the least decorrelation of two given polarimetric channels.

10) Minimum Rotation Angle $\theta_{|\gamma_{1-2}(\theta)|_{\text{min}}}$: it is defined as the rotation angle within the main range which produces the minimum coherence $|\gamma_{1-2}(\theta)|_{\text{min}}$, which is

$$\theta_{|\gamma_{1-2}(\theta)|_{\text{min}}}=\left\{\theta\,\middle|\,|\gamma_{1-2}(\theta)|=|\gamma_{1-2}(\theta)|_{\text{min}}\right\} \tag{2.4.15}$$

Minimum rotation angle indicates another specific state where the most serious decorrelation of two certain polarimetric channels occurs for a target in the polarimetric rotation domain.

2.4.3 Interpretation and Discussion

From the previous characterization, ten features can be derived for each polarimetric coherence pattern of two given polarimetric channels. According to the definitions, the physical meanings of these features are straightforward. These derived features are mainly dedicated to representing target scattering diversity between two polarimetric channels in the polarimetric rotation domain. More specifically, the original coherence $|\gamma_{1-2}(\theta)|_{\text{org}}$ indicates the coherence of two given polarimetric channels under the imaging geometry between the polarimetric radar and target without orientation adjustment. The coherence degree $|\gamma_{1-2}(\theta)|_{\text{mean}}$ measures the average coherence in the polarimetric rotation domain, while the coherence fluctuation $|\gamma_{1-2}(\theta)|_{\text{std}}$ reflects the standard deviation of the coherence between the considered polarimetric channels. For polarimetric roll-invariant scatterers, the coherence degree $|\gamma_{1-2}(\theta)|_{\text{mean}}$ will be close to the original coherence $|\gamma_{1-2}(\theta)|_{\text{org}}$, and the coherence fluctuation $|\gamma_{1-2}(\theta)|_{\text{std}}$ will approach zero. However, for polarimetric roll-variant scatterers, the coherence degree $|\gamma_{1-2}(\theta)|_{\text{mean}}$ will be apart from the original coherence $|\gamma_{1-2}(\theta)|_{\text{org}}$, while the coherence fluctuation $|\gamma_{1-2}(\theta)|_{\text{std}}$ will be significantly enhanced. In this sense, the combination of these three features has the potential to describe the polarimetric roll-variant degree of scatterers. The maximum coherence $|\gamma_{1-2}(\theta)|_{\max}$ and the minimum coherence $|\gamma_{1-2}(\theta)|_{\min}$ respectively represent the highest and the lowest coherence values of two polarimetric channels in the polarimetric coherence pattern. The secondary features of the coherence contrast $|\gamma_{1-2}(\theta)|_{\text{contrast}}$ and the coherence anisotropy $|\gamma_{1-2}(\theta)|_{\text{A}}$ can be derived thereafter. For polarimetric roll-invariant scatterers, the maximum coherence $|\gamma_{1-2}(\theta)|_{\max}$ and the minimum coherence $|\gamma_{1-2}(\theta)|_{\min}$ will be close to each other. In this vein, both the coherence contrast $|\gamma_{1-2}(\theta)|_{\text{contrast}}$ and the coherence anisotropy $|\gamma_{1-2}(\theta)|_{\text{A}}$ will be very minor. On the other aspect, for polarimetric roll-variant scatterers which are more common in reality, these features have the potential to reflect the hidden differences among different targets which can assist target detection and classification. The concept of the coherence beamwidth $|\gamma_{1-2}(\theta)|_{\text{bw0.95}}$ is transferred from the 3-dB beamwidth of the antenna radiation pattern. It is the index to identify the rotation angle range where the polarimetric coherence values are no less than 0.95 times $|\gamma_{1-2}(\theta)|_{\max}$. In other words, $|\gamma_{1-2}(\theta)|_{\text{bw0.95}}$ indicates the stability range of certain polarimetric coherence for scatterers. The maximum rotation angle $\theta_{|\gamma_{1-2}(\theta)|_{\max}}$ and the minimum rotation angle $\theta_{|\gamma_{1-2}(\theta)|_{\min}}$ are the specific rotation angles which lead to the maximum coherence $|\gamma_{1-2}(\theta)|_{\max}$ and the minimum coherence $|\gamma_{1-2}(\theta)|_{\min}$ in the polarimetric rotation domain, respectively. These two polarimetric angle features highly rely on target types and their relative orientations to polarimetric radar. The diversities of these two polarimetric angle features have the possibility to differentiate land covers with different structures. Meanwhile, these features can also discriminate targets with the same type but with different orientations. For example, for buildings parallel to the flight pass, $|\gamma_{1-2}(\theta)|_{\max}$ is close to the original coherence $|\gamma_{1-2}(\theta)|_{\text{org}}$. Meanwhile, for oriented buildings, $|\gamma_{1-2}(\theta)|_{\max}$ can obviously be enhanced in the polarimetric rotation domain.

Considering the four independent polarimetric coherence patterns ($|\gamma_{\text{HH-VV}}(\theta)|$, $|\gamma_{\text{HH-HV}}(\theta)|$, $|\gamma_{\text{(HH+VV)-(HH-VV)}}(\theta)|$ and $|\gamma_{\text{(HH-VV)-(HV)}}(\theta)|$, 40 hidden features in total can be derived. These new polarimetric features completely describe the hidden information embedded in polarimetric coherence patterns for given polarimetric channels.

Finally, it should be pointed out that although the expression of the polarimetric coherence pattern shown in (2.4.3) is compact, the analytic expressions of the derived features are not always available. Without any loss, it is recommended to obtain these features numerically.

2.4.4 Demonstration and Investigation

Providing timely information for large-scale agricultural monitoring has become one of the main PolSAR applications in the civil field. At present, X- and C-band polarimetric radar data have been used in existing literature to monitor and estimate rice phenology [2, 74]. These studies show that PolSAR can identify the phenological

TABLE 2.4.1
Two-Dimension Polarimetric Coherence Pattern Features

	Polarimetric Coherence Pattern			
Polarimetric Feature	$\left\|\gamma_{\mathrm{HH\text{-}HV}}(\theta)\right\|$	$\left\|\gamma_{\mathrm{HH\text{-}VV}}(\theta)\right\|$	$\left\|\gamma_{\mathrm{(HH+VV)\text{-}(HH\text{-}VV)}}(\theta)\right\|$	$\left\|\gamma_{\mathrm{(HH\text{-}VV)\text{-}(HV)}}(\theta)\right\|$
Original coherence	$\left\|\gamma_{\mathrm{HH\text{-}HV}}(\theta)\right\|_{\mathrm{org}}$	$\left\|\gamma_{\mathrm{HH\text{-}VV}}(\theta)\right\|_{\mathrm{org}}$	$\left\|\gamma_{\mathrm{(HH+VV)\text{-}(HH\text{-}VV)}}(\theta)\right\|_{\mathrm{org}}$	$\left\|\gamma_{\mathrm{(HH\text{-}VV)\text{-}(HV)}}(\theta)\right\|_{\mathrm{org}}$
Coherence degree	$\left\|\gamma_{\mathrm{HH\text{-}HV}}(\theta)\right\|_{\mathrm{mean}}$	$\left\|\gamma_{\mathrm{HH\text{-}VV}}(\theta)\right\|_{\mathrm{mean}}$	$\left\|\gamma_{\mathrm{(HH+VV)\text{-}(HH\text{-}VV)}}(\theta)\right\|_{\mathrm{mean}}$	$\left\|\gamma_{\mathrm{(HH\text{-}VV)\text{-}(HV)}}(\theta)\right\|_{\mathrm{mean}}$
Coherence fluctuation	$\left\|\gamma_{\mathrm{HH\text{-}HV}}(\theta)\right\|_{\mathrm{std}}$	$\left\|\gamma_{\mathrm{HH\text{-}VV}}(\theta)\right\|_{\mathrm{std}}$	$\left\|\gamma_{\mathrm{(HH+VV)\text{-}(HH\text{-}VV)}}(\theta)\right\|_{\mathrm{std}}$	$\left\|\gamma_{\mathrm{(HH\text{-}VV)\text{-}(HV)}}(\theta)\right\|_{\mathrm{std}}$
Maximum coherence	$\left\|\gamma_{\mathrm{HH\text{-}HV}}(\theta)\right\|_{\max}$	$\left\|\gamma_{\mathrm{HH\text{-}VV}}(\theta)\right\|_{\max}$	$\left\|\gamma_{\mathrm{(HH+VV)\text{-}(HH\text{-}VV)}}(\theta)\right\|_{\max}$	$\left\|\gamma_{\mathrm{(HH\text{-}VV)\text{-}(HV)}}(\theta)\right\|_{\max}$
Minimum coherence	$\left\|\gamma_{\mathrm{HH\text{-}HV}}(\theta)\right\|_{\min}$	$\left\|\gamma_{\mathrm{HH\text{-}VV}}(\theta)\right\|_{\min}$	$\left\|\gamma_{\mathrm{(HH+VV)\text{-}(HH\text{-}VV)}}(\theta)\right\|_{\min}$	$\left\|\gamma_{\mathrm{(HH\text{-}VV)\text{-}(HV)}}(\theta)\right\|_{\min}$
Coherence contrast	$\left\|\gamma_{\mathrm{HH\text{-}HV}}(\theta)\right\|_{\mathrm{contrast}}$	$\left\|\gamma_{\mathrm{HH\text{-}VV}}(\theta)\right\|_{\mathrm{contrast}}$	$\left\|\gamma_{\mathrm{(HH+VV)\text{-}(HH\text{-}VV)}}(\theta)\right\|_{\mathrm{contrast}}$	$\left\|\gamma_{\mathrm{(HH\text{-}VV)\text{-}(HV)}}(\theta)\right\|_{\mathrm{contrast}}$
Coherence anisotropy	$\left\|\gamma_{\mathrm{HH\text{-}HV}}(\theta)\right\|_{\mathrm{A}}$	$\left\|\gamma_{\mathrm{HH\text{-}VV}}(\theta)\right\|_{\mathrm{A}}$	$\left\|\gamma_{\mathrm{(HH+VV)\text{-}(HH\text{-}VV)}}(\theta)\right\|_{\mathrm{A}}$	$\left\|\gamma_{\mathrm{(HH\text{-}VV)\text{-}(HV)}}(\theta)\right\|_{\mathrm{A}}$
Coherence beamwidth	$\theta_{\left\|\gamma_{\mathrm{HH\text{-}HV}}(\theta)\right\|_{\mathrm{bw0.95}}}$	$\theta_{\left\|\gamma_{\mathrm{HH\text{-}VV}}(\theta)\right\|_{\mathrm{bw0.95}}}$	$\theta_{\left\|\gamma_{\mathrm{(HH+VV)\text{-}(HH\text{-}VV)}}(\theta)\right\|_{\mathrm{bw0.95}}}$	$\theta_{\left\|\gamma_{\mathrm{(HH\text{-}VV)\text{-}(HV)}}(\theta)\right\|_{\mathrm{bw0.95}}}$
Maximum rotation angle	$\theta_{\left\|\gamma_{\mathrm{HH\text{-}HV}}(\theta)\right\|_{\max}}$	$\theta_{\left\|\gamma_{\mathrm{HH\text{-}VV}}(\theta)\right\|_{\max}}$	$\theta_{\left\|\gamma_{\mathrm{(HH+VV)\text{-}(HH\text{-}VV)}}(\theta)\right\|_{\max}}$	$\theta_{\left\|\gamma_{\mathrm{(HH\text{-}VV)\text{-}(HV)}}(\theta)\right\|_{\max}}$
Minimum rotation angle	$\theta_{\left\|\gamma_{\mathrm{HH\text{-}HV}}(\theta)\right\|_{\min}}$	$\theta_{\left\|\gamma_{\mathrm{HH\text{-}VV}}(\theta)\right\|_{\min}}$	$\theta_{\left\|\gamma_{\mathrm{(HH+VV)\text{-}(HH\text{-}VV)}}(\theta)\right\|_{\min}}$	$\theta_{\left\|\gamma_{\mathrm{(HH\text{-}VV)\text{-}(HV)}}(\theta)\right\|_{\min}}$

stages of rice. For most agricultural monitoring systems, crop identification is a key step [75, 76]. Crop recognition ability is of great value to crop cultivation, management, and yield estimation. This section mainly introduces the application of polarimetric coherence patterns for crop recognition.

2.4.4.1 UAVSAR PolSAR Data Description

The PolSAR data over southern Manitoba, Canada, acquired by the JPL airborne L-band UAVSAR [77], are adopted for demonstration of polarimetric coherence patterns and the derived features. The provided data have already been 3-look processed in the range and 12-look processed in the azimuth. The range and azimuth resolutions are respectively 5 and 7 meters. A region containing various land covers, mainly including broadleaf, forage crops, soybeans, corn, wheat, rapeseed, and oats, is selected for experimental studies. The SimiTest speckle filter [72], especially suitable for polarimetric coherence estimation, is adopted for speckle reduction. The SimiTest filter has been applied using a 15 × 15 moving window, and the filtered PolSAR RGB composite image with a Pauli basis is shown in Figure 2.4.2 (a). A ground-truth map for the seven known land covers is shown in Figure 2.4.2 (b).

2.4.4.2 Polarimetric Coherence Pattern Visualization and Investigation

For each of the seven known land covers, one sample pixel is randomly selected, and their polarimetric coherence patterns of $\left|\gamma_{\text{HH-VV}}(\theta)\right|$, $\left|\gamma_{\text{HH-HV}}(\theta)\right|$, $\left|\gamma_{\text{(HH+VV)-(HH-VV)}}(\theta)\right|$ and $\left|\gamma_{\text{(HH-VV)-(HV)}}(\theta)\right|$ are generated and shown in Figure 2.4.3. It is observed that polarimetric coherence patterns from different polarization combinations show quite various characteristics. For example, polarimetric coherence patterns $\left|\gamma_{\text{(HH+VV)-(HH-VV)}}(\theta)\right|$ and $\left|\gamma_{\text{(HH-VV)-(HV)}}(\theta)\right|$ do not show a sidelobe-like effect for all these land covers. Their extreme maximum values correspond to the maximum values, while their extreme minimum values correspond to the minimum values. Since the periods of $\left|\gamma_{\text{(HH+VV)-(HH-VV)}}(\theta)\right|$ and $\left|\gamma_{\text{(HH-VV)-(HV)}}(\theta)\right|$ are respectively $\pi/2$ and $\pi/4$, their polarimetric coherence patterns generally show four-lobed and eight-lobed shapes in the full range of $[-\pi,\pi)$. In comparison, the polarimetric

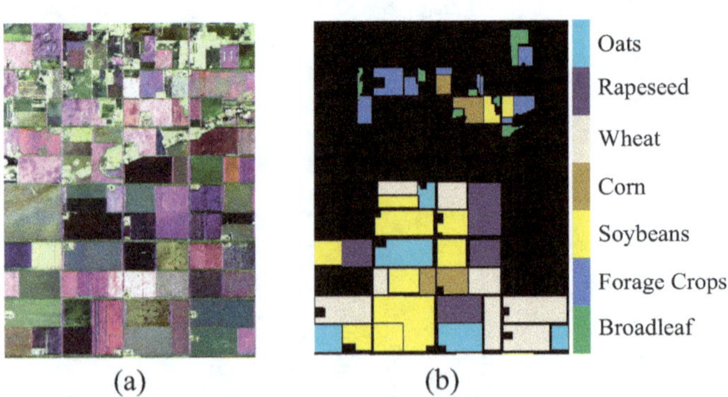

FIGURE 2.4.2 UAVSAR PolSAR data. (a) PolSAR Pauli image, (b) ground-truth.

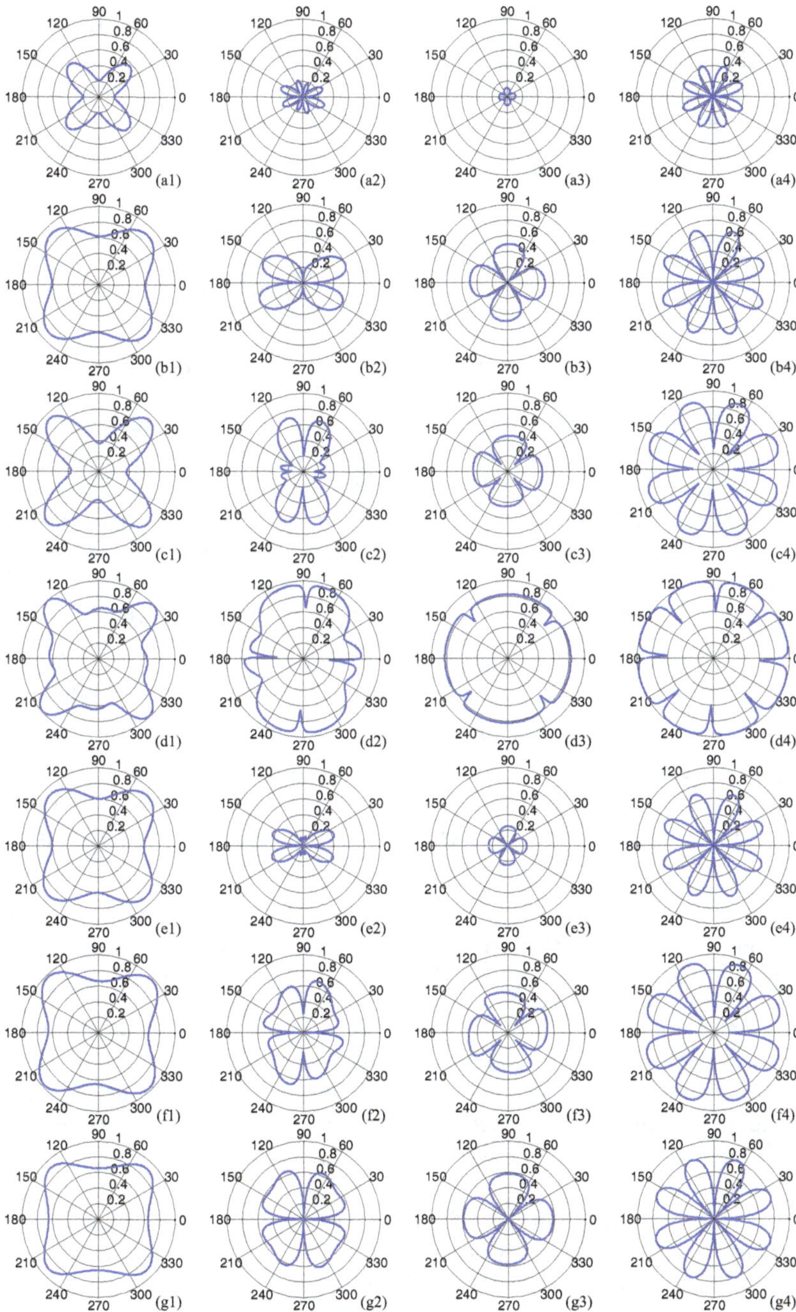

FIGURE 2.4.3 Polarimetric coherence patterns of the seven land covers. (a1)–(a4) Broadleaf, (b1)–(b4) forage crops, (c1)–(c4) soybeans, (d1)–(d4) corn, (e1)–(e4) wheat, (f1)–(f4) rapeseed, and (g1)–(g4) oats. The numbers 1–4 indicate $\left|\gamma_{\text{HH-VV}}(\theta)\right|$, $\left|\gamma_{\text{HH-HV}}(\theta)\right|$, $\left|\gamma_{(\text{HH+VV})\text{-}(\text{HH-VV})}(\theta)\right|$, and $\left|\gamma_{(\text{HH-VV})\text{-}(\text{HV})}(\theta)\right|$, respectively.

coherence patterns of $\left|\gamma_{\text{HH-VV}}(\theta)\right|$ and $\left|\gamma_{\text{HH-HV}}(\theta)\right|$ show more complicated shapes since a sidelobe-like effect presents for several land covers, such as the corn, shown in Figures 2.4.3 (d1) and 2.4.3 (d2). In addition, for a given polarization combination, the polarimetric coherence patterns from different land covers also show various responses, especially for $\left|\gamma_{\text{HH-VV}}(\theta)\right|$ and $\left|\gamma_{\text{HH-HV}}(\theta)\right|$, which have quite different shapes. Even for $\left|\gamma_{(\text{HH+VV})\text{-}(\text{HH-VV})}(\theta)\right|$ and $\left|\gamma_{(\text{HH-VV})\text{-}(\text{HV})}(\theta)\right|$, with a relatively certain shape, detailed characters such as the maximum and minimum coherence values differ among these land covers. From these seven land cover samples, it is also observed that the maximum rotation angle $\theta_{\left|\gamma_{\text{HH-VV}}(\theta)\right|_{\max}}$ approaches $\pm\frac{\pi}{4}$, while the maximum rotation angle $\theta_{\left|\gamma_{(\text{HH-VV})\text{-}(\text{HV})}(\theta)\right|_{\max}}$ approaches $\pm\frac{\pi}{8}$. Although these sample pixels are randomly selected, these phenomena are also generally valid for other samples.

Furthermore, the combinations of these polarimetric coherence patterns show great potential for land cover discrimination and classification. For the seven known classes of land covers, there are 21 class pairs in all. With visual inspection, the majority of these class pairs can be easily discriminated by these polarimetric coherence patterns. Taking the class pair of broadleaf and forage crops as an example, the shapes of polarimetric coherence patterns of $\left|\gamma_{\text{HH-VV}}(\theta)\right|$, $\left|\gamma_{\text{HH-HV}}(\theta)\right|$, and $\left|\gamma_{(\text{HH+VV})\text{-}(\text{HH-VV})}(\theta)\right|$ are obviously different, shown in Figures 2.4.3 (a1)–(a3) and 2.4.3 (b1)–(b3). Even though the shapes of polarimetric coherence pattern $\left|\gamma_{(\text{HH-VV})\text{-}(\text{HV})}(\theta)\right|$ are similar for this class pair, $\left|\gamma_{(\text{HH-VV})\text{-}(\text{HV})}(\theta)\right|_{\max}$ are respectively 0.4148 and 0.6948, which can successfully discriminate them. Only for two class pairs of forage crops and wheat, rapeseed and oats, each class pair share similar shapes of all four polarimetric coherence patterns, even though classes in these two pairs can also be separated with a detailed investigation of the derived features. Quantitatively, the $\left|\gamma_{\text{HH-HV}}(\theta)\right|_{\max}$ and $\left|\gamma_{(\text{HH+VV})\text{-}(\text{HH-VV})}(\theta)\right|_{\max}$ for forage crops are 0.6047 and 0.4922, which are much higher than those of wheat, with values of 0.4309 and 0.2475, accordingly. In addition, $\left|\gamma_{(\text{HH+VV})\text{-}(\text{HH-VV})}(\theta)\right|_{\min}$ and $\left|\gamma_{(\text{HH-VV})\text{-}(\text{HV})}(\theta)\right|_{\min}$ for rapeseed are 0.1344 and 0.2149, while those of oats approach 0, with values of 0.0061 and 0.0355, respectively. Therefore, the two class pairs of forage crops and wheat, and rapeseed and oats can also be well differentiated by the derived features. As a result, polarimetric coherence patterns provide powerful tools to reflect even minor differences among land cover, and the derived features are suitable for land cover classification.

The derived features from the full-scene will be investigated next. Principally, these phenomena are intrinsically determined by the various scatterer characteristics in the polarimetric rotation domain, which are also the foundations for further development of land cover classification.

2.4.4.3 Derived Features from Polarimetric Coherence Patterns

The derived features from polarimetric coherence patterns of $\left|\gamma_{\text{HH-VV}}(\theta)\right|$, $\left|\gamma_{\text{HH-HV}}(\theta)\right|$, $\left|\gamma_{(\text{HH+VV})\text{-}(\text{HH-VV})}(\theta)\right|$, and $\left|\gamma_{(\text{HH-VV})\text{-}(\text{HV})}(\theta)\right|$ for the full-scene are calculated and shown in Figures 2.4.4–2.4.7, respectively. As expected, these features contain rich information of local land covers which is hidden in the polarimetric rotation domain. These features also show great sensitivity to various land covers. For polarimetric coherence pattern $\left|\gamma_{\text{HH-VV}}(\theta)\right|$, compared with the original coherence $\left|\gamma_{\text{HH-VV}}(\theta)\right|_{\text{org}}$, the maximum coherence $\left|\gamma_{\text{HH-VV}}(\theta)\right|_{\text{max}}$ is obviously enhanced, while the minimum coherence $\left|\gamma_{\text{HH-VV}}(\theta)\right|_{\text{min}}$ shows relatively similar characteristics to $\left|\gamma_{\text{HH-VV}}(\theta)\right|_{\text{org}}$. For land covers with similar $\left|\gamma_{\text{HH-VV}}(\theta)\right|_{\text{org}}$ and $\left|\gamma_{\text{HH-VV}}(\theta)\right|_{\text{min}}$, the values of the minimum rotation angle $\theta_{\left|\gamma_{\text{HH-VV}}(\theta)\right|_{\text{min}}}$ approach 0. Meanwhile, the majority of values of the maximum rotation angle $\theta_{\left|\gamma_{\text{HH-VV}}(\theta)\right|_{\text{min}}}$ are close to $\pm\frac{\pi}{4}$, which marks $\theta_{\left|\gamma_{\text{HH-VV}}(\theta)\right|_{\text{min}}}$ with the least sensitivity to these land covers. The remaining features of $\left|\gamma_{\text{HH-VV}}(\theta)\right|_{\text{mean}}$, $\left|\gamma_{\text{HH-VV}}(\theta)\right|_{\text{std}}$, $\left|\gamma_{\text{HH-VV}}(\theta)\right|_{\text{max}}$, $\left|\gamma_{\text{HH-VV}}(\theta)\right|_{\text{A}}$, and $\left|\gamma_{\text{HH-VV}}(\theta)\right|_{\text{bw0.95}}$ also show good sensitivity and added information to these local land covers compared with the original coherence $\left|\gamma_{\text{HH-VV}}(\theta)\right|_{\text{org}}$. For polarimetric coherence pattern $\left|\gamma_{\text{HH-HV}}(\theta)\right|$, the original coherence $\left|\gamma_{\text{HH-HV}}(\theta)\right|_{\text{org}}$ is clearly enhanced and weakened by the maximum coherence $\left|\gamma_{\text{HH-HV}}(\theta)\right|_{\text{max}}$ and the minimum coherence $\left|\gamma_{\text{HH-HV}}(\theta)\right|_{\text{min}}$, respectively. Visually, the features of $\left|\gamma_{\text{HH-HV}}(\theta)\right|_{\text{mean}}$, $\left|\gamma_{\text{HH-HV}}(\theta)\right|_{\text{std}}$, $\left|\gamma_{\text{HH-HV}}(\theta)\right|_{\text{max}}$, $\left|\gamma_{\text{HH-HV}}(\theta)\right|_{\text{A}}$, and $\theta_{\left|\gamma_{\text{HH-HV}}(\theta)\right|_{\text{max}}}$ have better sensitivity to these various land covers than

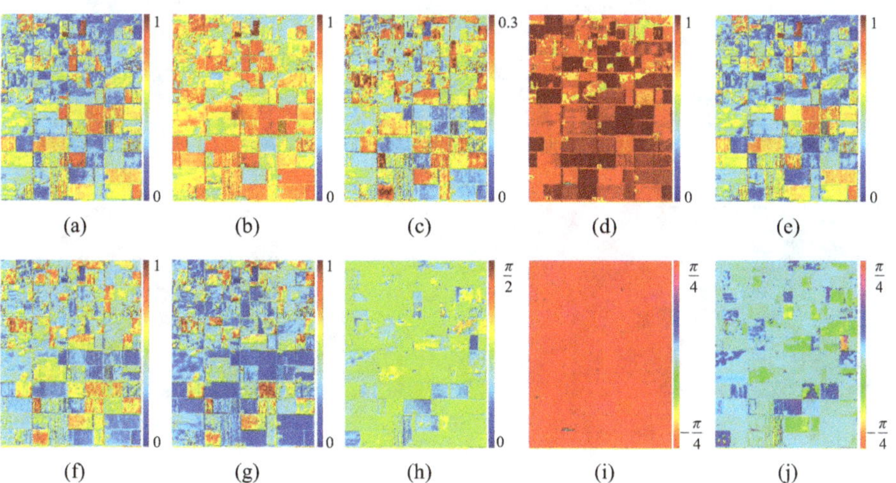

FIGURE 2.4.4 Features derived from polarimetric coherence pattern $\left|\gamma_{\text{HH-VV}}(\theta)\right|$. (a) Original coherence $\left|\gamma_{\text{HH-VV}}(\theta)\right|_{\text{org}}$, (b) coherence degree $\left|\gamma_{\text{HH-VV}}(\theta)\right|_{\text{mean}}$, (c) coherence fluctuation $\left|\gamma_{\text{HH-VV}}(\theta)\right|_{\text{std}}$, (d) maximum coherence $\left|\gamma_{\text{HH-VV}}(\theta)\right|_{\text{max}}$, (e) minimum coherence $\left|\gamma_{\text{HH-VV}}(\theta)\right|_{\text{min}}$, (f) coherence contrast $\left|\gamma_{\text{HH-VV}}(\theta)\right|_{\text{contrast}}$, (g) coherence anisotropy $\left|\gamma_{\text{HH-VV}}(\theta)\right|_{\text{A}}$, (h) coherence beamwidth $\theta_{\left|\gamma_{\text{HH-HV}}(\theta)\right|_{\text{bw0.95}}}$, (i) maximum rotation angle $\theta_{\left|\gamma_{\text{HH-VV}}(\theta)\right|_{\text{max}}}$, and (j) minimum rotation angle $\theta_{\left|\gamma_{\text{HH-VV}}(\theta)\right|_{\text{min}}}$.

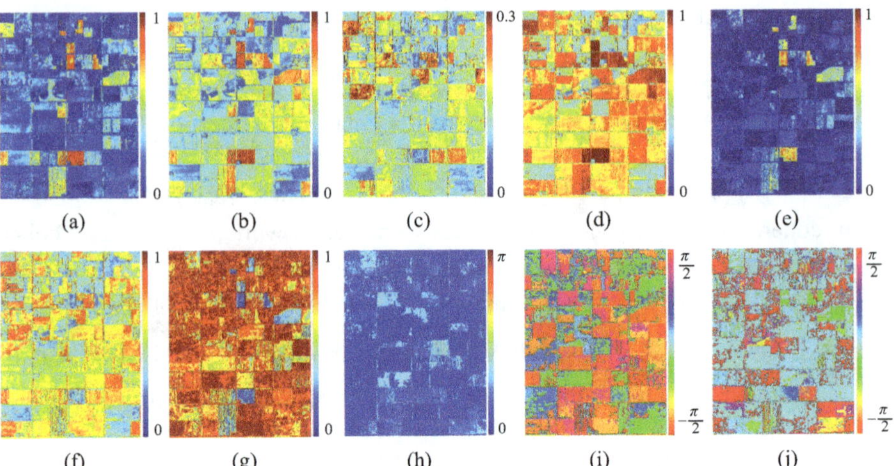

FIGURE 2.4.5 Features derived from polarimetric coherence pattern $\left|\gamma_{\text{HH-HV}}(\theta)\right|$. (a) Original coherence $\left|\gamma_{\text{HH-HV}}(\theta)\right|_{\text{org}}$, (b) coherence degree $\left|\gamma_{\text{HH-HV}}(\theta)\right|_{\text{mean}}$, (c) coherence fluctuation $\left|\gamma_{\text{HH-HV}}(\theta)\right|_{\text{std}}$, (d) maximum coherence $\left|\gamma_{\text{HH-HV}}(\theta)\right|_{\max}$, (e) minimum coherence $\left|\gamma_{\text{HH-HV}}(\theta)\right|_{\min}$, (f) coherence contrast $\left|\gamma_{\text{HH-HV}}(\theta)\right|_{\text{contrast}}$, (g) coherence anisotropy $\left|\gamma_{\text{HH-HV}}(\theta)\right|_{\text{A}}$, (h) coherence beamwidth $\theta_{\left|\gamma_{\text{HH-VV}}(\theta)\right|_{\text{bw0.95}}}$, (i) maximum rotation angle $\theta_{\left|\gamma_{\text{HH-HV}}(\theta)\right|_{\max}}$, and (j) minimum rotation angle $\theta_{\left|\gamma_{\text{HH-HV}}(\theta)\right|_{\min}}$.

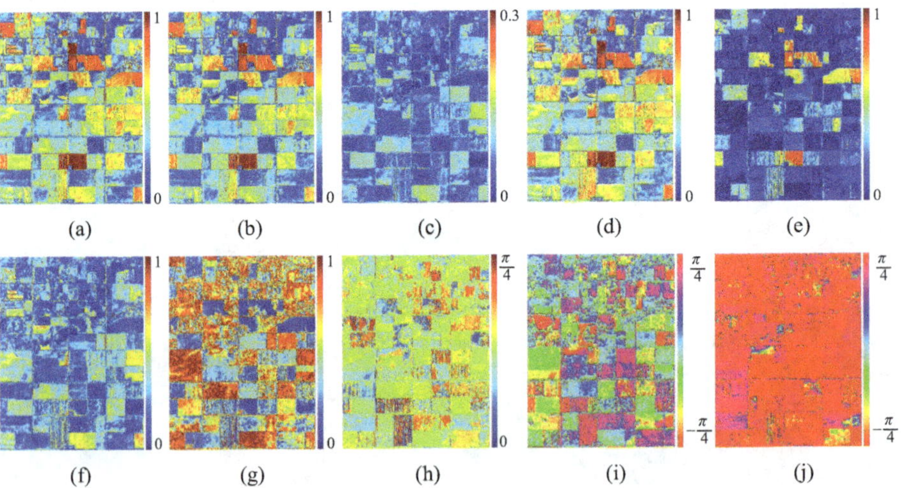

FIGURE 2.4.6 Features derived from polarimetric coherence pattern $\left|\gamma_{\text{(HH+VV)-(HH-VV)}}(\theta)\right|$. (a) Original coherence $\left|\gamma_{\text{(HH+VV)-(HH-VV)}}(\theta)\right|_{\text{org}}$, (b) coherence degree $\left|\gamma_{\text{(HH+VV)-(HH-VV)}}(\theta)\right|_{\text{mean}}$, (c) coherence fluctuation $\left|\gamma_{\text{(HH+VV)-(HH-VV)}}(\theta)\right|_{\text{std}}$, (d) maximum coherence $\left|\gamma_{\text{(HH+VV)-(HH-VV)}}(\theta)\right|_{\max}$, (e) minimum coherence $\left|\gamma_{\text{(HH+VV)-(HH-VV)}}(\theta)\right|_{\min}$, (f) coherence contrast $\left|\gamma_{\text{(HH+VV)-(HH-VV)}}(\theta)\right|_{\text{contrast}}$, (g) coherence anisotropy $\left|\gamma_{\text{(HH+VV)-(HH-VV)}}(\theta)\right|_{\text{A}}$, (h) coherence beamwidth $\theta_{\left|\gamma_{\text{(HH+VV)-(HH-VV)}}(\theta)\right|_{\text{bw0.95}}}$, (i) maximum rotation angle $\theta_{\left|\gamma_{\text{(HH+VV)-(HH-VV)}}(\theta)\right|_{\max}}$, and (j) minimum rotation angle $\theta_{\left|\gamma_{\text{(HH+VV)-(HH-VV)}}(\theta)\right|_{\min}}$.

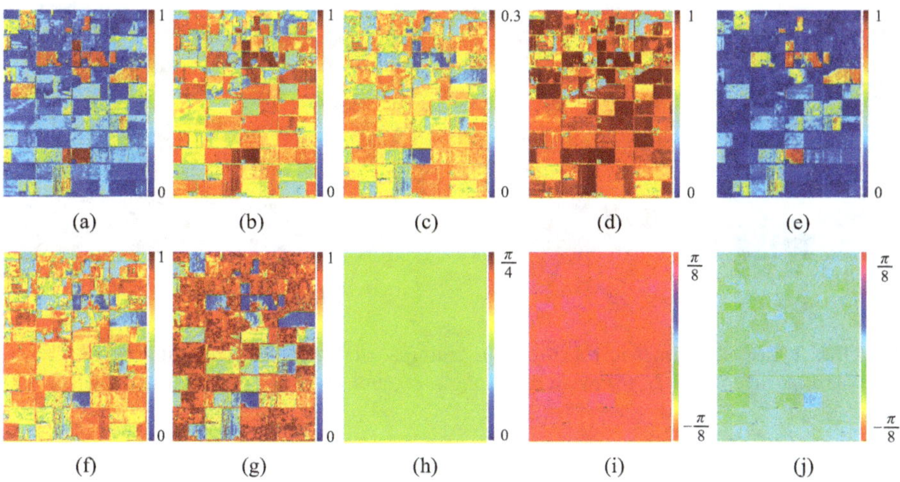

FIGURE 2.4.7 Features derived from polarimetric coherence pattern $\left|\gamma_{(\text{HH-VV})\text{-}(\text{HV})}(\theta)\right|$. (a) Original coherence $\left|\gamma_{(\text{HH-VV})\text{-}(\text{HV})}(\theta)\right|_{\text{org}}$, (b) coherence degree $\left|\gamma_{(\text{HH-VV})\text{-}(\text{HV})}(\theta)\right|_{\text{mean}}$, (c) coherence fluctuation $\left|\gamma_{(\text{HH-VV})\text{-}(\text{HV})}(\theta)\right|_{\text{std}}$, (d) maximum coherence $\left|\gamma_{(\text{HH-VV})\text{-}(\text{HV})}(\theta)\right|_{\max}$, (e) minimum coherence $\left|\gamma_{(\text{HH-VV})\text{-}(\text{HV})}(\theta)\right|_{\min}$, (f) coherence contrast $\left|\gamma_{(\text{HH-VV})\text{-}(\text{HV})}(\theta)\right|_{\text{contrast}}$, (g) coherence anisotropy $\left|\gamma_{(\text{HH-VV})\text{-}(\text{HV})}(\theta)\right|_{\text{A}}$, (h) coherence beamwidth $\theta_{\left|\gamma_{(\text{HH-VV})\text{-}(\text{HV})}(\theta)\right|_{\text{bw0.95}}}$, (i) maximum rotation angle $\theta_{\left|\gamma_{(\text{HH-VV})\text{-}(\text{HV})}(\theta)\right|_{\max}}$, and (j) minimum rotation angle $\theta_{\left|\gamma_{(\text{HH-VV})\text{-}(\text{HV})}(\theta)\right|_{\min}}$.

the remaining features. For polarimetric coherence pattern $\left|\gamma_{(\text{HH+VV})\text{-}(\text{HH-VV})}(\theta)\right|$, it is observed that the features of $\left|\gamma_{(\text{HH+VV})\text{-}(\text{HH-VV})}(\theta)\right|_{\text{org}}$, $\left|\gamma_{(\text{HH+VV})\text{-}(\text{HH-VV})}(\theta)\right|_{\text{mean}}$, and $\left|\gamma_{(\text{HH+VV})\text{-}(\text{HH-VV})}(\theta)\right|_{\max}$ are very close. Although the original coherence is obviously weakened by the minimum coherence $\left|\gamma_{(\text{HH+VV})\text{-}(\text{HH-VV})}(\theta)\right|_{\min}$, the polarimetric coherence fluctuation $\left|\gamma_{(\text{HH+VV})\text{-}(\text{HH-VV})}(\theta)\right|_{\text{std}}$ remains very minor for the majority of these samples, especially compared with the other three polarimetric coherence patterns. This can be further validated by the coherence beamwidth $\theta_{\left|\gamma_{(\text{HH+VV})\text{-}(\text{HH-VV})}(\theta)\right|_{\text{bw0.95}}}$, since the majority of these samples have a large stability range of polarimetric coherence. Furthermore, most values of the minimum rotation angle $\theta_{\left|\gamma_{(\text{HH+VV})\text{-}(\text{HH-VV})}(\theta)\right|_{\min}}$ are close to $\pm\frac{\pi}{4}$, which marks $\theta_{\left|\gamma_{(\text{HH+VV})\text{-}(\text{HH-VV})}(\theta)\right|_{\min}}$ with the least sensitivity to these land covers. For polarimetric coherence pattern $\left|\gamma_{(\text{HH-VV})\text{-}(\text{HV})}(\theta)\right|$, the enhancement and weakening of the original coherence $\left|\gamma_{(\text{HH-VV})\text{-}(\text{HV})}(\theta)\right|_{\text{org}}$ are both significant by the maximum coherence $\left|\gamma_{(\text{HH-VV})\text{-}(\text{HV})}(\theta)\right|_{\max}$ and the minimum coherence $\left|\gamma_{(\text{HH-VV})\text{-}(\text{HV})}(\theta)\right|_{\min}$. Meanwhile, as also observed from Figure 2.4.3, values of the

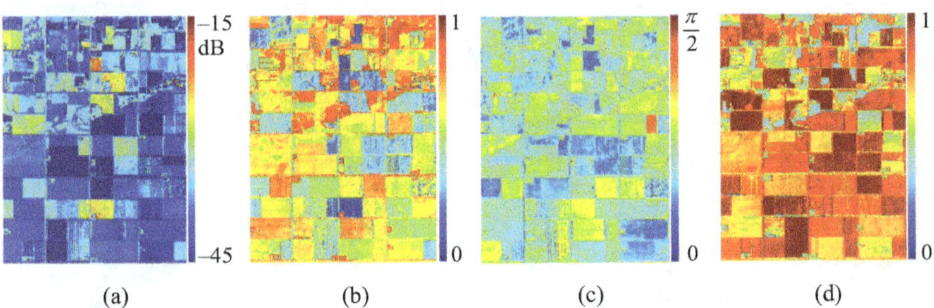

FIGURE 2.4.8 Polarimetric roll-invariant features. (a) Total backscattering power SPAN, (b) polarimetric entropy *H*, (c) mean alpha angle $\bar{\alpha}$, and (d) polarimetric anisotropy *Ani*.

maximum rotation angle $\theta_{\left|\gamma_{\text{(HH-VV)-(HV)}}(\theta)\right|_{\max}}$ are close to $\pm\frac{\pi}{8}$, while those of the minimum rotation angle $\theta_{\left|\gamma_{\text{(HH-VV)-(HV)}}(\theta)\right|_{\min}}$ are close to 0. Furthermore, the coherence beamwidth $\theta_{\left|\gamma_{\text{(HH-VV)-(HV)}}(\theta)\right|_{\text{bw0.95}}}$ has an almost constant value of about $\frac{\pi}{8}$, which is the half period of $\left|\gamma_{\text{(HH-VV)-(HV)}}(\theta)\right|$. In this vein, generally, these three features are not suitable for land cover discrimination. In comparison, the remaining features derived from $\left|\gamma_{\text{(HH-VV)-(HV)}}(\theta)\right|$ show good sensitivity performance.

In addition, the commonly used polarimetric roll-invariant features of total backscattering power $SPAN$, polarimetric entropy H, mean alpha angle $\bar{\alpha}$, and polarimetric anisotropy Ani are displayed in Figure 2.4.8 for comparison. Basically, these two kinds of features show good complementary potential to fully understand the scattering mechanism of various land covers. For quantitative investigation of these derived features, target classification will be adopted for demonstration in the following section.

2.4.4.4 Polarimetric Coherence Maximization

Polarimetric coherence features strongly relate to the types and orientations of the local scatterers. This section describes how to enhance the polarimetric coherence of land cover in the polarimetric rotation domain and provides demonstrations with AIRSAR P-, L-, and C-band PolSAR data over Flevoland, the Netherlands. The details of these multi-frequency datasets are described in Section 2.3.6. The original coherence $\left|\gamma_{\text{HH-VV}}(\theta)\right|_{\text{org}}$, $\left|\gamma_{\text{HH-HV}}(\theta)\right|_{\text{org}}$, $\left|\gamma_{\text{(HH+VV)-(HV)}}(\theta)\right|_{\text{org}}$, and $\left|\gamma_{\text{(HH-VV)-(HV)}}(\theta)\right|_{\text{org}}$ and the maximum coherence $\left|\gamma_{\text{HH-VV}}(\theta)\right|_{\max}$, $\left|\gamma_{\text{HH-HV}}(\theta)\right|_{\max}$, $\left|\gamma_{\text{(HH+VV)-(HV)}}(\theta)\right|_{\max}$, and $\left|\gamma_{\text{(HH-VV)-(HV)}}(\theta)\right|_{\max}$ of four independent polarimetric coherence patterns are calculated and shown in Figures 2.4.9–2.4.11 for the P-, L-, and C-band datasets, respectively. Meanwhile, the histograms of these polarimetric coherence features are

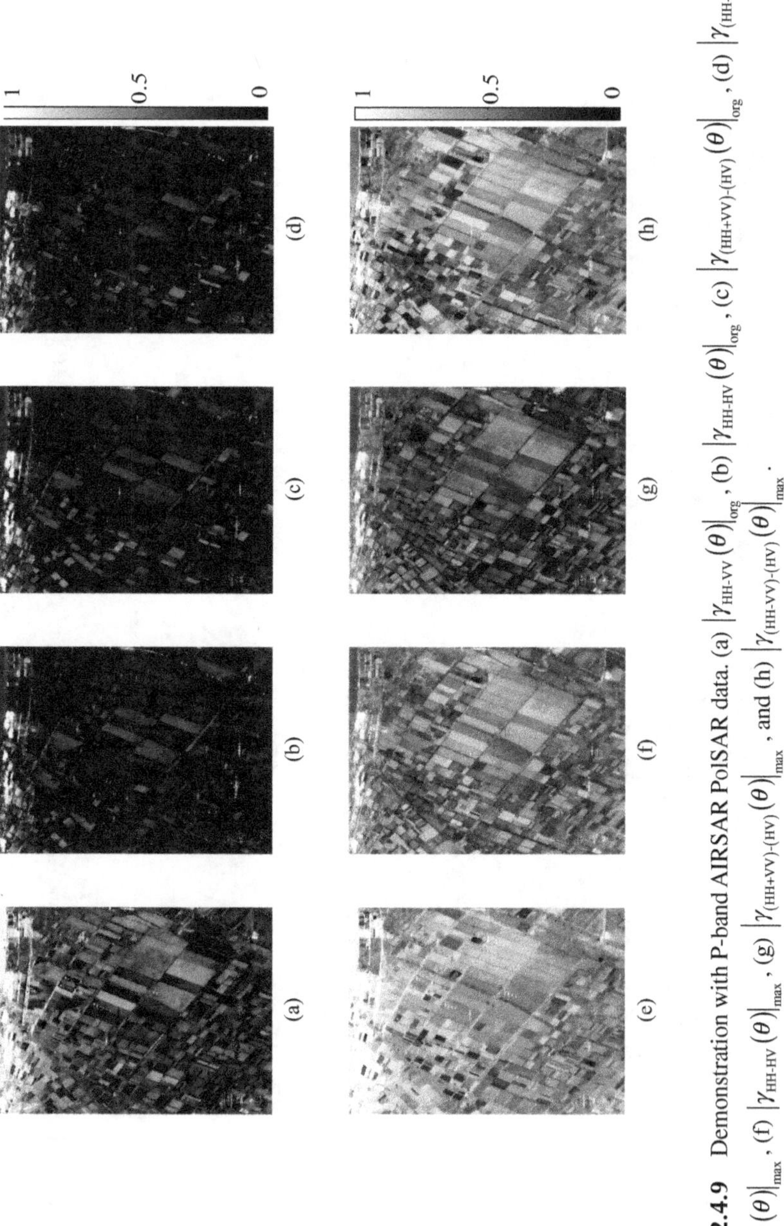

FIGURE 2.4.9 Demonstration with P-band AIRSAR PolSAR data. (a) $\left|\gamma_{\text{HH-VV}}(\theta)\right|_{\text{org}}$, (b) $\left|\gamma_{\text{HH-HV}}(\theta)\right|_{\text{org}}$, (c) $\left|\gamma_{(\text{HH+VV})\text{-}(\text{HV})}(\theta)\right|_{\text{org}}$, (d) $\left|\gamma_{(\text{HH-VV})\text{-}(\text{HV})}(\theta)\right|_{\text{org}}$, (e) $\left|\gamma_{\text{HH-VV}}(\theta)\right|_{\max}$, (f) $\left|\gamma_{\text{HH-HV}}(\theta)\right|_{\max}$, (g) $\left|\gamma_{(\text{HH+VV})\text{-}(\text{HV})}(\theta)\right|_{\max}$, and (h) $\left|\gamma_{(\text{HH-VV})\text{-}(\text{HV})}(\theta)\right|_{\max}$.

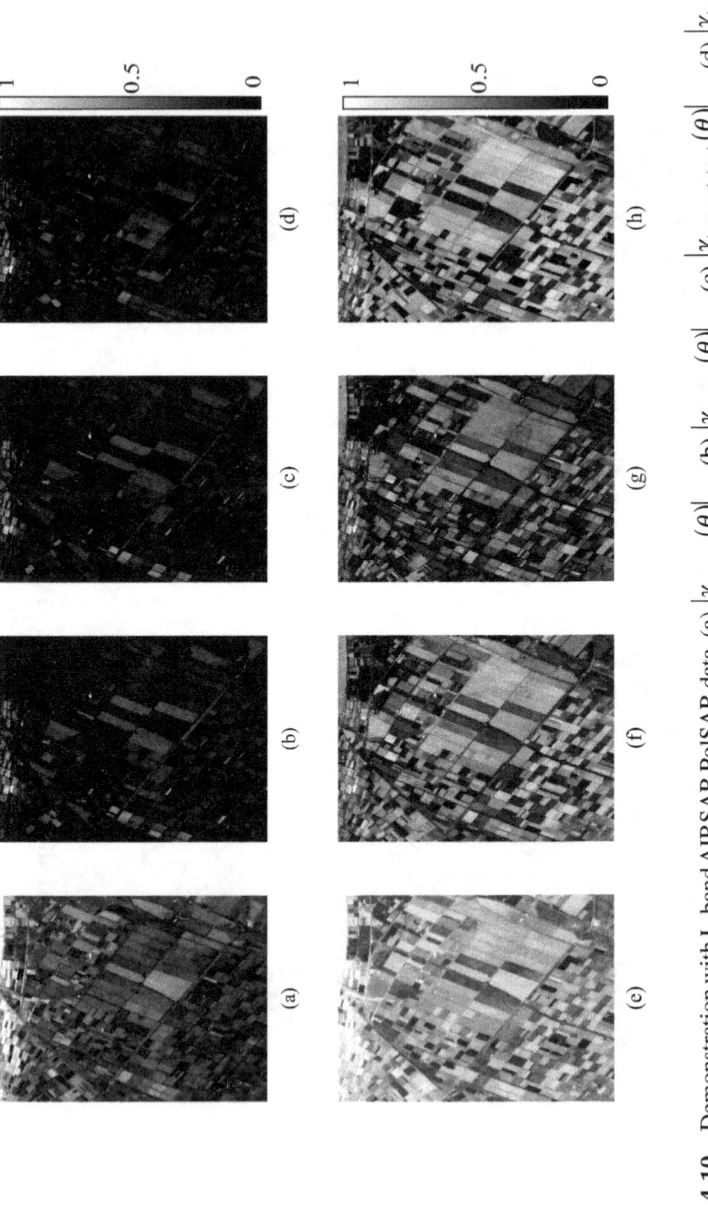

FIGURE 2.4.10 Demonstration with L-band AIRSAR PolSAR data. (a) $\left|\gamma_{\text{HH-VV}}(\theta)\right|_{\text{org}}$, (b) $\left|\gamma_{\text{HH-HV}}(\theta)\right|_{\text{org}}$, (c) $\left|\gamma_{\text{(HH+VV)-(HV)}}(\theta)\right|_{\text{org}}$, (d) $\left|\gamma_{\text{(HH-VV)-(HV)}}(\theta)\right|_{\text{org}}$, (e) $\left|\gamma_{\text{HH-VV}}(\theta)\right|_{\max}$, (f) $\left|\gamma_{\text{HH-HV}}(\theta)\right|_{\max}$, (g) $\left|\gamma_{\text{(HH+VV)-(HV)}}(\theta)\right|_{\max}$, and (h) $\left|\gamma_{\text{(HH-VV)-(HV)}}(\theta)\right|_{\max}$.

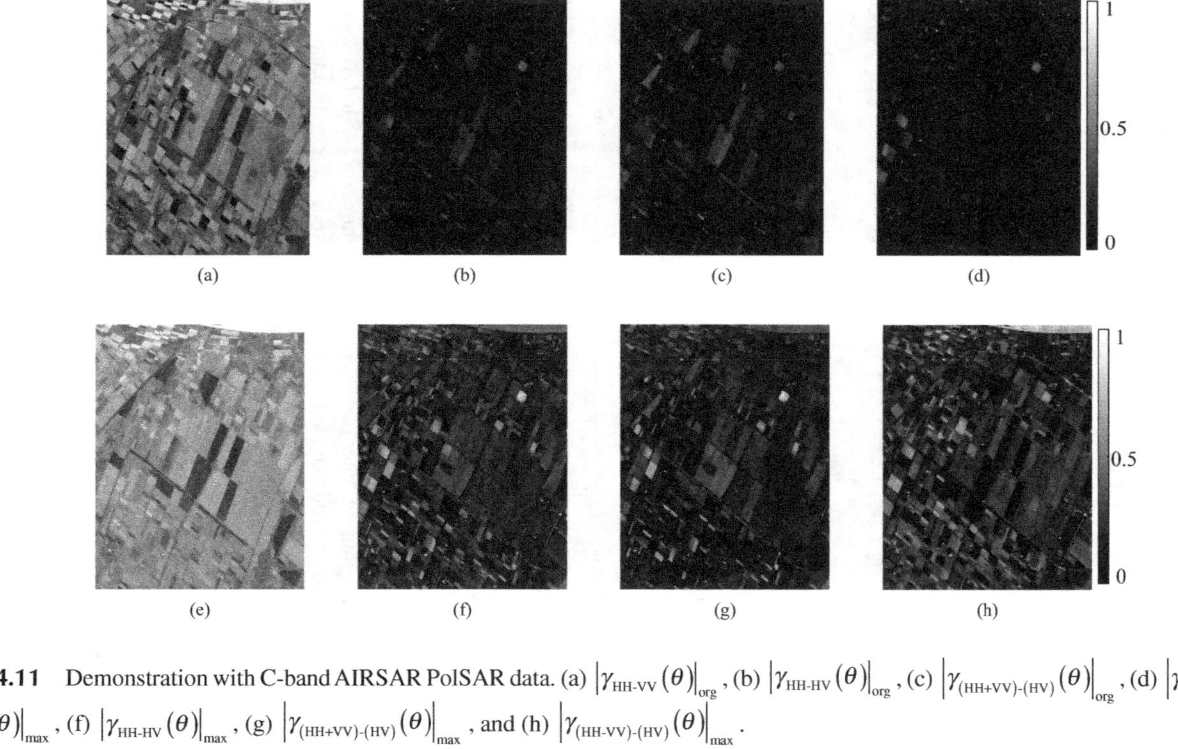

FIGURE 2.4.11 Demonstration with C-band AIRSAR PolSAR data. (a) $\left|\gamma_{\text{HH-VV}}(\theta)\right|_{\text{org}}$, (b) $\left|\gamma_{\text{HH-HV}}(\theta)\right|_{\text{org}}$, (c) $\left|\gamma_{(\text{HH+VV})\text{-}(\text{HV})}(\theta)\right|_{\text{org}}$, (d) $\left|\gamma_{(\text{HH-VV})\text{-}(\text{HV})}(\theta)\right|_{\text{org}}$, (e) $\left|\gamma_{\text{HH-VV}}(\theta)\right|_{\text{max}}$, (f) $\left|\gamma_{\text{HH-HV}}(\theta)\right|_{\text{max}}$, (g) $\left|\gamma_{(\text{HH+VV})\text{-}(\text{HV})}(\theta)\right|_{\text{max}}$, and (h) $\left|\gamma_{(\text{HH-VV})\text{-}(\text{HV})}(\theta)\right|_{\text{max}}$.

shown in Figures 2.4.12–2.4.14, respectively. As expected, without the maximization processing, these crop areas obviously show very low polarimetric coherence values. The majority of these values are below 0.5. Among the four polarimetric coherence patterns, only the $\left|\gamma_{\text{HH-VV}}(\theta)\right|_{\text{org}}$ of co-polarization channels achieves relatively high values. This phenomenon can be explained from two aspects. The first is that crop fields usually satisfy the azimuth symmetry assumption with $\left\langle S_{\text{HH}}S_{\text{HV}}^{*}\right\rangle \approx 0$ and $\left\langle S_{\text{VV}}S_{\text{HV}}^{*}\right\rangle \approx 0$ and produce low values of cross-correlation between two polarimetric

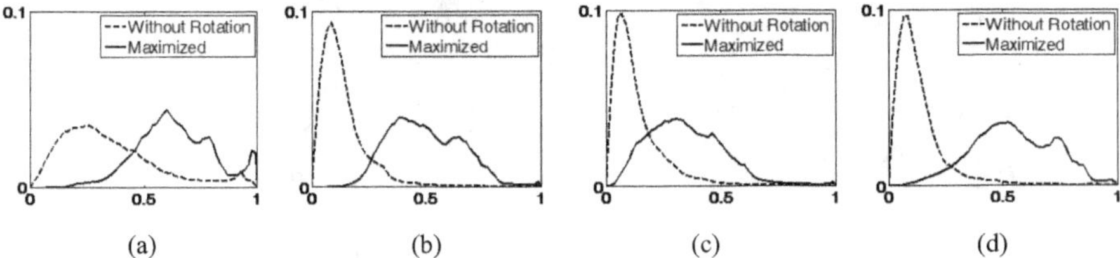

FIGURE 2.4.12 Histograms of polarimetric coherence features for P-band AIRSAR PolSAR data in terms of original coherence and maximum coherence for polarimetric coherence patterns of (a) $\left|\gamma_{\text{HH-VV}}(\theta)\right|$, (b) $\left|\gamma_{\text{HH-HV}}(\theta)\right|$, (c) $\left|\gamma_{\text{(HH+VV)-(HV)}}(\theta)\right|$, and (d) $\left|\gamma_{\text{(HH-VV)-(HV)}}(\theta)\right|$.

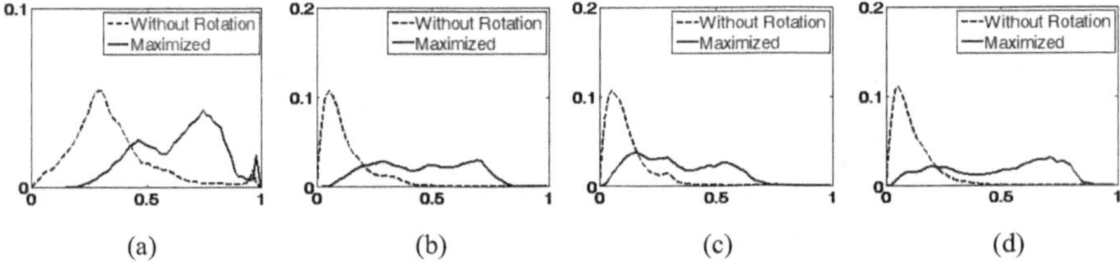

FIGURE 2.4.13 Histograms of polarimetric coherence features for L-band AIRSAR PolSAR data in terms of original coherence and maximum coherence for polarimetric coherence patterns of (a) $\left|\gamma_{\text{HH-VV}}(\theta)\right|$, (b) $\left|\gamma_{\text{HH-HV}}(\theta)\right|$, (c) $\left|\gamma_{\text{(HH+VV)-(HV)}}(\theta)\right|$, and (d) $\left|\gamma_{\text{(HH-VV)-(HV)}}(\theta)\right|$.

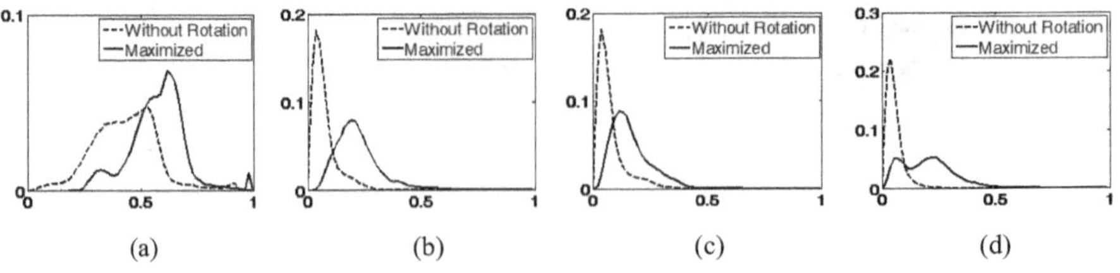

FIGURE 2.4.14 Histograms of polarimetric coherence features for C-band AIRSAR PolSAR data in terms of original coherence and maximum coherence for polarimetric coherence patterns of (a) $\left|\gamma_{\text{HH-VV}}(\theta)\right|$, (b) $\left|\gamma_{\text{HH-HV}}(\theta)\right|$, (c) $\left|\gamma_{\text{(HH+VV)-(HV)}}(\theta)\right|$, and (d) $\left|\gamma_{\text{(HH-VV)-(HV)}}(\theta)\right|$.

channels. The second is that crop areas usually exhibit a large portion of volume scattering mechanism, and the volume decorrelation effect leads to low polarimetric correlation. In this vein, polarimetric coherence features are seldom investigated and utilized for crop fields in the literature.

From Figure 2.4.9–2.4.14, with the maximization processing in the polarimetric rotation domain, the original polarimetric coherence features are significantly enhanced for all three frequencies and four combinations of polarimetric channels. It is observed that for a certain polarimetric coherence feature, the corresponding values from the P- and L-bands are higher than those from the C-band both without and with the maximization processing, respectively. This phenomenon can be interpreted by the different penetration depths among P-, L-, and C-band frequencies. P- and L-band microwaves can penetrate deeper than C-band. The backscattering from the ground and stem can form double-bounce scattering and be less affected by temporal decorrelation and volume scattering decorrelation. Besides, surface scattering directly from the soil surface can also be significant at the P- and L-bands. Another phenomenon is observed that for certain frequency bands, the original coherence $|\gamma_{\text{HH-VV}}(\theta)|_{\text{org}}$ and the maximum coherence $|\gamma_{\text{HH-VV}}(\theta)|_{\max}$ of polarimetric coherence pattern $|\gamma_{\text{HH-VV}}(\theta)|$ generally exhibit the highest values among those from the other three polarimetric coherence patterns. The main reason lies in that they are not affected by the reflection symmetry effect, which can greatly decrease the coherence degree. Another possible reason is that both HH and VV are co-polarization channels with a higher signal to noise ratio. Therefore, the noise decorrelation effect is not as obvious as for other channels. In addition, results from Figure 2.4.9–2.4.11 also show that both the original coherence and the maximum coherence are strongly polarimetric dependent and target dependent. This phenomenon becomes more obvious after the maximization and the local land covers are more distinguishable. These properties are the basis and evidence for crop discrimination demonstrated in the following section.

In addition, quantitative comparisons in terms of mean values of the original coherence and the maximum coherence are summarized in Tables 2.4.2–2.4.4 for the P-, L-, and C-band full-scene data, respectively. For the P-band, the mean values of the original coherence of $|\gamma_{\text{HH-VV}}(\theta)|$, $|\gamma_{\text{HH-HV}}(\theta)|$, $|\gamma_{\text{(HH+VV)-(HV)}}(\theta)|$, and $|\gamma_{\text{(HH-VV)-(HV)}}(\theta)|$ are respectively 0.35, 0.15, 0.14, and 0.14, while they are enhanced up to 0.64, 0.50, 0.35, and 0.54 accordingly by the maximization processing. The corresponding enhanced percentages are 82.86%, 233.33%, 150.00%, and 285.71%. The situations in the L-band are very close to those in the P-band. The corresponding mean values of the original coherence of $|\gamma_{\text{HH-VV}}(\theta)|$, $|\gamma_{\text{HH-HV}}(\theta)|$, $|\gamma_{\text{(HH+VV)-(HV)}}(\theta)|$, and $|\gamma_{\text{(HH-VV)-(HV)}}(\theta)|$ are respectively 0.35, 0.13, 0.12, and 0.11, while they are maximized up to 0.64, 0.45, 0.33, and 0.48 accordingly. The corresponding enhanced percentages are 82.86%, 246.15%, 175.00%, and 336.36%, respectively. Compared with the P- and L-bands, except for $|\gamma_{\text{HH-VV}}(\theta)|$, the coherence features from C-band data show even lower values. The mean values are 0.43, 0.07, 0.08, and 0.06 for original coherences of $|\gamma_{\text{HH-VV}}(\theta)|$, $|\gamma_{\text{HH-HV}}(\theta)|$, $|\gamma_{\text{(HH+VV)-(HV)}}(\theta)|$, and $|\gamma_{\text{(HH-VV)-(HV)}}(\theta)|$. The corresponding mean values of maximized coherence features are 0.57, 0.22, 0.17, and 0.22, with enhanced percentages of 32.56%, 214.29%, 112.50%, and 266.67%. As a

TABLE 2.4.2
Comparison of Polarimetric Coherence without and with Maximization in the Polarimetric Rotation Domain for AIRSAR P-Band Full-Scene Data

Polarimetric Coherence Pattern	Means of Original Coherence	Means of Maximum Coherence	Enhanced Percentage (%)
$\lvert\gamma_{\text{HH-VV}}(\theta)\rvert$	0.35	0.64	82.86
$\lvert\gamma_{\text{HH-HV}}(\theta)\rvert$	0.15	0.50	233.33
$\lvert\gamma_{\text{(HH+VV)-(HV)}}(\theta)\rvert$	0.14	0.35	150.00
$\lvert\gamma_{\text{(HH-VV)-(HV)}}(\theta)\rvert$	0.14	0.54	285.71

TABLE 2.4.3
Comparison of Polarimetric Coherence without and with Maximization in the Polarimetric Rotation Domain for AIRSAR L-Band Full-Scene Data

Polarimetric Coherence Pattern	Means of Original Coherence	Means of Maximum Coherence	Enhanced Percentage (%)
$\lvert\gamma_{\text{HH-VV}}(\theta)\rvert$	0.35	0.64	82.86
$\lvert\gamma_{\text{HH-HV}}(\theta)\rvert$	0.13	0.45	246.15
$\lvert\gamma_{\text{(HH+VV)-(HV)}}(\theta)\rvert$	0.12	0.33	175.00
$\lvert\gamma_{\text{(HH-VV)-(HV)}}(\theta)\rvert$	0.11	0.48	336.36

TABLE 2.4.4
Comparison of Polarimetric Coherence without and with Maximization in the Polarimetric Rotation Domain for AIRSAR C-Band Full-Scene Sata

Polarimetric Coherence Pattern	Means of Original Coherence	Means of Maximum Coherence	Enhanced Percentage (%)
$\lvert\gamma_{\text{HH-VV}}(\theta)\rvert$	0.43	0.57	32.56
$\lvert\gamma_{\text{HH-HV}}(\theta)\rvert$	0.07	0.22	214.29
$\lvert\gamma_{\text{(HH+VV)-(HV)}}(\theta)\rvert$	0.08	0.17	112.50
$\lvert\gamma_{\text{(HH-VV)-(HV)}}(\theta)\rvert$	0.06	0.22	266.67

result, the polarimetric coherence maximization processing is effective to enhance the polarization coherence of land covers for multi-frequency data, various polarization channel combinations, and various land covers.

2.5 TWO-DIMENSION POLARIMETRIC CORRELATION PATTERN INTERPRETATION TOOL

2.5.1 Definition and Characterization of Two-Dimension Polarimetric Correlation Patterns

For polarimetric radar, the correlation of two polarization channels also contains rich information. Inspired by the polarimetric coherence pattern [26–29], the correlation of two polarization channels at a fixed angle can also be extended to the polarimetric rotation domain. The two-dimension polarimetric correlation pattern (polarimetric correlation pattern for short) [30, 31] is obtained as

$$\left|\hat{\gamma}_{1\text{-}2}(\theta)\right| = \left|\left\langle s_1(\theta) s_2^*(\theta)\right\rangle\right| \qquad \theta \in [-\pi, \pi) \tag{2.5.1}$$

where the value range of $\left|\hat{\gamma}_{1\text{-}2}(\theta)\right|$ is $[0, +\infty)$.

Considering the reciprocal condition, six typical polarimetric correlation patterns are available based on the Lexicographic vector $\mathbf{k}_\text{L} = [S_\text{HH} \quad \sqrt{2}S_\text{HV} \quad S_\text{VV}]^\text{T}$ and the Pauli vector $\mathbf{k}_\text{P} = \frac{1}{\sqrt{2}}[S_\text{HH} + S_\text{VV} \quad S_\text{HH} - S_\text{VV} \quad 2S_\text{HV}]^\text{T}$: $\left|\hat{\gamma}_\text{HH-HV}(\theta)\right|$, $\left|\hat{\gamma}_\text{HH-VV}(\theta)\right|$, $\left|\hat{\gamma}_\text{VV-HV}\right|$, $\left|\hat{\gamma}_\text{(HH+VV)-(HH-VV)}(\theta)\right|$, $\left|\hat{\gamma}_\text{(HH+VV)-(HV)}\right|$, and $\left|\hat{\gamma}_\text{(HH-VV)-(HV)}(\theta)\right|$. Among them, there are two groups of polarimetric correlation patterns that are respectively equivalent:

$$\left|\hat{\gamma}_\text{(HH+VV)-(HH-VV)}(\theta)\right| = \left|\hat{\gamma}_\text{(HH+VV)-(HV)}(\theta + \pi/4)\right| \tag{2.5.2}$$

$$\left|\hat{\gamma}_\text{HH-VV}(\theta)\right| = \left|\hat{\gamma}_\text{VV-HV}(\theta + \pi/2)\right| \tag{2.5.3}$$

Thus, only four independent polarimetric correlation patterns, $\left|\hat{\gamma}_\text{HH-HV}(\theta)\right|$, $\left|\hat{\gamma}_\text{HH-VV}(\theta)\right|$, $\left|\hat{\gamma}_\text{(HH+VV)-(HH-VV)}(\theta)\right|$, and $\left|\hat{\gamma}_\text{(HH-VV)-(HV)}(\theta)\right|$, are considered thereinafter. An example of polarimetric correlation patterns is shown in Figure 2.5.1, which indicates a ship pixel and a sea clutter pixel, respectively. The polarimetric correlation variations are quite different between them. Similar to the polarimetric coherence pattern interpretation tool, ten polarimetric features can also be derived to describe the characteristics of the polarimetric correlation pattern.

Using $\left|\hat{\gamma}_{1\text{-}2}(\theta)\right|$ as an example, the definitions of these features are given as follows

1) Original Correlation $\left|\hat{\gamma}_{1\text{-}2}(\theta)\right|_\text{org}$: it is the correlation without any rotation processing:

$$\left|\hat{\gamma}_{1\text{-}2}(\theta)\right|_\text{org} = \left|\hat{\gamma}_{1\text{-}2}(0)\right| = \left|\hat{\gamma}_{1\text{-}2}\right| \tag{2.5.4}$$

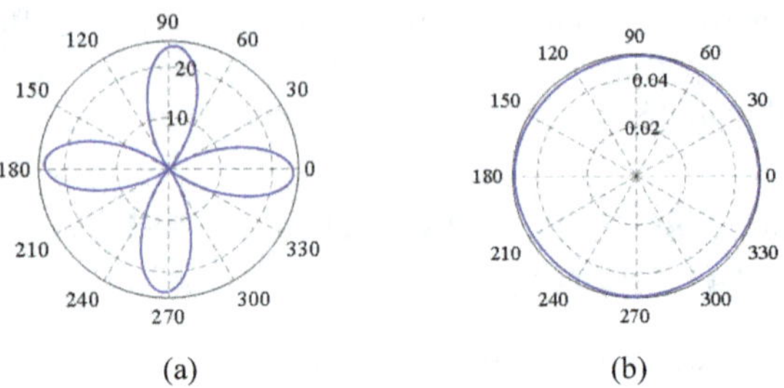

FIGURE 2.5.1 Illustration of the two-dimension polarimetric correlation pattern examples. (a) Ship pixel. (b) Sea clutter pixel.

This is the normally used polarimetric correlation feature in the literature and indicates a target decorrelation effect for two polarization channels under the original imaging geometry.

2) Correlation Degree $\left|\hat{\gamma}_{1-2}(\theta)\right|_{\text{mean}}$: it is the mean value of correlation in the polarimetric rotation domain and indicates the general correlation level:

$$\left|\hat{\gamma}_{1-2}(\theta)\right|_{\text{mean}} = \frac{1}{2\pi}\int_{-\pi}^{\pi}\left|\hat{\gamma}_{1-2}(\theta)\right|\mathrm{d}\theta \tag{2.5.5}$$

Correlation degree $\left|\hat{\gamma}_{1-2}(\theta)\right|_{\text{mean}}$ is a measure for targets' average decorrelation effect in the polarimetric rotation domain. The higher the correlation degree, the lower the decorrelation phenomenon.

3) Correlation Fluctuation $\left|\hat{\gamma}_{1-2}(\theta)\right|_{\text{std}}$: it is the standard deviation of correlation in the polarimetric rotation domain and indicates the correlation fluctuation:

$$\left|\hat{\gamma}_{1-2}(\theta)\right|_{\text{std}} = \sqrt{\frac{1}{2\pi}\int_{-\pi}^{\pi}\left(\left|\hat{\gamma}_{1-2}(\theta)\right| - \left|\hat{\gamma}_{1-2}(\theta)\right|_{\text{mean}}\right)^2\mathrm{d}\theta} \tag{2.5.6}$$

Correlation fluctuation $\left|\hat{\gamma}_{1-2}(\theta)\right|_{\text{std}}$ can identify target scattering diversity. Generally, the higher the correlation fluctuation, the more obvious the target scattering diversity in the polarimetric rotation domain. For polarimetric roll-invariant scatterers, correlation fluctuation $\left|\hat{\gamma}_{1-2}(\theta)\right|_{\text{std}}$ will shrink to zero.

4) Maximum Correlation $\left|\hat{\gamma}_{1-2}(\theta)\right|_{\max}$: it is the maximum correlation in the polarimetric rotation domain:

$$\left|\hat{\gamma}_{1-2}(\theta)\right|_{\max} = \max_{\theta\in[-\pi,\pi)}\left(\left|\hat{\gamma}_{1-2}(\theta)\right|\right) \tag{2.5.7}$$

It indicates the upper limit of polarimetric correlation of two polarization channels via orientation adjustment.

5) Minimum Correlation $\left|\hat{\gamma}_{1-2}(\theta)\right|_{\min}$: it is the minimum correlation in the polarimetric rotation domain:

$$\left|\hat{\gamma}_{1-2}(\theta)\right|_{\min} = \min_{\theta\in[-\pi,\pi)}\left(\left|\hat{\gamma}_{1-2}(\theta)\right|\right) \tag{2.5.8}$$

The minimum correlation $\left|\hat{\gamma}_{1-2}(\theta)\right|_{\min}$ gives the lower limit of polarimetric correlation through rotation processing in the polarimetric rotation domain.

6) Correlation Contrast $\left|\hat{\gamma}_{1-2}(\theta)\right|_{\text{contrast}}$: it is defined as

$$\left|\hat{\gamma}_{1-2}(\theta)\right|_{\text{contrast}} = \left|\hat{\gamma}_{1-2}(\theta)\right|_{\max} - \left|\hat{\gamma}_{1-2}(\theta)\right|_{\min} \tag{2.5.9}$$

It reflects the absolute contrast in the polarimetric rotation domain. In addition, correlation contrast $\left|\hat{\gamma}_{1-2}(\theta)\right|_{\text{contrast}}$ can also partially reflect target scattering diversity in the polarimetric rotation domain. For polarimetric roll-invariant targets without orientation dependency, correlation contrast $\left|\hat{\gamma}_{1-2}(\theta)\right|_{\text{contrast}}$ will become zero.

7) Correlation Anisotropy $\left|\hat{\gamma}_{1-2}(\theta)\right|_{\text{A}}$: it is defined as

$$\left|\hat{\gamma}_{1-2}(\theta)\right|_{\text{A}} = \frac{\left|\hat{\gamma}_{1-2}(\theta)\right|_{\max} - \left|\hat{\gamma}_{1-2}(\theta)\right|_{\min}}{\left|\hat{\gamma}_{1-2}(\theta)\right|_{\max} + \left|\hat{\gamma}_{1-2}(\theta)\right|_{\min}} \tag{2.5.10}$$

It reflects the relative contrast in the polarimetric rotation domain. Correlation anisotropy $\left|\hat{\gamma}_{1-2}(\theta)\right|_{\text{A}}$ is a complementary feature to correlation contrast $\left|\hat{\gamma}_{1-2}(\theta)\right|_{\text{contrast}}$. For polarimetric roll-variant targets with relatively low correlation degree $\left|\hat{\gamma}_{1-2}(\theta)\right|_{\text{contrast}}$, correlation anisotropy $\left|\gamma_{1-2}(\theta)\right|_{\text{A}}$ can further enhance the contrast and show better discriminant performance compared with correlation contrast $\left|\hat{\gamma}_{1-2}(\theta)\right|_{\text{contrast}}$.

8) Correlation Beamwidth $\left|\hat{\gamma}_{1-2}(\theta)\right|_{\text{bw0.95}}$: it is defined as the rotation angle range within which the correlation values are no less than $0.95\times\left|\hat{\gamma}_{1-2}(\theta)\right|_{\max}$, which is

$$\left|\hat{\gamma}_{1-2}(\theta)\right|_{\text{bw0.95}} = \left\{\theta \,\middle|\, \left|\hat{\gamma}_{1-2}(\theta)\right|_{\max} \geq \left|\hat{\gamma}_{1-2}(\theta)\right| \geq 0.95\times\left|\hat{\gamma}_{1-2}(\theta)\right|_{\max}\right\} \tag{2.5.11}$$

Correlation beamwidth is an index reflecting the sensitivity of the orientation dependency effect. The smaller $\left|\hat{\gamma}_{1-2}(\theta)\right|_{\text{bw0.95}}$ is, the greater the decorrelation effect and orientation dependency are.

9) Maximum Rotation Angle $\theta_{\left|\hat{\gamma}_{1-2}(\theta)\right|_{\max}}$: it is defined as the rotation angle within the main range which produces the maximum correlation $\left|\hat{\gamma}_{1-2}(\theta)\right|_{\max}$, which is

$$\theta_{|\hat{\gamma}_{1-2}(\theta)|_{\max}} = \left\{\theta \middle| |\hat{\gamma}_{1-2}(\theta)| = |\hat{\gamma}_{1-2}(\theta)|_{\max}\right\} \tag{2.5.12}$$

Maximum rotation angle indicates the most suitable state for a target in the polarimetric rotation domain, which leads to the least decorrelation of two given polarimetric channels.

10) Minimum Rotation Angle $\theta_{|\hat{\gamma}_{1-2}(\theta)|_{\min}}$: it is defined as the rotation angle within the main range which produces the minimum correlation $|\hat{\gamma}_{1-2}(\theta)|_{\min}$, which is

$$\theta_{|\hat{\gamma}_{1-2}(\theta)|_{\min}} = \left\{\theta \middle| |\hat{\gamma}_{1-2}(\theta)| = |\hat{\gamma}_{1-2}(\theta)|_{\min}\right\} \tag{2.5.13}$$

Minimum rotation angle indicates another specific state where the most serious decorrelation of two certain polarimetric channels occurs for a target in the polarimetric rotation domain.

The polarimetric correlation pattern features from $|\hat{\gamma}_{\text{HH-HV}}(\theta)|$, $|\hat{\gamma}_{\text{HH-VV}}(\theta)|$, $|\hat{\gamma}_{\text{(HH+VV)-(HH-VV)}}(\theta)|$, and $|\hat{\gamma}_{\text{(HH-VV)-(HV)}}(\theta)|$ are shown in Table 2.5.1.

2.5.2 Demonstration and Investigation

For canonical structures of plate, dihedral, dipole, cylinder, narrow dihedral, quarter device, and helix, the polarimetric scattering matrices and polarimetric correlation patterns are shown in Table 2.5.2. Note that the polarimetric correlation patterns have been normalized with the maximum correlation value for better visualization. It can be seen that the polarimetric correlation patterns are quite different among different structures. Taking $|\hat{\gamma}_{\text{HH-HV}}(\theta)|$ for example, the plate has zero correlation values among all rotation angles, while the helix has the same non-zero value among all the rotation angles. The dipole cylinder, and quarter devices all have four-lobed shapes; however, the beamwidths of these three structures become larger in order. The dihedral and narrow dihedral both have eight-lobed shapes, but the lobes in vertical direction are not the same. As for $|\hat{\gamma}_{\text{HH-VV}}(\theta)|$, the polarimetric correlation patterns are also various for different canonical structures. The dihedral, cylinder, quarter device, and helix show four-lobe, square, cross, and circle shapes, respectively. The dihedral and the dipole both have four-lobed shapes, but their maximum and minimum rotation angles are different. In this vein, the polarimetric correlation pattern can effectively depict the scattering characteristics of canonical structures, which provides support for manmade target structure recognition.

Apart from the analysis of canonical structures, polarimetric correlation pattern performances have been validated with real PolSAR datasets. A GaoFen-3 C band spaceborne PolSAR dataset, collected on March 30, 2017, over the ocean area, Hong Kong, China, is selected for demonstration. The image size is 1000 × 1000 pixels, and the resolution is 8 × 8 m. The Pauli image and labeled ground-truth are

TABLE 2.5.1
Two-Dimension Polarimetric Correlation Pattern Features

	Polarimetric Correlation Pattern			
Polarimetric Feature	$\lvert\hat{\gamma}_{\text{HH-HV}}(\theta)\rvert$	$\lvert\hat{\gamma}_{\text{HH-VV}}(\theta)\rvert$	$\lvert\hat{\gamma}_{\text{(HH+VV)-(HH-VV)}}(\theta)\rvert$	$\lvert\hat{\gamma}_{\text{(HH-VV)-(HV)}}(\theta)\rvert$
Original correlation	$\lvert\hat{\gamma}_{\text{HH-HV}}(\theta)\rvert_{\text{org}}$	$\lvert\hat{\gamma}_{\text{HH-VV}}(\theta)\rvert_{\text{org}}$	$\lvert\hat{\gamma}_{\text{(HH+VV)-(HH-VV)}}(\theta)\rvert_{\text{org}}$	$\lvert\hat{\gamma}_{\text{(HH-VV)-(HV)}}(\theta)\rvert_{\text{org}}$
Correlation degree	$\lvert\hat{\gamma}_{\text{HH-HV}}(\theta)\rvert_{\text{mean}}$	$\lvert\hat{\gamma}_{\text{HH-VV}}(\theta)\rvert_{\text{mean}}$	$\lvert\hat{\gamma}_{\text{(HH+VV)-(HH-VV)}}(\theta)\rvert_{\text{mean}}$	$\lvert\hat{\gamma}_{\text{(HH-VV)-(HV)}}(\theta)\rvert_{\text{mean}}$
Correlation fluctuation	$\lvert\hat{\gamma}_{\text{HH-HV}}(\theta)\rvert_{\text{std}}$	$\lvert\hat{\gamma}_{\text{HH-VV}}(\theta)\rvert_{\text{std}}$	$\lvert\hat{\gamma}_{\text{(HH+VV)-(HH-VV)}}(\theta)\rvert_{\text{std}}$	$\lvert\hat{\gamma}_{\text{(HH-VV)-(HV)}}(\theta)\rvert_{\text{std}}$
Maximum correlation	$\lvert\hat{\gamma}_{\text{HH-HV}}(\theta)\rvert_{\text{max}}$	$\lvert\hat{\gamma}_{\text{HH-VV}}(\theta)\rvert_{\text{max}}$	$\lvert\hat{\gamma}_{\text{(HH+VV)-(HH-VV)}}(\theta)\rvert_{\text{max}}$	$\lvert\hat{\gamma}_{\text{(HH-VV)-(HV)}}(\theta)\rvert_{\text{max}}$
Minimum correlation	$\lvert\hat{\gamma}_{\text{HH-HV}}(\theta)\rvert_{\text{min}}$	$\lvert\hat{\gamma}_{\text{HH-VV}}(\theta)\rvert_{\text{min}}$	$\lvert\hat{\gamma}_{\text{(HH+VV)-(HH-VV)}}(\theta)\rvert_{\text{min}}$	$\lvert\hat{\gamma}_{\text{(HH-VV)-(HV)}}(\theta)\rvert_{\text{min}}$
Correlation contrast	$\lvert\hat{\gamma}_{\text{HH-HV}}(\theta)\rvert_{\text{contrast}}$	$\lvert\hat{\gamma}_{\text{HH-VV}}(\theta)\rvert_{\text{contrast}}$	$\lvert\hat{\gamma}_{\text{(HH+VV)-(HH-VV)}}(\theta)\rvert_{\text{contrast}}$	$\lvert\hat{\gamma}_{\text{(HH-VV)-(HV)}}(\theta)\rvert_{\text{contrast}}$
Correlation anisotropy	$\lvert\hat{\gamma}_{\text{HH-HV}}(\theta)\rvert_{\text{A}}$	$\lvert\hat{\gamma}_{\text{HH-VV}}(\theta)\rvert_{\text{A}}$	$\lvert\hat{\gamma}_{\text{(HH+VV)-(HH-VV)}}(\theta)\rvert_{\text{A}}$	$\lvert\hat{\gamma}_{\text{(HH-VV)-(HV)}}(\theta)\rvert_{\text{A}}$
Maximum rotation angle	$\theta_{\lvert\hat{\gamma}_{\text{HH-HV}}(\theta)\rvert_{\text{bw0.95}}}$	$\theta_{\lvert\hat{\gamma}_{\text{HH-VV}}(\theta)\rvert_{\text{bw0.95}}}$	$\theta_{\lvert\hat{\gamma}_{\text{(HH+VV)-(HH-VV)}}(\theta)\rvert_{\text{bw0.95}}}$	$\theta_{\lvert\hat{\gamma}_{\text{(HH-VV)-(HV)}}(\theta)\rvert_{\text{bw0.95}}}$
Minimum rotation angle	$\theta_{\lvert\hat{\gamma}_{\text{HH-HV}}(\theta)\rvert_{\text{max}}}$	$\theta_{\lvert\hat{\gamma}_{\text{HH-VV}}(\theta)\rvert_{\text{max}}}$	$\theta_{\lvert\hat{\gamma}_{\text{(HH+VV)-(HH-VV)}}(\theta)\rvert_{\text{max}}}$	$\theta_{\lvert\hat{\gamma}_{\text{(HH-VV)-(HV)}}(\theta)\rvert_{\text{max}}}$
Correlation beamwidth	$\theta_{\lvert\hat{\gamma}_{\text{HH-HV}}(\theta)\rvert_{\text{min}}}$	$\theta_{\lvert\hat{\gamma}_{\text{HH-VV}}(\theta)\rvert_{\text{min}}}$	$\theta_{\lvert\hat{\gamma}_{\text{(HH+VV)-(HH-VV)}}(\theta)\rvert_{\text{min}}}$	$\theta_{\lvert\hat{\gamma}_{\text{(HH-VV)-(HV)}}(\theta)\rvert_{\text{min}}}$

TABLE 2.5.2
Polarimetric Correlation Patterns of Canonical Structures

Canonical Structure	Polarimetric Scattering Matrix	$\left\|\hat{\gamma}_{\text{HH-HV}}(\theta)\right\|$	$\left\|\hat{\gamma}_{\text{HH-VV}}(\theta)\right\|$	$\left\|\hat{\gamma}_{(\text{HH+VV})-(\text{HH HV})}(\theta)\right\|$	$\left\|\hat{\gamma}_{(\text{HH-VV})-(\text{HV})}(\theta)\right\|$
Plate	$\begin{bmatrix} 1 & 0 \\ 0 & 1 \end{bmatrix}$				
Dihedral	$\begin{bmatrix} 1 & 0 \\ 0 & -1 \end{bmatrix}$				
Dipole	$\begin{bmatrix} 1 & 0 \\ 0 & 0 \end{bmatrix}$				

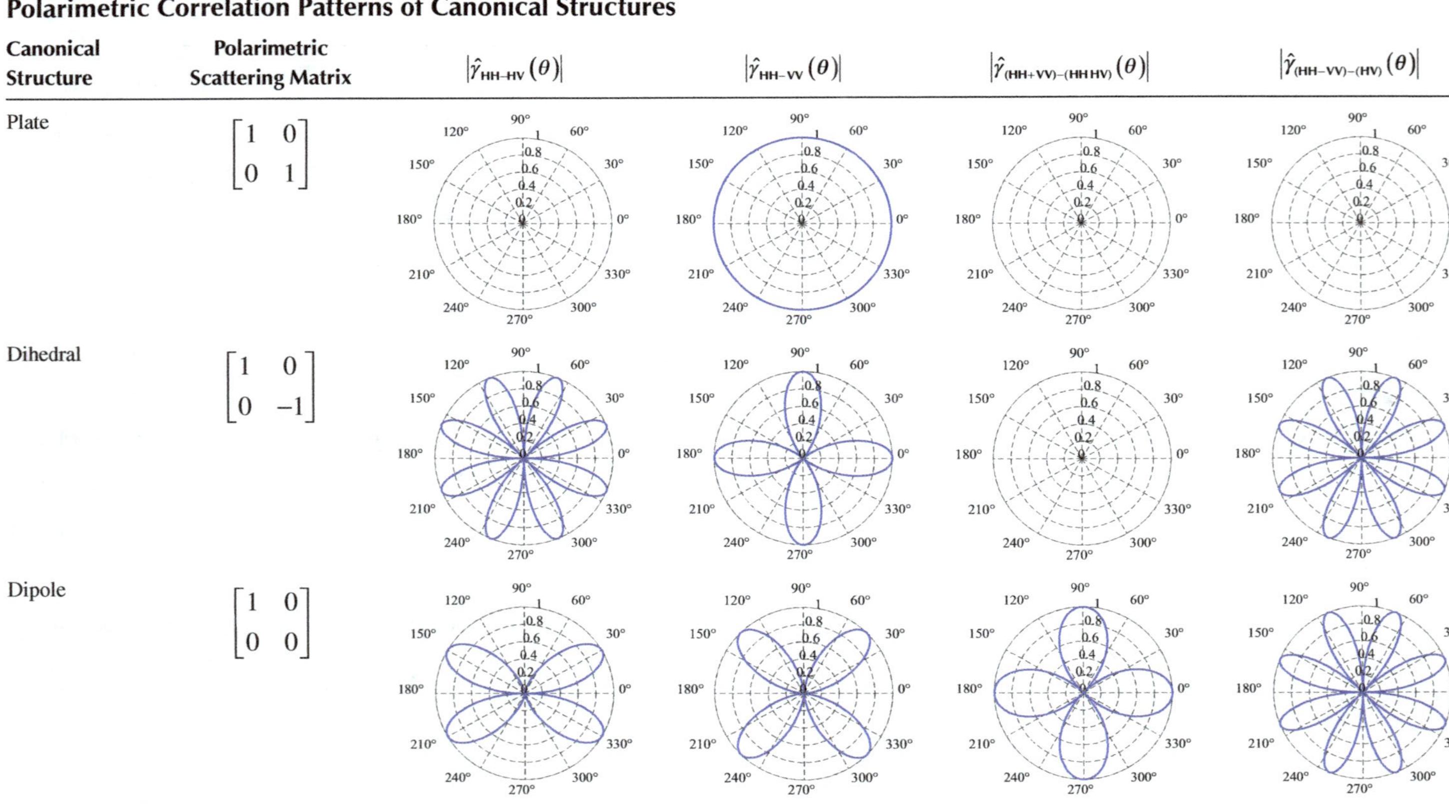

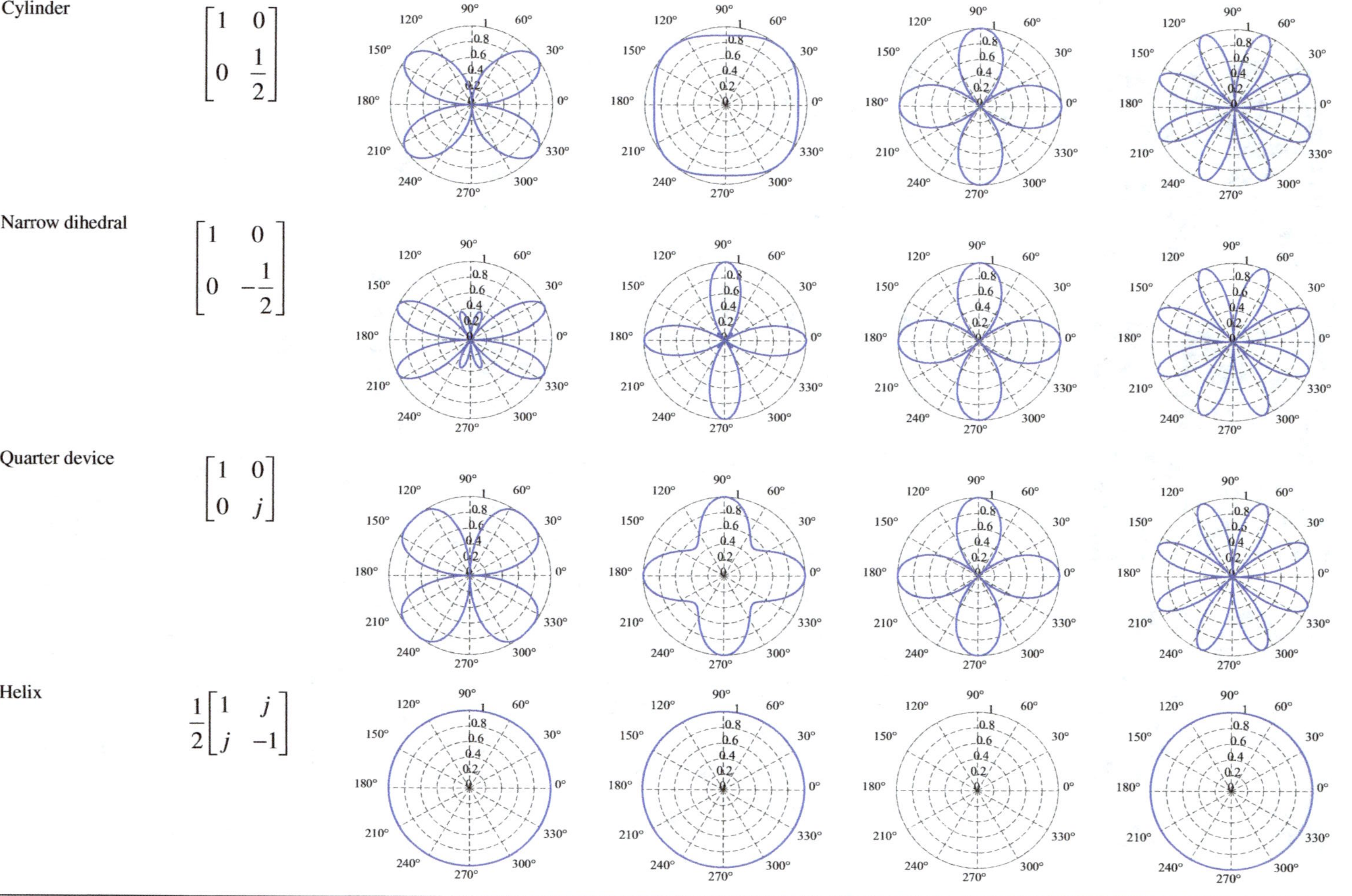
Cylinder
$\begin{bmatrix} 1 & 0 \\ 0 & \frac{1}{2} \end{bmatrix}$
Narrow dihedral
$\begin{bmatrix} 1 & 0 \\ 0 & -\frac{1}{2} \end{bmatrix}$
Quarter device
$\begin{bmatrix} 1 & 0 \\ 0 & j \end{bmatrix}$
Helix
$\frac{1}{2}\begin{bmatrix} 1 & j \\ j & -1 \end{bmatrix}$

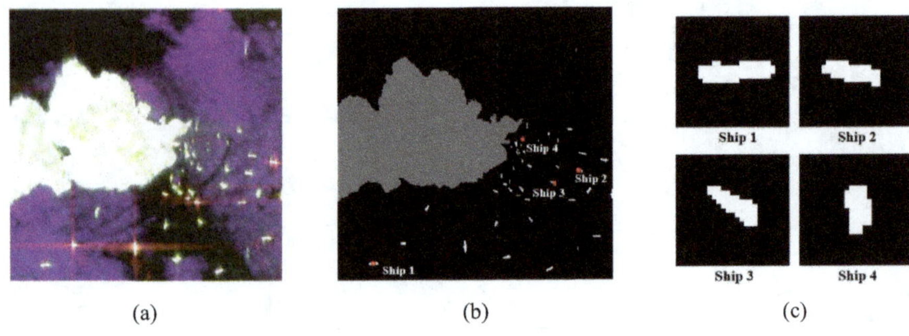

FIGURE 2.5.2 GaoFen-3 PolSAR data. (a) Pauli image, (b) ground-truth, (c) selected ship regions.

shown in Figure 2.5.2(a) and (b). Four ship regions with different ship orientations are extracted and shown in Figure 2.5.2 (c).

Taking $\left|\hat{\gamma}_{\text{HH-VV}}(\theta)\right|$ for example, the polarimetric correlation patterns for each pixel of these four ship regions are displayed in Figure 2.5.3. The red polarimetric correlation patterns indicate the ship pixels, while the blue polarimetric correlation patterns are the sea clutter pixels. It can be seen that the sea clutter pixels usually have a circular shape, corresponding to the plate structures. However, the polarimetric correlation patterns of ship pixels are almost four-lobed shapes, since the ships contain some dihedral structures. The polarimetric correlation patterns from real PolSAR datasets are consistent with the theoretical analysis summarized in Table 2.5.2. Therefore, the polarimetric correlation pattern interpretation tool has the capability to well discriminate ships and sea clutter, which provides a solid foundation for polarimetric rotation domain ship detection.

2.6 THREE-DIMENSION POLARIMETRIC CORRELATION PATTERN INTERPRETATION TOOL

Polarimetric radar can acquire rich polarization information and is widely used in many applications. In radar polarization information processing, target scattering responses are affected by both polarization orientation angle and polarization ellipticity angle. Extending polarimetric rotation domain techniques to the polarization ellipticity angle dimension has the capability to explore and utilize more complete target scattering diversity. In this section, a three-dimension polarimetric correlation pattern interpretation tool is introduced [32,33], which can visualize and exhibit targets' polarimetric rotation domain properties in terms of both the polarization orientation and ellipticity angles.

2.6.1 Polarization Basis Transformation

With (2.2.1), the polarimetric scattering matrix from an arbitrary polarization basis (X,Y) can be derived by the unitary transformation

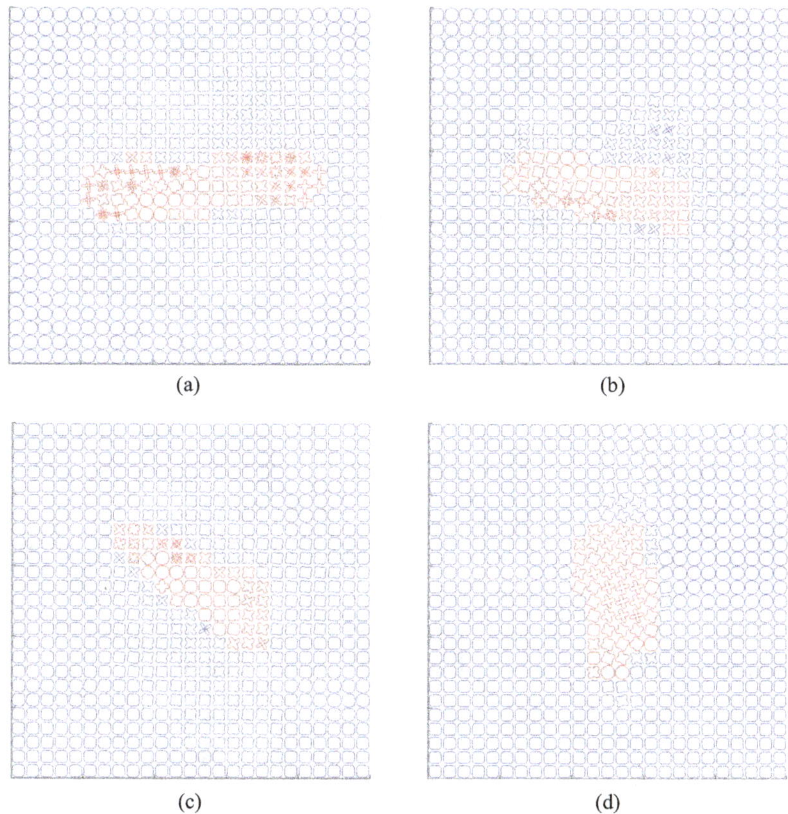

FIGURE 2.5.3 Two-dimension Polarimetric correlation patterns of four ship regions. (a) ship 1, (b) ship 2, (c) ship 3, (d) ship 4.

$$\mathbf{S}_{(\mathrm{X,Y})} = \mathbf{U}_2^{\mathrm{T}}(\theta,\tau)\mathbf{S}_{(\mathrm{H,V})}\mathbf{U}_2(\theta,\tau) \tag{2.6.1}$$

where θ is the polarization orientation angle (POA) and τ is the polarization ellipticity angle (PEA). $\mathbf{U}_2(\theta,\tau)$ stands for the unitary matrix and is the product of the POA unitary matrix $\mathbf{U}_2(\theta)$ and the PEA unitary matrix $\mathbf{U}_2(\tau)$:

$$\begin{aligned}\mathbf{U}_2(\theta,\tau) &= \mathbf{U}_2(\theta)\mathbf{U}_2(\tau) \\ &= \begin{bmatrix} \cos\theta & -\sin\theta \\ \sin\theta & \cos\theta \end{bmatrix}\begin{bmatrix} \cos\tau & \mathrm{j}\sin\tau \\ \mathrm{j}\sin\tau & \cos\tau \end{bmatrix}\end{aligned} \tag{2.6.2}$$

With a $(\mathrm{H,V})$ polarization basis, the Pauli vector $\mathbf{k}_{\mathrm{P}}$ and Lexicographic vector $\mathbf{k}_{\mathrm{L}}$ can be represented as

$$\mathbf{k}_{\mathrm{P}_{(\mathrm{H,V})}} = \frac{1}{\sqrt{2}}\left[S_{\mathrm{HH}} + S_{\mathrm{VV}} \quad S_{\mathrm{HH}} - S_{\mathrm{VV}} \quad 2S_{\mathrm{HV}}\right]^{\mathrm{T}} \tag{2.6.3}$$

$$\mathbf{k}_{\mathrm{L}_{(\mathrm{H,V})}} = \left[S_{\mathrm{HH}} \quad \sqrt{2}S_{\mathrm{HV}} \quad S_{\mathrm{HV}}\right]^{\mathrm{T}} \tag{2.6.4}$$

Utilizing the components of $\mathbf{S}_{(\mathrm{X,Y})}$ in (2.6.1), the Pauli and Lexicographic vectors with arbitrary polarization basis can be achieved. Taking Pauli vector for instance, the $\mathbf{k}_{\mathrm{P}_{(\mathrm{H,V})}}(\theta,\tau)$ is

$$\begin{aligned}
&\mathbf{k}_{\mathrm{P}_{(\mathrm{H,V})}}(\theta,\tau) \\
&= \frac{1}{\sqrt{2}}\left(S_{\mathrm{HH}}(\theta,\tau) + S_{\mathrm{VV}}(\theta,\tau) \quad S_{\mathrm{HH}}(\theta,\tau) - S_{\mathrm{VV}}(\theta,\tau) \quad 2S_{\mathrm{HV}}(\theta,\tau)\right) \\
&= \frac{1}{\sqrt{2}}\begin{bmatrix}
\cos(2\tau)(S_{\mathrm{HH}}+S_{\mathrm{VV}}) - \mathrm{j}\sin(2\tau)\sin(2\theta)(S_{\mathrm{HH}}-S_{\mathrm{VV}}) + 2\mathrm{j}\cos(2\theta)\sin(2\tau)S_{\mathrm{HV}} \\
\cos(2\theta)(S_{\mathrm{HH}}-S_{\mathrm{VV}}) + 2\sin(2\theta)S_{\mathrm{HV}} \\
\mathrm{j}\sin(2\tau)(S_{\mathrm{HH}}+S_{\mathrm{VV}}) - \cos(2\tau)\sin(2\theta)(S_{\mathrm{HH}}-S_{\mathrm{VV}}) + 2\cos(2\tau)\cos(2\theta)S_{\mathrm{HV}}
\end{bmatrix}
\end{aligned} \tag{2.6.5}$$

Furthermore, the relationship of $\mathbf{k}_{\mathrm{P}_{(\mathrm{X,Y})}}(\theta,\tau)$ and $\mathbf{k}_{\mathrm{P}_{(\mathrm{H,V})}}$ can be linked by the transformation matrix $\mathbf{U}_{\mathrm{P}_{(\mathrm{H,V})}2\mathrm{P}_{(\mathrm{X,Y})}}(\theta,\tau)$

$$\mathbf{k}_{\mathrm{P}_{(\mathrm{X,Y})}}(\theta,\tau) = \mathbf{U}_{\mathrm{P}_{(\mathrm{H,V})}2\mathrm{P}_{(\mathrm{X,Y})}}(\theta,\tau)\mathbf{k}_{\mathrm{P}_{(\mathrm{H,V})}} \tag{2.6.6}$$

where the transformation matrix $\mathbf{U}_{\mathrm{P}_{(\mathrm{H,V})}2\mathrm{P}_{(\mathrm{X,Y})}}(\theta,\tau)$ is

$$\begin{aligned}
\mathbf{U}_{\mathrm{P}_{(\mathrm{H,V})}2\mathrm{P}_{(\mathrm{X,Y})}}(\theta,\tau) &= \mathbf{U}_{\mathrm{P}_{(\mathrm{H,V})}2\mathrm{P}_{(\mathrm{X,Y})}}(\tau)\mathbf{U}_{\mathrm{P}_{(\mathrm{H,V})}2\mathrm{P}_{(\mathrm{X,Y})}}(\theta) \\
&= \begin{bmatrix} \cos 2\tau & 0 & \mathrm{j}\sin 2\tau \\ 0 & 1 & 0 \\ \mathrm{j}\sin 2\tau & 0 & \cos 2\tau \end{bmatrix}\begin{bmatrix} 1 & 0 & 0 \\ 0 & \cos 2\theta & \sin 2\theta \\ 0 & -\sin 2\theta & \cos 2\theta \end{bmatrix} \\
&= \begin{bmatrix} \cos(2\tau) & -\mathrm{j}\sin(2\tau)\sin(2\theta) & \mathrm{j}\cos(2\theta)\sin(2\tau) \\ 0 & \cos(2\theta) & \sin(2\theta) \\ \mathrm{j}\sin(2\tau) & -\cos(2\tau)\sin(2\theta) & \cos(2\tau)\cos(2\theta) \end{bmatrix}
\end{aligned} \tag{2.6.7}$$

$\mathbf{T}(\theta,\tau)$ from an arbitrary polarization basis $(\mathrm{X,Y})$ is

$$\mathbf{T}(\theta,\tau) = \mathbf{U}_{\mathrm{P}_{(\mathrm{H,V})}2\mathrm{P}_{(\mathrm{X,Y})}}(\theta,\tau)\mathbf{T}\mathbf{U}^{\mathrm{H}}_{\mathrm{P}_{(\mathrm{H,V})}2\mathrm{P}_{(\mathrm{X,Y})}}(\theta,\tau) \tag{2.6.8}$$

Note that the transformation matrix $\mathbf{T}(\theta,\tau)$ is different from that given in [1]. From [1], the transformation matrix is $\tilde{\mathbf{U}}_{\mathrm{P}_{(\mathrm{H,V})}2\mathrm{P}_{(\mathrm{X,Y})}}(\theta,\tau)=\mathbf{U}_{\mathrm{P}_{(\mathrm{H,V})}2\mathrm{P}_{(\mathrm{X,Y})}}(\theta)\mathbf{U}_{\mathrm{P}_{(\mathrm{H,V})}2\mathrm{P}_{(\mathrm{X,Y})}}(\tau)$, where the order of the two matrixes $\mathbf{U}_{\mathrm{P}_{(\mathrm{H,V})}2\mathrm{P}_{(\mathrm{X,Y})}}(\theta)$ and $\mathbf{U}_{\mathrm{P}_{(\mathrm{H,V})}2\mathrm{P}_{(\mathrm{X,Y})}}(\tau)$ is different. In fact, the polarimetric coherency matrix can also be derived from the polarimetric scattering matrix. For the polarization basis $(\mathrm{X,Y})$, the polarimetric coherency matrix $\mathbf{T}(\theta,\tau)$ derived from the polarimetric scattering matrix $\mathbf{S}_{(\mathrm{X,Y})}$ is consistent with the transformation matrix $\mathbf{U}_{\mathrm{P}_{(\mathrm{H,V})}2\mathrm{P}_{(\mathrm{X,Y})}}(\theta,\tau)$ shown in (2.6.7). Therefore, it is used as the polarization basis transformation matrix for the polarimetric coherency matrix in this book.

According to the similar transformation relationship shown in (2.3.22), the polarimetric covariance matrix $\mathbf{C}(\theta,\tau)$ from an arbitrary polarization basis $(\mathrm{X,Y})$ is obtained as

$$\mathbf{C}(\theta,\tau)=\mathbf{U}_{\mathrm{P}_{(\mathrm{H,V})}2\mathrm{L}_{(\mathrm{H,V})}}\mathbf{T}(\theta,\tau)\mathbf{U}^{\mathrm{H}}_{\mathrm{P}_{(\mathrm{H,V})}2\mathrm{L}_{(\mathrm{H,V})}} \tag{2.6.9}$$

where the transformation matrix $\mathbf{U}_{\mathrm{P}_{(\mathrm{H,V})}2\mathrm{L}_{(\mathrm{H,V})}}$ is

$$\mathbf{U}_{\mathrm{P}_{(\mathrm{H,V})}2\mathrm{L}_{(\mathrm{H,V})}}=\frac{1}{\sqrt{2}}\begin{bmatrix}1 & 1 & 0\\ 0 & 0 & \sqrt{2}\\ 1 & -1 & 0\end{bmatrix} \tag{2.6.10}$$

2.6.2 Three-Dimension Polarimetric Correlation Pattern

The two-dimension polarimetric correlation pattern $|\hat{\gamma}_{1-2}(\theta)|$ can be extended to the PEA dimension to obtain the three-dimension polarimetric correlation pattern:

$$|\hat{\gamma}_{1-2}(\theta,\tau)|=\left|\langle s_1(\theta,\tau)s_2^*(\theta,\tau)\rangle\right| \qquad \theta\in[-\pi,\pi),\tau\in[-\pi/4,\pi/4) \tag{2.6.11}$$

Similarly, six typical three-dimension polarimetric correlation patterns of $|\hat{\gamma}_{\mathrm{HH\text{-}HV}}(\theta,\tau)|$, $|\hat{\gamma}_{\mathrm{HH\text{-}VV}}(\theta,\tau)|$, $|\hat{\gamma}_{\mathrm{VV\text{-}HV}}(\theta,\tau)|$, $|\hat{\gamma}_{(\mathrm{HH+VV})\text{-}(\mathrm{HH\text{-}VV})}(\theta,\tau)|$, $|\hat{\gamma}_{(\mathrm{HH\text{-}VV})\text{-}(\mathrm{HV})}(\theta,\tau)|$, and $|\hat{\gamma}_{(\mathrm{HH+VV})\text{-}(\mathrm{HV})}(\theta,\tau)|$ are available. Furthermore, three equivalence relationships can be obtained:

$$|\hat{\gamma}_{(\mathrm{HH+VV})-(\mathrm{HH\text{-}VV})}(\theta,\tau)|=|\hat{\gamma}_{(\mathrm{HH+VV})\text{-}(\mathrm{HV})}(\theta+\pi/4,\tau)| \tag{2.6.12}$$

$$|\hat{\gamma}_{\mathrm{HH\text{-}HV}}(\theta,\tau)|=|\hat{\gamma}_{\mathrm{VV\text{-}HV}}(\theta+\pi/2,\tau)| \tag{2.6.13}$$

$$|\hat{\gamma}_{(\mathrm{HH\text{-}VV})-(\mathrm{HV})}(\theta,\tau)|=|\hat{\gamma}_{(\mathrm{HH+VV})-(\mathrm{HH\text{-}VV})}(\theta,\tau+\pi/4)| \tag{2.6.14}$$

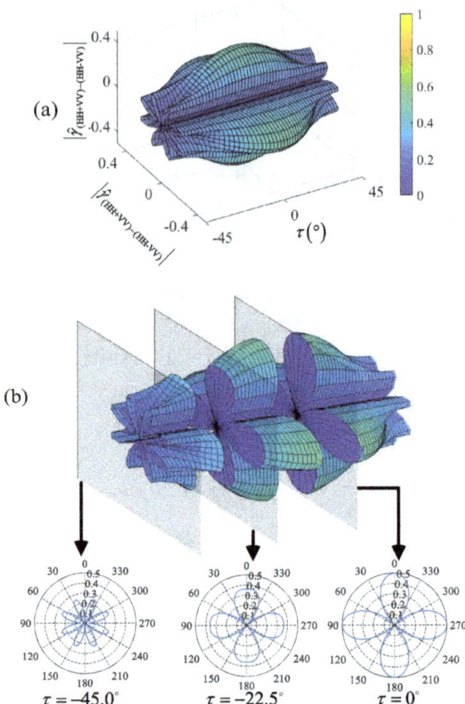

FIGURE 2.6.1 (a) Illustration of dipole's three-dimension polarimetric correlation pattern $\left|\hat{\gamma}_{(\mathrm{HH+VV})-(\mathrm{HH-VV})}(\theta,\tau)\right|$, (b) sectional views on different PEAs.

Thus, only the three independent three-dimension polarimetric correlation patterns of $\left|\hat{\gamma}_{\mathrm{HH\text{-}HV}}(\theta,\tau)\right|$, $\left|\hat{\gamma}_{\mathrm{HH\text{-}VV}}(\theta,\tau)\right|$, and $\left|\hat{\gamma}_{(\mathrm{HH+VV})\text{-}(\mathrm{HH\text{-}VV})}(\theta,\tau)\right|$ are considered thereafter.

In order to visualize the three-dimension polarimetric correlation pattern, the cylindrical coordinate system is adopted. An example of three-dimension polarimetric correlation pattern $\left|\hat{\gamma}_{(\mathrm{HH+VV})-(\mathrm{HH-VV})}(\theta,\tau)\right|$ from a dipole is shown in Figure 2.6.1 (a). In the coordinate, one dimension stands for the PEA. With a given PEA, the other two dimensions become the two-dimension polarimetric correlation pattern. The surface color is relative to the correlation value. The lighter the surface color, the higher the correlation value. The sectional views on different PEAs are displayed in Figure 2.6.1 (b), which are defined as the polarimetric correlation pattern planes. The difference of the polarimetric correlation patterns in terms of different PEAs can be observed, and the confidential information is analyzed in the next section.

2.6.3 Three-Dimension Polarimetric Correlation Pattern Investigation of Canonical Structures

Canonical structures, including the plane, dihedral, dipole, and helix, are considered for three-dimension polarimetric correlation pattern investigation. The corresponding polarimetric scattering matrix are

$$\mathbf{S}_{\text{Plate}} = \begin{bmatrix} 1 & 0 \\ 0 & 1 \end{bmatrix} \tag{2.6.15}$$

$$\mathbf{S}_{\text{Dihedral}} = \begin{bmatrix} 1 & 0 \\ 0 & -1 \end{bmatrix} \tag{2.6.16}$$

$$\mathbf{S}_{\text{Dipole}} = \begin{bmatrix} 1 & 0 \\ 0 & 0 \end{bmatrix} \tag{2.6.17}$$

$$\mathbf{S}_{\text{Helix}} = \frac{1}{2}\begin{bmatrix} 1 & \mathrm{j} \\ \mathrm{j} & -1 \end{bmatrix} \tag{2.6.18}$$

The three-dimension polarimetric correlation patterns of four canonical scattering structures are shown in Figure 2.6.2. It can be seen that there are significant differences of three-dimension polarimetric correlation patterns among different scattering structures. Meanwhile, it is noted that $\left[-45^\circ, 45^\circ\right]$ is a main period in the PEA τ direction in most cases except $\left|\hat{\gamma}_{\text{HH-HV}}(\theta,\tau)\right|$ for helix structure. In this vein, the manifolds are displayed for $\tau \in \left[-45^\circ, 45^\circ\right]$.

2.6.3.1 Periodicity and Symmetry Analyzation

The three-dimension polarimetric correlation pattern is usually periodic with the given PEA. The following investigation takes the three-dimension polarimetric correlation pattern $\left|\hat{\gamma}_{\text{HH-HV}}(\theta,\tau)\right|$ as an example for analysis. For planes and helices, the three-dimension polarimetric correlation pattern $\left|\hat{\gamma}_{\text{HH-HV}}(\theta,\tau)\right|$ maintains a circular shape in terms of the PEA, which shows the polarimetric roll-invariant characteristic. For dihedrals, the periodicity of the three-dimension polarimetric correlation pattern $\left|\hat{\gamma}_{\text{HH-HV}}(\theta,\tau)\right|$ is 45° when PEA $\tau = 0$, while it is 90° for the other PEAs. For dipoles, the three-dimension polarimetric correlation pattern $\left|\hat{\gamma}_{\text{HH-HV}}(\theta,\tau)\right|$ is polarimetric roll-invariant when PEA $\tau = \pm 45^\circ$, while the periodicity is 90° for the other PEAs. The three-dimension polarimetric correlation patterns of plane, dipole, and dihedral structures are symmetric in the range of $\tau \in \left[-45^\circ, 45^\circ\right]$, while that of the helix structure is symmetry in the range of $\tau \in \left[-45^\circ, 135^\circ\right]$, shown in Figure 2.6.3.

2.6.3.2 Distribution of Extreme Values

There are also significant differences in the extreme values distribution of the three-dimension polarimetric correlation pattern for different scattering structures. Taking the three-dimension polarimetric correlation pattern $\left|\hat{\gamma}_{\text{HH-HV}}(\theta,\tau)\right|$ as an example, the analytical expressions of the three-dimension polarimetric correlation pattern for different scattering structures are shown in Table 2.6.1. Note that the following analysis is in the main ranges of $\theta \in [-90^\circ, 90^\circ]$ and $\tau \in [-45^\circ, 45^\circ]$. For plane structure, the maximum value points distribute circularly at the three-dimension polarimetric correlation pattern slices when $\tau = \pm 22.5^\circ$, while the minimum value points are located at the slices when $\tau = 0, \pm 45^\circ$. For dihedral structure, the

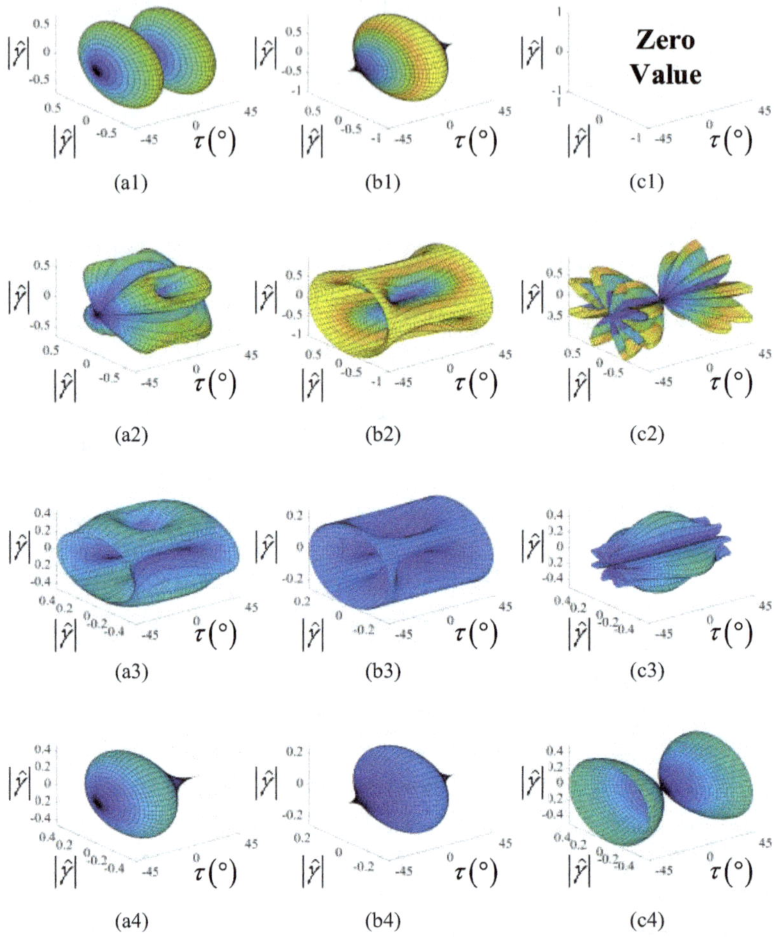

FIGURE 2.6.2 Three-dimension polarimetric correlation patterns for canonical structures. The letters a–c in each column indicate $\left|\hat{\gamma}_{\text{HH-HV}}(\theta,\tau)\right|$, $\left|\hat{\gamma}_{\text{HH-VV}}(\theta,\tau)\right|$, and $\left|\hat{\gamma}_{(\text{HH+VV})\text{-}(\text{HH-VV})}(\theta,\tau)\right|$, respectively. The numbers 1–4 in each row indicate canonical structures of the plane, dihedral, dipole, and helix structures, respectively.

analytical expression of the curve where the maximum value points are located is hard to give straightforwardly. Nevertheless, the maximum value points locate on an elliptic arc, observed from Figure 2.6.2 (a2). The minimum value points locate at the three-dimension polarimetric correlation pattern slices when $\tau = \pm 45^\circ$ for any θ and $\tau = 0^\circ$ for $\theta = 0^\circ, \pm 45^\circ, \pm 90^\circ$. For dipole structure, the analytical expression of the curve where the maximum value points are located is also not straightforward. From Figure 2.6.2 (a3), it is observed that the maximum value points also locate

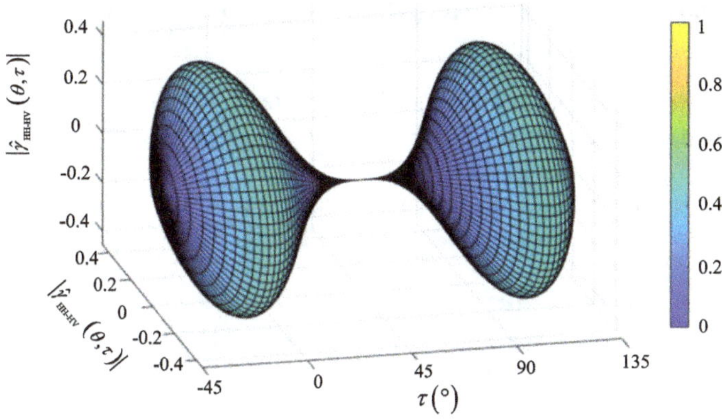

FIGURE 2.6.3 Illustration of the three-dimension polarimetric correlation pattern $\left|\hat{\gamma}_{\text{HH-HV}}(\theta,\tau)\right|$ for a helix with the τ range of $\left[-45^\circ,135^\circ\right]$. The periodicity along the τ direction is also valid.

TABLE 2.6.1

The Three-Dimension Polarimetric Correlation Pattern $\left|\hat{\gamma}_{\text{HH-HV}}(\theta,\tau)\right|$ Expressions of Canonical Structures

Canonical Structure	Analytical Expression
Plane	$\frac{1}{2}\left\|\sin(4\tau)\right\|$
Dihedral	$\left\|\cos(2\tau)\sin(2\theta)\sqrt{\cos^2(2\theta)+\sin^2(2\theta)\sin^2(2\tau)}\right\|$
Dipole	$\frac{1}{2}\left[\sin^2(\tau)\sin^2(\theta)+\cos^2(\theta)\cos^2(\tau)\right]\sqrt{\cos^2(2\tau)\sin^2(2\theta)+\sin^2(2\tau)}$
Helix	$\frac{\left\|\cos(2\tau)\left[\sin(2\tau)-1\right]\right\|}{4}$

at an elliptic arc. The minimum value points locate at the three-dimension polarimetric correlation pattern slices when $\tau=0^\circ$ for $\theta=0,\pm 90^\circ$. For helix structure, the maximum value points locate at the three-dimension polarimetric correlation pattern slices when $\tau=-15^\circ$, while the minimum value points locate at the three-dimension polarimetric correlation pattern slices when $\tau=\pm 45^\circ$. In summary, the distributions of the maximum and minimum points of different structures are not the same. It can be noted that when $\tau=\pm 45^\circ$, except for the dipole structure, the values of $\left|\hat{\gamma}_{\text{HH-HV}}(\theta,\tau)\right|$ from other structures are always zero. On the condition of $\tau=0^\circ$, the $\left|\hat{\gamma}_{\text{HH-HV}}(\theta,\tau)\right|$ values from the plane structure are irregular to the polarimetric orientation angle and always maintain zero.

2.6.4 Characterization of Three-Dimension Polarimetric Correlation Patterns

The three-dimension polarimetric correlation pattern provides another visualization tool to fully depict the changes of correlation values in terms of the PEA and POA. For quantitative characterization and description, three new polarimetric feature sets, including global features, local features, and mutual features, are introduced as follows.

2.6.4.1 The Concept of Curvature

In differential geometry, curvature is an important theoretical tool to describe the characteristics of space surfaces. There are three canonical surface curvatures: principal curvature, mean curvature, and Gaussian curvature. For one point over the surface, there are countless curves. Each curve can obtain a curvature value. The maximum and minimum curvature are denoted as κ_1 and κ_2, respectively, which are both named the principal curvature. The principal curvature can represent the maximum bending degree on the surface. The mean value of κ_1 and κ_2 is defined as the mean curvature, which can also describe the local bending situation on the surface. As long as one point on the surface bends in one direction, there is non-zero mean curvature, except for the saddle point, which satisfies $\kappa_1 = -\kappa_2$, and the mean curvature is zero. The product of κ_1 and κ_2 is defined as the Gaussian curvature, which can measure the overall bending degree in a local range and whether the surface has bending in all directions.

The calculation is as follows. For surface curvature, the coordinate vector of Euclidean space is defined as $\begin{bmatrix} x & y & z \end{bmatrix}^{\mathrm{T}}$, and the normal vector to the surface $\mathcal{S}$ can be expressed as

$$\mathbf{n} = \left[\frac{\partial \mathcal{S}}{\partial x} \times \frac{\partial \mathcal{S}}{\partial y}\right] \Big/ \left|\frac{\partial \mathcal{S}}{\partial x} \times \frac{\partial \mathcal{S}}{\partial y}\right| \tag{2.6.19}$$

where $\times$ is the cross product.

The first fundamental forms $\mathbf{I}(\mathcal{S})$ and the second fundamental forms $\mathbf{II}\ (\mathcal{S})$ of the surface are thus defined as [78]

$$\mathbf{I}(\mathcal{S}) = \begin{bmatrix} \dfrac{\partial \mathcal{S}}{\partial x}\dfrac{\partial \mathcal{S}}{\partial x}, & \dfrac{\partial \mathcal{S}}{\partial x}\dfrac{\partial \mathcal{S}}{\partial y} \\ \dfrac{\partial \mathcal{S}}{\partial y}\dfrac{\partial \mathcal{S}}{\partial x}, & \dfrac{\partial \mathcal{S}}{\partial y}\dfrac{\partial \mathcal{S}}{\partial y} \end{bmatrix} \tag{2.6.20}$$

$$\mathbf{II}(\mathcal{S}) = \begin{bmatrix} \dfrac{\partial^2 \mathcal{S}}{\partial x^2}\mathbf{n}, \dfrac{\partial^2 \mathcal{S}}{\partial x \partial y}\mathbf{n} \\ \dfrac{\partial^2 \mathcal{S}}{\partial y \partial x}\mathbf{n}, \dfrac{\partial^2 \mathcal{S}}{\partial y^2}\mathbf{n} \end{bmatrix} \tag{2.6.21}$$

The Gaussian curvature GC and the mean curvature MC are [78]

$$\mathrm{GC} = \kappa_1 \kappa_2 = \frac{\mathrm{Det}(\mathbf{II})}{\mathrm{Det}(\mathbf{I})} \tag{2.6.22}$$

$$\mathrm{MC} = \frac{1}{2}(\kappa_1 + \kappa_2) = \frac{1}{2}\mathrm{Tr}\left(\frac{\boldsymbol{II}}{\boldsymbol{I}}\right) \tag{2.6.23}$$

In this vein, the maximum principal curvature κ_1 and the minimum principal curvature κ_2 can be calculated as

$$\kappa_1 = \mathrm{MC} + \sqrt{\mathrm{MC}^2 - \mathrm{GC}} \tag{2.6.24}$$

$$\kappa_2 = \mathrm{MC} - \sqrt{\mathrm{MC}^2 - \mathrm{GC}} \tag{2.6.25}$$

For one curve L on the surface section, the curve curvature CC can indicate the bending degree for this two-dimension situation, and its expression is [78]

$$\mathrm{CC} = \frac{\left|L^{(2)}\right|}{\left[1 + \left(L^{(1)}\right)^2\right]^{\frac{3}{2}}} \tag{2.6.26}$$

where superscripts $^{(1)}$ and $^{(2)}$ mean the first and second derivatives.

2.6.4.2 Surface Curvature for Canonical Structures

Observed from Figure 2.6.2, the surface of the three-dimension polarimetric correlation pattern is fluctuating. In some regions of the three-dimension polarimetric correlation pattern, both valley structures and sharp structures are formed. This indicates that the curvature features can effectively characterize the spatial variations of a three-dimension polarimetric correlation pattern along the PEA and POA directions. For four canonical structures, the Gaussian curvature, mean curvature, and the maximum value of two principal curvature results κ_1 and κ_2 of the three-dimension polarimetric correlation pattern $\left|\hat{\gamma}_{\text{HH-HV}}(\theta,\tau)\right|$ are displayed in Figures 2.6.4, 2.6.5, and 2.6.6, respectively.

In Figure 2.6.4, for plane structure, the Gaussian curvature value is always zero, which is independent of the PEA and POA. This conclusion can also be seen from Figure 2.6.2 (a1). With the change of the POA, the values of the three-dimension polarimetric correlation pattern $\left|\hat{\gamma}_{\text{HH-HV}}(\theta,\tau)\right|$ are always constant at the PEA-plane for the plane structure, and its Gaussian curvature feature is zero. For a dihedral structure, its Gaussian curvature feature has two peaks located at $\{\theta = \pm 45°, \tau = 0°\}$, shown in Figure 2.6.4, which are the minimum value points of the three-dimension polarimetric correlation pattern $\left|\hat{\gamma}_{\text{HH-HV}}(\theta,\tau)\right|$. In addition, compared with the plane and helix structures, the three-dimension polarimetric correlation pattern

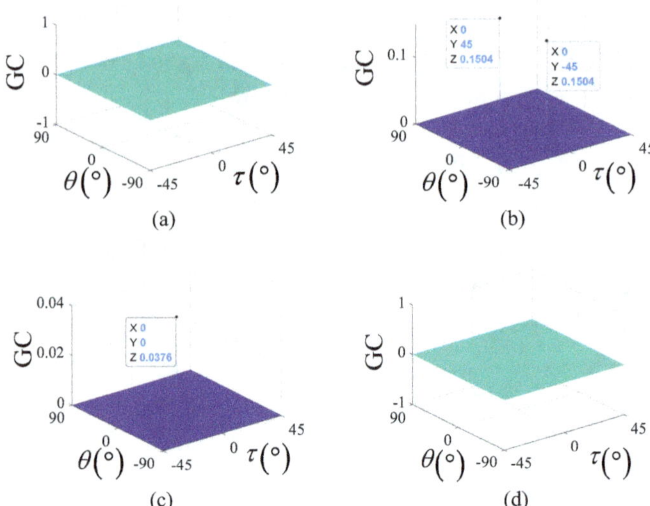

FIGURE 2.6.4 Gaussian curvature results from three-dimension polarimetric correlation pattern $\left|\hat{\gamma}_{\text{HH-HV}}(\theta,\tau)\right|$ for four canonical structures. (a) Plane, (b) dihedral, (c) dipole, and (d) helix.

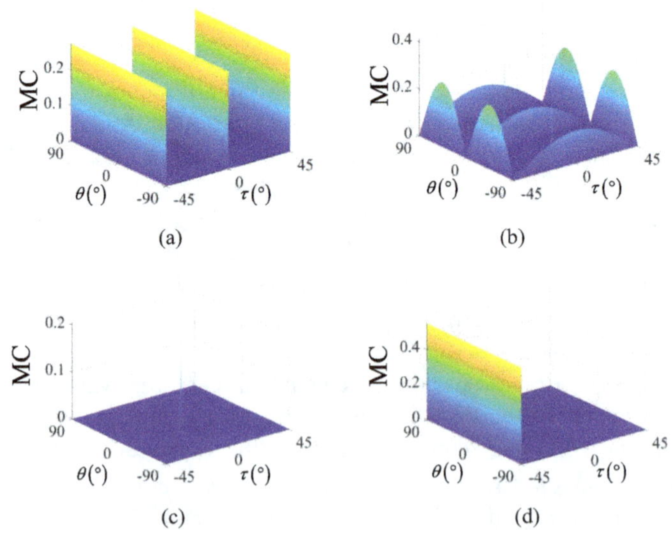

FIGURE 2.6.5 Mean curvature results from three-dimension polarimetric correlation pattern $\left|\hat{\gamma}_{\text{HH-HV}}(\theta,\tau)\right|$ for four canonical structures. (a) Plane, (b) dihedral, (c) dipole, and (d) helix.

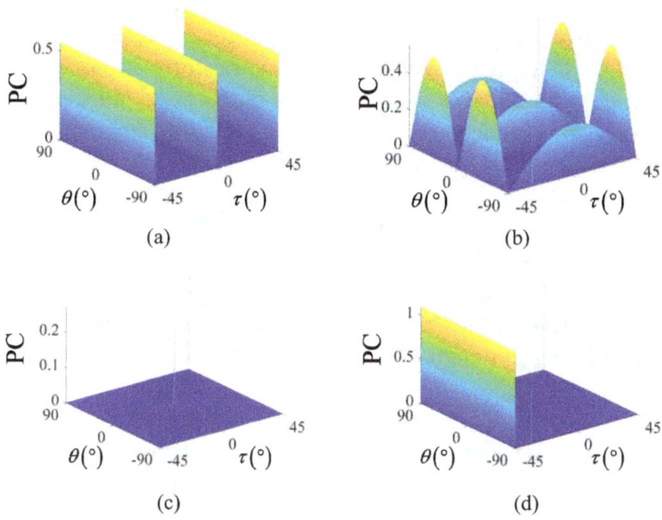

FIGURE 2.6.6 Principal curvature results from three-dimension polarimetric correlation pattern $\left|\hat{\gamma}_{\text{HH-HV}}(\theta,\tau)\right|$ for four canonical structures. (a) Plane, (b) dihedral, (c) dipole, and (d) helix.

values are greatly changing at $\{\theta = \pm 45°, \tau = 0°\}$ from the dihedral structure and at $\{\theta = 0, \tau = 0°\}$ from the dipole structure. In this vein, both the maximum value of the Gaussian curvature feature and its location provide the possibility to distinguish different scattering mechanisms.

For mean curvature features shown in Figure 2.6.5, planar bulges and pulses can be observed. Taking the plane structure as an example, its mean curvature feature shows three planar bulges respectively at $\tau = 0°$ and $\tau = \pm 45°$. Two peaks are observed at $\{\theta = \pm 45°, \tau = 0°\}$ for the dihedral structure, while the dipole structure has one peak at $\{\theta = 0°, \tau = 0°\}$, which stands for the extreme point accordingly. Except for the absolute feature values, the shapes of the principal curvature feature maps are closely similar to those of the corresponding mean curvature feature maps, as shown in Figures 2.6.5 and 2.6.6.

In conclusion, compared with the mean curvature and principal curvature results, the values of the Gaussian curvature feature are more pure, since its non-zero values are only available when surface bending in all directions is satisfied. Gaussian curvature can present the overall change degree of a three-dimension polarimetric correlation pattern in terms of various PEAs and POAs. The higher the Gaussian curvature values, the more intense changes appear to be in both the PEA-plane and POA-plane. The mean curvature represents the average change degree of a three-dimension polarimetric correlation pattern in terms of various PEAs and POAs, while the principal curvature characterizes the maximum bending degree for surface points of a three-dimension polarimetric correlation pattern. In addition, the curve curvature has the capability to indicate the bending degree for curve points from the PEA-plane or POA-plane of a three-dimension polarimetric correlation pattern. Therefore, three surface curvature features and one curve curvature feature provide effective solutions to represent, characterize, and utilize the rich information embedded in the three-dimension polarimetric correlation pattern.

2.6.4.3 Characterization of Three-Dimension Polarimetric Correlation Pattern

In order to quantitatively characterize and represent a three-dimension polarimetric correlation pattern, three kinds of new polarimetric feature sets, including global features, local features, and mutual features, are defined.

1) **Global Features:** global features are derived from the three-dimension manifold of a three-dimension polarimetric correlation pattern. The global features mainly include the maximum correlation, minimum correlation, and surface curvature features (such as the maximum Gaussian curvature, maximum mean curvature, and the maximum principal curvature). The schematic diagram of global features is shown in Figure 2.6.7. The definitions of these features are as follows:

   a) Maximum Correlation $\left|\hat{\gamma}_{1-2}(\theta,\tau)\right|_{\text{all-max}}$: It is the maximum correlation of a three-dimension polarimetric correlation pattern:

$$\left|\hat{\gamma}_{1-2}(\theta,\tau)\right|_{\text{all-max}} = \max_{\theta\in[-\pi,\pi),\tau\in[-\pi/4,\pi/4)}\left(\left|\hat{\gamma}_{1-2}(\theta,\tau)\right|\right) \tag{2.6.27}$$

   It indicates the upper limit of polarimetric correlation of two polarization channels via both POA and PEA adjustment.

   b) Minimum Correlation $\left|\hat{\gamma}_{1-2}(\theta,\tau)\right|_{\text{all-min}}$: It is the minimum correlation of a three-dimension polarimetric correlation pattern:

$$\left|\hat{\gamma}_{1-2}(\theta,\tau)\right|_{\text{all-min}} = \min_{\theta\in[-\pi,\pi),\tau\in[-\pi/4,\pi/4)}\left(\left|\hat{\gamma}_{1-2}(\theta,\tau)\right|\right) \tag{2.6.28}$$

   It indicates the lower limit of polarimetric correlation of two polarization channels via both POA and PEA adjustment.

   c) Maximum Gaussian Curvature $\left|\hat{\gamma}_{1-2}(\theta,\tau)\right|_{\text{MGC}}$: It is defined as the maximum value of the Gaussian curvatures:

$$\left|\hat{\gamma}_{1-2}(\theta,\tau)\right|_{\text{MGC}} = \max_{\theta\in[-\pi,\pi),\tau\in[-\pi/4,\pi/4)}\left(\left|\text{GC}(\theta,\tau)\right|\right) \tag{2.6.29}$$

   It indicates the maximum bending degree of a three-dimension polarimetric correlation pattern. Note that Gaussian curvature is the essential

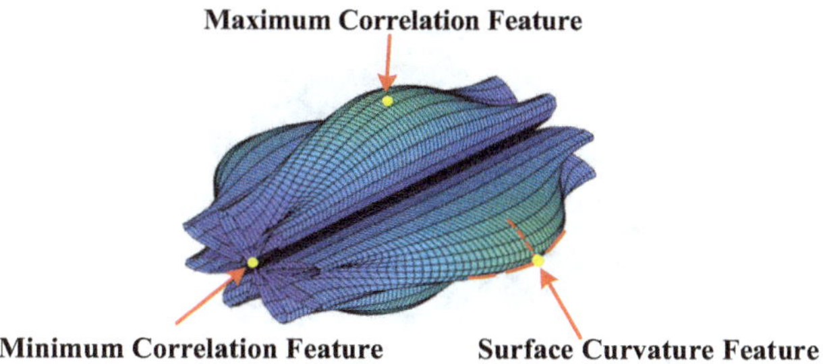

FIGURE 2.6.7 Schematic diagram of global features.

attribute of a surface and will not change in terms of any non-stretching transformation to the surface.

d) Maximum Mean Curvature $\left|\hat{\gamma}_{1-2}(\theta,\tau)\right|_{\text{MMC}}$: It is defined as the maximum value of the mean curvatures:

$$\left|\hat{\gamma}_{1-2}(\theta,\tau)\right|_{\text{MMC}} = \max_{\theta\in[-\pi,\pi),\tau\in[-\pi/4,\pi/4)}\left(\left|\text{MC}(\theta,\tau)\right|\right) \tag{2.6.30}$$

It indicates the maximum mean bending degree of a three-dimension polarimetric correlation pattern.

e) Maximum Principal Curvature $\left|\hat{\gamma}_{1-2}(\theta,\tau)\right|_{\text{MPC}}$: It is defined as the maximum value of two kinds of the principal curvatures:

$$\left|\hat{\gamma}_{1-2}(\theta,\tau)\right|_{\text{MPC}} = \max\left[\max_{\theta\in[-\pi,\pi),\tau\in[-\pi/4,\pi/4)}\left(\left|\kappa_1(\theta,\tau)\right|\right), \max_{\theta\in[-\pi,\pi),\tau\in[-\pi/4,\pi/4)}\left(\left|\kappa_2(\theta,\tau)\right|\right)\right] \tag{2.6.31}$$

It indicates the maximum bending degree of a three-dimension polarimetric correlation pattern.

2) **Local Features:** local features are derived from the two-dimension planes of a three-dimension polarimetric correlation pattern. Given a PEA or a POA, a two-dimension plane of the three-dimension polarimetric correlation pattern can be obtained, and a set of local features can be defined thereafter. Examples of the PEA-plane and POA-plane are shown in Figures 2.6.8 and 2.6.9.

First, with a given PEA, a PEA-plane, shown in Figure 2.6.8, can be obtained. For each given PEA, seven local features can be respectively defined, as follows

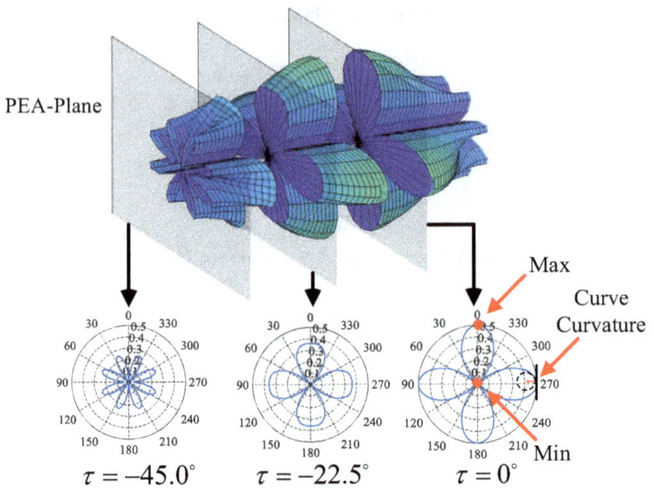

FIGURE 2.6.8 Schematic diagram of PEA-planes and local features.

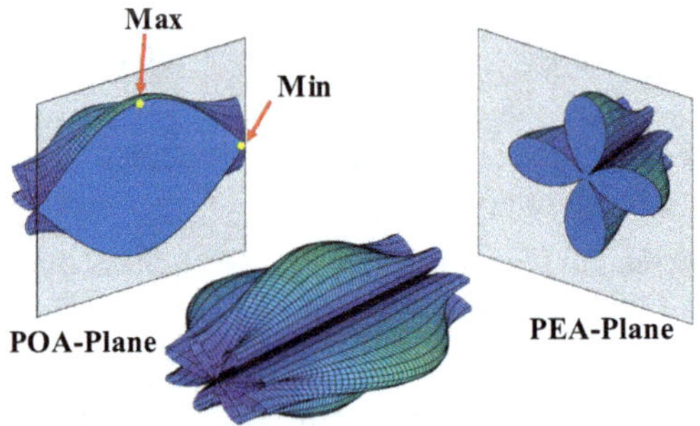

FIGURE 2.6.9 Schematic diagram of POA-plane and PEA-plane.

a) Slice Original Correlation $\left|\hat{\gamma}_{1-2}(\theta,\tau)\right|_{\text{org}}^{\tau=\alpha}$: It is the polarimetric correlation value with POA $\theta=0$ in the PEA-plane of $\tau=\alpha$:

$$\left|\hat{\gamma}_{1-2}(\theta,\tau)\right|_{\text{org}}^{\tau=\alpha} = \left|\hat{\gamma}_{1-2}(0,\tau=\alpha)\right| \tag{2.6.32}$$

b) Slice Correlation Degree $\left|\hat{\gamma}_{1-2}(\theta,\tau)\right|_{\text{mean}}^{\tau=\alpha}$: It is the mean of polarimetric correlation values in the PEA-plane of $\tau=\alpha$:

$$\left|\hat{\gamma}_{1-2}(\theta,\tau)\right|_{\text{mean}}^{\tau=\alpha} = \frac{1}{2\pi}\int_{-\pi}^{\pi}\left|\hat{\gamma}_{1-2}(\theta,\tau)\right|^{\tau=\alpha}\mathrm{d}\theta \tag{2.6.33}$$

c) Slice Correlation Fluctuation $\left|\hat{\gamma}_{1-2}(\theta,\tau)\right|_{\text{std}}^{\tau=\alpha}$: It is the standard deviation of polarimetric correlation values in the PEA-plane of $\tau=\alpha$:

$$\left|\hat{\gamma}_{1-2}(\theta,\tau)\right|_{\text{std}}^{\tau=\alpha} = \sqrt{\frac{1}{2\pi}\int_{-\pi}^{\pi}\left(\left|\hat{\gamma}_{1-2}(\theta,\tau)\right|^{\tau=\alpha} - \left|\hat{\gamma}_{1-2}(\theta,\tau)\right|_{\text{mean}}^{\tau=\alpha}\right)^2 \mathrm{d}\theta} \tag{2.6.34}$$

d) Slice Maximum Correlation $\left|\hat{\gamma}_{1-2}(\theta,\tau)\right|_{\max}^{\tau=\alpha}$: It is the maximum polarimetric correlation value in the PEA-plane of $\tau=\alpha$:

$$\left|\hat{\gamma}_{1-2}(\theta,\tau)\right|_{\max}^{\tau=\alpha} = \max_{\theta\in[-\pi,\pi)}\left(\left|\hat{\gamma}_{1-2}(\theta,\tau)\right|^{\tau=\alpha}\right) \tag{2.6.35}$$

e) Slice Minimum Correlation $\left|\hat{\gamma}_{1-2}(\theta,\tau)\right|_{\min}^{\tau=\alpha}$: It is the minimum polarimetric correlation value in the PEA-plane of $\tau=\alpha$:

$$\left|\hat{\gamma}_{1-2}(\theta,\tau)\right|_{\min}^{\tau=\alpha} = \min_{\theta\in[-\pi,\pi)}\left(\left|\hat{\gamma}_{1-2}(\theta,\tau)\right|^{\tau=\alpha}\right) \tag{2.6.36}$$

f) Slice Correlation Contrast $\left|\hat{\gamma}_{1-2}(\theta,\tau)\right|_{\text{contrast}}^{\tau=\alpha}$: It is defined as the difference between the slice maximum correlation and slice minimum correlation in the PEA-plane of $\tau=\alpha$:

$$\left|\hat{\gamma}_{1-2}(\theta,\tau)\right|_{\text{contrast}}^{\tau=\alpha} = \left|\hat{\gamma}_{1-2}(\theta,\tau)\right|_{\max}^{\tau=\alpha} - \left|\hat{\gamma}_{1-2}(\theta,\tau)\right|_{\min}^{\tau=\alpha} \tag{2.6.37}$$

It partially indicates the target scattering diversity in the PEA-plane of $\tau=\alpha$.

g) Slice Maximum Curve Curvature $\left|\hat{\gamma}_{1-2}(\theta,\tau)\right|_{\text{MCC}}^{\tau=\alpha}$: It is defined as the maximum value of curve curvatures in the PEA-plane of $\tau=\alpha$:

$$\left|\hat{\gamma}_{1-2}(\theta,\tau)\right|_{\text{MCC}}^{\tau=\alpha} = \max_{\theta\in[-\pi,\pi)}\left(\left|\text{CC}(\theta)\right|\right) \tag{2.6.38}$$

It indicates the maximum change degree with the POA. Curve curvature CC is defined in (2.6.26).

Second, for a certain POA, a POA-plane can be derived shown in Figure 2.6.9. The PEA-plane is also shown for comparison. Similarly, seven local features can be defined:

a) Slice Original Correlation $\left|\hat{\gamma}_{1-2}(\theta,\tau)\right|_{\text{org}}^{\theta=\beta}$: It is the polarimetric correlation value with PEA $\tau=0$ in the POA-plane of $\theta=\beta$:

$$\left|\hat{\gamma}_{1-2}(\theta,\tau)\right|_{\text{org}}^{\theta=\beta} = \left|\hat{\gamma}_{1-2}(\theta=\beta,0)\right| \tag{2.6.39}$$

b) Slice Correlation Degree $\left|\hat{\gamma}_{1-2}(\theta,\tau)\right|_{\text{mean}}^{\theta=\beta}$: It is the mean of polarimetric correlation values in the POA-plane of $\theta=\beta$:

$$\left|\hat{\gamma}_{1-2}(\theta,\tau)\right|_{\text{mean}}^{\theta=\beta}=\frac{1}{2\pi}\int_{-\pi/4}^{\pi/4}\left|\hat{\gamma}_{1-2}(\theta,\tau)\right|^{\theta=\beta}\mathrm{d}\tau \tag{2.6.40}$$

c) Slice Correlation Fluctuation $\left|\hat{\gamma}_{1-2}(\theta,\tau)\right|_{\text{std}}^{\theta=\beta}$: It is the standard deviation of polarimetric correlation values in the POA-plane of $\theta=\beta$:

$$\left|\hat{\gamma}_{1-2}(\theta,\tau)\right|_{\text{std}}^{\theta=\beta}=\sqrt{\frac{1}{2\pi}\int_{-\pi/4}^{\pi/4}\left(\left|\hat{\gamma}_{1-2}(\theta,\tau)\right|^{\theta=\beta}-\left|\hat{\gamma}_{1-2}(\theta,\tau)\right|_{\text{mean}}^{\theta=\beta}\right)^{2}\mathrm{d}\tau} \tag{2.6.41}$$

d) Slice Maximum Correlation $\left|\hat{\gamma}_{1-2}(\theta,\tau)\right|_{\text{max}}^{\theta=\beta}$: It is the maximum polarimetric correlation value in the POA-plane of $\theta=\beta$:

$$\left|\hat{\gamma}_{1-2}(\theta,\tau)\right|_{\text{max}}^{\theta=\beta}=\max_{\tau\in[-\pi/4,\pi/4)}\left(\left|\hat{\gamma}_{1-2}(\theta,\tau)\right|^{\theta=\beta}\right) \tag{2.6.42}$$

e) Slice Minimum Correlation $\left|\hat{\gamma}_{1-2}(\theta,\tau)\right|_{\text{min}}^{\theta=\beta}$: It is the minimum polarimetric correlation value in the POA-plane of $\theta=\beta$:

$$\left|\hat{\gamma}_{1-2}(\theta,\tau)\right|_{\text{min}}^{\theta=\beta}=\min_{\tau\in[-\pi/4,\pi/4)}\left(\left|\hat{\gamma}_{1-2}(\theta,\tau)\right|^{\theta=\beta}\right) \tag{2.6.43}$$

f) Slice Correlation Contrast $\left|\hat{\gamma}_{1-2}(\theta,\tau)\right|_{\text{contrast}}^{\theta=\beta}$: It is defined as the difference between the slice maximum correlation and slice minimum correlation in the POA-plane of $\theta=\beta$:

$$\left|\hat{\gamma}_{1-2}(\theta,\tau)\right|_{\text{contrast}}^{\theta=\beta}=\left|\hat{\gamma}_{1-2}(\theta,\tau)\right|_{\text{max}}^{\theta=\beta}-\left|\hat{\gamma}_{1-2}(\theta,\tau)\right|_{\text{min}}^{\theta=\beta} \tag{2.6.44}$$

g) Slice Maximum Curve Curvature $\left|\hat{\gamma}_{1-2}(\theta,\tau)\right|_{\text{MCC}}^{\theta=\beta}$: It is defined as the maximum curve curvature in the POA-plane of $\theta=\beta$:

$$\left|\hat{\gamma}_{1-2}(\theta,\tau)\right|_{\text{MCC}}^{\theta=\beta}=\max_{\tau\in[-\pi/4,\pi/4)}\left(\left|\text{CC}(\tau)\right|\right) \tag{2.6.45}$$

In general, this section focuses on four cases, including the PEA of $\tau\in\{0°,45°\}$ and the POA of $\theta\in\{0°,45°\}$. In this vein, for each three-dimension polarimetric correlation pattern, 28 local features can be extracted in total.

3) **Mutual Features:** mutual features are deduced from two different planes and indicate the interrelationships between them.

a) Correlation Difference $\left|\hat{\gamma}_{1-2}(\theta,\tau)\right|_{\text{diff}}$: It is the difference between the maximum values of two PEA-planes:

$$\left|\hat{\gamma}_{1-2}(\theta,\tau)\right|_{\text{diff}} = \left|\left|\hat{\gamma}_{1-2}(\theta,\tau)\right|_{\max}^{\tau=\alpha_1} - \left|\hat{\gamma}_{1-2}(\theta,\tau)\right|_{\max}^{\tau=\alpha_2}\right| \tag{2.6.46}$$

where α_1 and α_2 are two different PEAs. In this book, $\alpha_1 = 0°$ and $\alpha_2 = 45°$ are mainly considered.

2.6.4.4 Interpretation and Analysis

Based on the quantitative characterization of the three-dimension polarimetric correlation pattern and the four typical PEA-plane and POA-plane ($\tau \in \{0°, 45°\}$ and $\tau \in \{0°, 45°\}$), 5 global features, 28 local features and 1 mutual feature can be obtained. Therefore, 34 new features can be extracted from three independent three-dimension polarimetric correlation patterns $\left|\hat{\gamma}_{\text{HH-HV}}(\theta,\tau)\right|$, $\left|\hat{\gamma}_{\text{HH-VV}}(\theta,\tau)\right|$ and $\left|\hat{\gamma}_{\text{(HH+VV)-(HH-VV)}}(\theta,\tau)\right|$, respectively, which are shown in Table 2.6.2.

For the global features, the most intuitive difference between three-dimension polarimetric correlation patterns is the degree of surface bending. The curvature features can well descript this bending degree. In detail, the maximum Gaussian curvature represents the overall maximum bending degree and has a non-zero value on the condition of the surface bending in all directions. Besides, the higher the maximum Gaussian curvature value is, the more intense the change is in both the PEA and POA planes. For the local features, taking the PEA-plane of $\tau = \alpha$ as an example, the slice original correlation $\left|\hat{\gamma}_{1-2}(\theta,\tau)\right|_{\text{org}}^{\tau=\alpha}$ indicates the target decorrelation effect for two polarization channels at a certain PEA. The slice correlation degree $\left|\hat{\gamma}_{1-2}(\theta,\tau)\right|_{\text{mean}}^{\tau=\alpha}$ is the average value of polarimetric correlation values, while the slice correlation fluctuation $\left|\hat{\gamma}_{1-2}(\theta,\tau)\right|_{\text{std}}^{\tau=\alpha}$ indicates the standard deviation of polarimetric correlation values. For the slice correlation contrast $\left|\hat{\gamma}_{1-2}(\theta,\tau)\right|_{\text{contrast}}^{\tau=\alpha}$, it reveals the target scattering diversity in the PEA-plane of $\tau = \alpha$. The slice maximum correlation value $\left|\hat{\gamma}_{1-2}(\theta,\tau)\right|_{\max}^{\tau=\alpha}$ and the slice minimum correlation value $\left|\hat{\gamma}_{1-2}(\theta,\tau)\right|_{\min}^{\tau=\alpha}$ represent the highest and lowest polarimetric correlation values in the PEA-plane of $\tau = \alpha$, respectively. Specifically, on the condition of $\tau = 0°$, the slice polarimetric features from the PEA-plane are equivalent to the two-dimension polarimetric correlation pattern features. Furthermore, the slice maximum curve curvature $\left|\hat{\gamma}_{1-2}(\theta,\tau)\right|_{\text{MCC}}^{\tau=\alpha}$ indicates the variation intensity of polarimetric correlation values in the PEA-plane of $\tau = \alpha$. For mutual features, the correlation difference $\left|\hat{\gamma}_{1-2}(\theta,\tau)\right|_{\text{diff}}$ indicates the interrelationships between different PEA-planes.

TABLE 2.6.2
Three-Dimension Polarimetric Correlation Pattern Features

Polarimetric Feature			Three-Dimension Polarimetric Correlation Pattern		
			$\lvert\hat{\gamma}_{\text{HH-HV}}(\theta,\tau)\rvert$	$\lvert\hat{\gamma}_{\text{HH-HV}}(\theta,\tau)\rvert$	$\lvert\hat{\gamma}_{\text{(HH+VV)-(HH-VV)}}(\theta,\tau)\rvert$
Global feature		Maximum correlation	$\lvert\hat{\gamma}_{\text{HH-HV}}(\theta,\tau)\rvert_{\text{all-max}}$	$\lvert\hat{\gamma}_{\text{HH-VV}}(\theta,\tau)\rvert_{\text{all-max}}$	$\lvert\hat{\gamma}_{\text{(HH+VV)-(HH-VV)}}(\theta,\tau)\rvert_{\text{all-max}}$
		Minimum correlation	$\lvert\hat{\gamma}_{\text{HH-HV}}(\theta,\tau)\rvert_{\text{all-min}}$	$\lvert\hat{\gamma}_{\text{HH-VV}}(\theta,\tau)\rvert_{\text{all-min}}$	$\lvert\hat{\gamma}_{\text{(HH+VV)-(HH-VV)}}(\theta,\tau)\rvert_{\text{all-min}}$
		Maximum Gaussian curvature	$\lvert\hat{\gamma}_{\text{HH-HV}}(\theta,\tau)\rvert_{\text{MGC}}$	$\lvert\hat{\gamma}_{\text{HH-VV}}(\theta,\tau)\rvert_{\text{MGC}}$	$\lvert\hat{\gamma}_{\text{(HH+VV)-(HH-VV)}}(\theta,\tau)\rvert_{\text{MGC}}$
		Maximum mean curvature	$\lvert\hat{\gamma}_{\text{HH-HV}}(\theta,\tau)\rvert_{\text{MMC}}$	$\lvert\hat{\gamma}_{\text{HH-VV}}(\theta,\tau)\rvert_{\text{MMC}}$	$\lvert\hat{\gamma}_{\text{(HH+VV)-(HH-VV)}}(\theta,\tau)\rvert_{\text{MMC}}$
		Maximum principle curvature	$\lvert\hat{\gamma}_{\text{HH-HV}}(\theta,\tau)\rvert_{\text{MPC}}$	$\lvert\hat{\gamma}_{\text{HH-VV}}(\theta,\tau)\rvert_{\text{MPC}}$	$\lvert\hat{\gamma}_{\text{(HH+VV)-(HH-VV)}}(\theta,\tau)\rvert_{\text{MPC}}$
Local feature	PEA-plane	Slice original correlation	$\lvert\hat{\gamma}_{\text{HH-HV}}(\theta,\tau)\rvert_{\text{org}}^{\tau=\alpha}$	$\lvert\hat{\gamma}_{\text{HH-VV}}(\theta,\tau)\rvert_{\text{org}}^{\tau=\alpha}$	$\lvert\hat{\gamma}_{\text{(HH+VV)-(HH-VV)}}(\theta,\tau)\rvert_{\text{org}}^{\tau=\alpha}$
		Slice correlation degree	$\lvert\hat{\gamma}_{\text{HH-HV}}(\theta,\tau)\rvert_{\text{mean}}^{\tau=\alpha}$	$\lvert\hat{\gamma}_{\text{HH-VV}}(\theta,\tau)\rvert_{\text{mean}}^{\tau=\alpha}$	$\lvert\hat{\gamma}_{\text{(HH+VV)-(HH-VV)}}(\theta,\tau)\rvert_{\text{mean}}^{\tau=\alpha}$
		Slice correlation fluctuation	$\lvert\hat{\gamma}_{\text{HH-HV}}(\theta,\tau)\rvert_{\text{std}}^{\tau=\alpha}$	$\lvert\hat{\gamma}_{\text{HH-VV}}(\theta,\tau)\rvert_{\text{std}}^{\tau=\alpha}$	$\lvert\hat{\gamma}_{\text{(HH+VV)-(HH-VV)}}(\theta,\tau)\rvert_{\text{std}}^{\tau=\alpha}$
		Slice maximum correlation	$\lvert\hat{\gamma}_{\text{HH-HV}}(\theta,\tau)\rvert_{\text{max}}^{\tau=\alpha}$	$\lvert\hat{\gamma}_{\text{HH-VV}}(\theta,\tau)\rvert_{\text{max}}^{\tau=\alpha}$	$\lvert\hat{\gamma}_{\text{(HH+VV)-(HH-VV)}}(\theta,\tau)\rvert_{\text{max}}^{\tau=\alpha}$
		Slice minimum correlation	$\lvert\hat{\gamma}_{\text{HH-HV}}(\theta,\tau)\rvert_{\text{min}}^{\tau=\alpha}$	$\lvert\hat{\gamma}_{\text{HH-VV}}(\theta,\tau)\rvert_{\text{min}}^{\tau=\alpha}$	$\lvert\hat{\gamma}_{\text{(HH+VV)-(HH-VV)}}(\theta,\tau)\rvert_{\text{min}}^{\tau=\alpha}$
		Slice correlation contrast	$\lvert\hat{\gamma}_{\text{HH-HV}}(\theta,\tau)\rvert_{\text{contrast}}^{\tau=\alpha}$	$\lvert\hat{\gamma}_{\text{HH-VV}}(\theta,\tau)\rvert_{\text{contrast}}^{\tau=\alpha}$	$\lvert\hat{\gamma}_{\text{(HH+VV)-(HH-VV)}}(\theta,\tau)\rvert_{\text{contrast}}^{\tau=\alpha}$
		Slice maximum curve curvature	$\lvert\hat{\gamma}_{\text{HH-HV}}(\theta,\tau)\rvert_{\text{MCC}}^{\tau=\alpha}$	$\lvert\hat{\gamma}_{\text{HH-VV}}(\theta,\tau)\rvert_{\text{MCC}}^{\tau=\alpha}$	$\lvert\hat{\gamma}_{\text{(HH+VV)-(HH-VV)}}(\theta,\tau)\rvert_{\text{MCC}}^{\tau=\alpha}$

	POA-plane	Slice original correlation	$\left\|\hat{\gamma}_{\text{HH-HV}}(\theta,\tau)\right\|_{\text{org}}^{\theta=\beta}$	$\left\|\hat{\gamma}_{\text{HH-VV}}(\theta,\tau)\right\|_{\text{org}}^{\theta=\beta}$	$\left\|\hat{\gamma}_{\text{(HH+VV)-(HH-VV)}}(\theta,\tau)\right\|_{\text{org}}^{\theta=\beta}$
		Slice correlation degree	$\left\|\hat{\gamma}_{\text{HH-HV}}(\theta,\tau)\right\|_{\text{mean}}^{\theta=\beta}$	$\left\|\hat{\gamma}_{\text{HH-VV}}(\theta,\tau)\right\|_{\text{mean}}^{\theta=\beta}$	$\left\|\hat{\gamma}_{\text{(HH+VV)-(HH-VV)}}(\theta,\tau)\right\|_{\text{mean}}^{\theta=\beta}$
		Slice correlation fluctuation	$\left\|\hat{\gamma}_{\text{HH-HV}}(\theta,\tau)\right\|_{\text{std}}^{\theta=\beta}$	$\left\|\hat{\gamma}_{\text{HH-VV}}(\theta,\tau)\right\|_{\text{std}}^{\theta=\beta}$	$\left\|\hat{\gamma}_{\text{(HH+VV)-(HH-VV)}}(\theta,\tau)\right\|_{\text{std}}^{\theta=\beta}$
		Slice maximum correlation	$\left\|\hat{\gamma}_{\text{HH-HV}}(\theta,\tau)\right\|_{\text{max}}^{\theta=\beta}$	$\left\|\hat{\gamma}_{\text{HH-VV}}(\theta,\tau)\right\|_{\text{max}}^{\theta=\beta}$	$\left\|\hat{\gamma}_{\text{(HH+VV)-(HH-VV)}}(\theta,\tau)\right\|_{\text{max}}^{\theta=\beta}$
		Slice minimum correlation	$\left\|\hat{\gamma}_{\text{HH-HV}}(\theta,\tau)\right\|_{\text{min}}^{\theta=\beta}$	$\left\|\hat{\gamma}_{\text{HH-VV}}(\theta,\tau)\right\|_{\text{min}}^{\theta=\beta}$	$\left\|\hat{\gamma}_{\text{(HH+VV)-(HH-VV)}}(\theta,\tau)\right\|_{\text{min}}^{\theta=\beta}$
		Slice correlation contrast	$\left\|\hat{\gamma}_{\text{HH-HV}}(\theta,\tau)\right\|_{\text{contrast}}^{\theta=\beta}$	$\left\|\hat{\gamma}_{\text{HH-VV}}(\theta,\tau)\right\|_{\text{contrast}}^{\theta=\beta}$	$\left\|\hat{\gamma}_{\text{(HH+VV)-(HH-VV)}}(\theta,\tau)\right\|_{\text{contrast}}^{\theta=\beta}$
		Slice maximum curve curvature	$\left\|\hat{\gamma}_{\text{HH-HV}}(\theta,\tau)\right\|_{\text{MCC}}^{\theta=\beta}$	$\left\|\hat{\gamma}_{\text{HH-VV}}(\theta,\tau)\right\|_{\text{MCC}}^{\theta=\beta}$	$\left\|\hat{\gamma}_{\text{(HH+VV)-(HH-VV)}}(\theta,\tau)\right\|_{\text{MCC}}^{\theta=\beta}$
Mutual feature		Correlation difference	$\left\|\hat{\gamma}_{\text{HH-HV}}(\theta,\tau)\right\|_{\text{diff}}$	$\left\|\hat{\gamma}_{\text{HH-VV}}(\theta,\tau)\right\|_{\text{diff}}$	$\left\|\hat{\gamma}_{\text{(HH+VV)-(HH-VV)}}(\theta,\tau)\right\|_{\text{diff}}$

2.7 SUMMARY

In this chapter, the core ideas of polarimetric rotation domain interpretation theory and the corresponding interpretation tools are introduced in detail, including uniform polarimetric matrix rotation theory, the two-dimension polarimetric coherence pattern interpretation tool, the two-dimension polarimetric correlation pattern interpretation tool, and the three-dimension polarimetric correlation pattern interpretation tool. A series of new polarimetric feature sets is derived. Combined with a variety of PolSAR data, the application potentials of various polarimetric rotation domain features are demonstrated. The polarimetric rotation domain interpretation theory provides new ideas and new methods for the characterization, understanding, and utilization of target scattering diversity effects. The subsequent chapters will introduce the application research results of the polarimetric rotation domain interpretation theory.

REFERENCES

1. J. S. Lee and E. Pottier, *Polarimetric Radar Imaging: From Basics to Applications*. Boca Raton: CRC Press, 2009.
2. J. M. Lopez Sanchez, F. Vicente Guijalba, J. D. Ballester Berman, and S. R. Cloude, "Polarimetric response of rice fields at C-band: Analysis and phenology retrieval," *IEEE Transactions on Geoscience and Remote Sensing,* vol. 52, no. 5, pp. 2977–2993, 2014.
3. C. Yonezawa, M. Negishi, K. Azuma, M. Watanabe, N. Ishitsuka, S. Ogawa, et al., "Growth monitoring and classification of rice fields using multitemporal RADARSAT-2 full-polarimetric data," *International Journal of Remote Sensing,* vol. 33, no. 18, pp. 5696–5711, 2012.
4. S. W. Chen, Y. Z. Li, and X. S. Wang, "Crop discrimination based on polarimetric correlation coefficients optimization for PolSAR data," *International Journal of Remote Sensing,* vol. 36, no. 16, pp. 4233–4249, 2015.
5. J. S. Lee, M. R. Grunes, T. L. Ainsworth, L. Du, D. L. Schuler, and S. R. Cloude, "Unsupervised classification using polarimetric decomposition and the complex Wishart classifier," *IEEE Transactions on Geoscience and Remote Sensing,* vol. 37, no. 5, pp. 2249–2258, 1999.
6. R. Paladini, L. F. Famil, P. Eric, M. Martorella, F. Berizzi, and E. Dalle Mese, "Point target classification via fast lossless and sufficient invariant decomposition of high-resolution and fully polarimetric SAR/ISAR data," *Proceedings of the IEEE,* vol. 101, no. 3, pp. 798–830, 2013.
7. H. Skriver, "Crop classification by multitemporal C- and L-band single- and dual-polarization and fully polarimetric SAR," *IEEE Transactions on Geoscience and Remote Sensing,* vol. 50, no. 6, pp. 2138–2149, 2012.
8. S. R. Cloude and E. Pottier, "An entropy based classification scheme for land applications of polarimetric SAR," *IEEE Transactions On Geoscience And Remote Sensing,* vol. 35, no. 1, pp. 68–78, 1997.
9. S. P. Xiao, S. W. Chen, Y. L. Chang, Y. Z. Li, and M. Sato, "Polarimetric coherence optimization and its application for manmade target extraction in PolSAR data," *IEICE Transactions on Electronics,* vol. 97, no. 6, pp. 566–574, 2014.
10. D. Xiang, T. Tang, Y. Ban, and Y. Su, "Man-made target detection from polarimetric SAR data via nonstationarity and asymmetry," *IEEE Journal of Selected Topics in Applied Earth Observations and Remote Sensing,* vol. 9, no. 4, pp. 1459–1469, 2016.

11. C. S. Tao, S. W. Chen, Y. Z. Li, and S. P. Xiao, "PolSAR land cover classification based on roll-invariant and selected hidden polarimetric features in the rotation domain," *Remote Sensing,* vol. 9, no. 7, p. 660, 2017.
12. Y. Yamaguchi, "Disaster monitoring by fully polarimetric SAR data acquired with ALOS-PALSAR," *Proceedings of the IEEE,* vol. 100, no. 10, pp. 2851–2860, 2012.
13. M. Sato, S. W. Chen, and M. Satake, "Polarimetric SAR analysis of tsunami damage following the March 11, 2011 East Japan earthquake," *Proceedings of the IEEE,* vol. 100, no. 10, pp. 2861–2875, 2012.
14. S. W. Chen and M. Sato, "Tsunami damage investigation of built-up areas using multi-temporal spaceborne full polarimetric SAR images," *IEEE Transactions on Geoscience and Remote Sensing,* vol. 51, no. 4, pp. 1985–1997, 2013.
15. S. W. Chen, X. S. Wang, and M. Sato, "Urban damage level mapping based on scattering mechanism investigation using fully polarimetric SAR data for the 3.11 East Japan earthquake," *IEEE Transactions on Geoscience and Remote Sensing,* vol. 54, no. 12, pp. 6919–6929, 2016.
16. L. Zhao, J. Yang, P. Li, L. Zhang, L. Shi, and F. Lang, "Damage assessment in urban areas using post-earthquake airborne PolSAR imagery," *International Journal of Remote Sensing,* vol. 34, no. 24, pp. 8952–8966, 2013.
17. J. S. Lee, D. L. Schuler, T. L. Ainsworth, E. Krogager, D. Kasilingam, and W. M. Boerner, "On the estimation of radar polarization orientation shifts induced by terrain slopes," *IEEE Transactions on Geoscience and Remote Sensing,* vol. 40, no. 1, pp. 30–41, 2002.
18. S. W. Chen, M. Ohki, M. Shimada, and M. Sato, "Deorientation effect investigation for model-based decomposition over oriented built-up areas," *IEEE Geoscience and Remote Sensing Letters,* vol. 10, no. 2, pp. 273–277, 2013.
19. S. W. Chen, Y. Z. Li, X. S. Wang, S. P. Xiao, and M. Sato, "Modeling and interpretation of scattering mechanisms in polarimetric synthetic aperture radar: Advances and perspectives," *IEEE Signal Processing Magazine,* vol. 31, no. 4, pp. 79–89, 2014.
20. R. Touzi, "Target scattering decomposition in terms of roll-invariant target parameters," *IEEE Transactions on Geoscience and Remote Sensing,* vol. 45, no. 1, pp. 73–84, 2007.
21. S. W. Chen, X. S. Wang, S. P. Xiao, and M. Sato, *Target Scattering Mechanism in Polarimetric Synthetic Aperture Radar-Interpretation and Application.* Singapore: Springer, 2018.
22. J. S. Lee and T. L. Ainsworth, "The effect of orientation angle compensation on coherency matrix and polarimetric target decompositions," *IEEE Transactions on Geoscience and Remote Sensing,* vol. 49, no. 1, pp. 53–64, 2011.
23. S. W. Chen, "Characterization and application of electromagnetic scattering in polarimetric imaging radar," Ph.D. dissertation, Tohoku University, Japan, 2012.
24. S. W. Chen, Y. Z. Li, D. H. Dai, X. S. Wang, S. P. Xiao, and M. Sato, "Uniform polarimetric matrix rotation theory," in *IEEE International Geoscience and Remote Sensing Symposium*, Melbourne, Australia, 2013, pp. 4166–4169.
25. S. W. Chen, X. S. Wang, and M. Sato, "Uniform polarimetric matrix rotation theory and its applications," *IEEE Transactions on Geoscience and Remote Sensing,* vol. 52, no. 8, pp. 4756–4770, 2014.
26. S. W. Chen, Y. Z. Li, X. S. Wang, and S. P. Xiao, "Polarimetric SAR target scattering interpretation in rotation domain: Theory and application," *Journal of Radars,* vol. 6, no. 5, pp. 442–455, 2017.
27. S. W. Chen, Y. Z. Li, and X. S. Wang, "A visualization tool for polarimetric SAR data investigation," in *The 11th European Synthetic Aperture Radar Conference*, Hamburg, Germany, 2016, pp. 579–582.
28. S. W. Chen and X. S. Wang, "Polarimetric coherence pattern: A visualization tool for PolSAR data investigation," in *IEEE International Geoscience and Remote Sensing Symposium*, Beijing, China, 2016, pp. 7509–7512.

29. S. W. Chen, "Polarimetric coherence pattern: A visualization and characterization tool for PolSAR data investigation," *IEEE Transactions on Geoscience and Remote Sensing,* vol. 56, no. 1, pp. 286–297, 2018.
30. X. C. Cui, C. S. Tao, S. W. Chen, and Y. Su, "PolSAR ship detection with polarimetric correlation pattern," in *Asia-Pacific Conference on Synthetic Aperture Radar*, Xiamen, Chinn, 2019, pp. 1–4.
31. X. C. Cui, C. S. Tao, Y. Su, and S. W. Chen, "PolSAR ship detection based on polarimetric correlation pattern," *IEEE Geoscience and Remote Sensing Letters,* vol. 18, no. 3, pp. 471–475, 2021.
32. M. D. Li, G. Q. Wu, S. P. Xiao, and S. W. Chen, "A three-dimension polarimetric correlation pattern interpretation tool and its application," in *CIE International Conference on Radar*, Haikou, China, 2021, pp. 582–585.
33. M. D. Li, S. P. Xiao, and S. W. Chen, "Three-dimension polarimetric correlation pattern interpretation tool and its application," *IEEE Transactions on Geoscience and Remote Sensing,* vol. 60, pp. 1–16, 2022.
34. S. W. Chen and C. S. Tao, "PolSAR image classification using polarimetric-feature–driven deep convolutional neural network," *IEEE Geoscience and Remote Sensing Letters,* vol. 15, no. 4, pp. 627–631, 2018.
35. S. W. Chen, C. S. Tao, X. S. Wang, and S. P. Xiao, "PolSAR target classification using polarimetric-feature–driven deep convolutional neural network," in *IEEE International Geoscience and Remote Sensing Symposium*, Valencia, Spain, 2018, pp. 4407–4410.
36. C. S. Tao, S. W. Chen, Y. Z. Li, and S. P. Xiao, "Polarimetric SAR terrain classification using polarimetric features derived from rotation domain," *Journal of Radars,* vol. 6, no. 5, pp. 524–532, 2016.
37. S. W. Chen, C. S. Tao, Y. Z. Li, X. S. Wang, and S. P. Xiao, "Polarimetric SAR image classification with polarimetric rotation domain features and deep convolutional neural network," in *The 4th China High Resolution Earth Observation Conference*, Wuhan, China, 2017.
38. C. S. Tao, S. W. Chen, Y. Z. Li, and S. P. Xiao, "PolSAR land cover classification based on the selected features in polarimetric rotation domain," in *The 3rd Advanced Symposium on Imaging Radar Earth Observation*, Changsha, China, 2016, pp. 310–319.
39. S. W. Chen, Y. Z. Li, X. S. Wang, and S. P. Xiao, "Manmade target extraction in PolSAR data using polarimetric coherence optimization," in *IET International Radar Conference*, Hangzhou, China, 2015, pp. 1–4.
40. S. W. Chen, C. S. Tao, X. S. Wang, and S. P. Xiao, "Polarimetric SAR targets detection and classification with deep convolutional neural network," in *Progress in Electromagnetics Research Symposium*, Toyama, Japan, 2018, pp. 2227–2234.
41. H. Zhang, Y. Z. Li, and S. W. Chen, "Compact-pol SAR urban area extraction with extended polarimetric correlation pattern," in *IEEE International Geoscience and Remote Sensing Symposium*, Kuala Lumpur, Malaysia, 2022.
42. X. C. Cui, Y. Su, and S. W. Chen, "Polarimetric SAR ship detection based on polarimetric rotation domain features and superpixel technique," *Journal of Radars,* vol. 10, no. 1, pp. 35–48, 2021.
43. C. L. Yang, M. D. Li, Q. X. Chi, and S. W. Chen, "Polarimetric SAR aircraft detection based on polarimetric scattering interpretation and superpixel technique," in *The 17th Chinese National Symposium on Radio Propagation*, Yan'an, China, 2022, pp. 1–4.
44. X. C. Cui, Y. Su, and S. W. Chen, "PolSAR ship detection based on polarimetric correlation pattern," in *The 20th National Conference on Image and Graphics*, Urumqi, China, 2020, pp. 1–7.
45. C. S. Tao, S. W. Chen, and S. P. Xiao, "PolSAR ship detection based on the selected polarimetric features and deep CNN," in *The 4th Advanced Symposium on Imaging Radar Earth Observation*, Shenzhen, China, 2018, pp. 75–77.

46. H. L. Li, M. D. Li, X. C. Cui, and S. W. Chen, "Man-made target structure recognition with polarimetric correlation pattern and roll-invariant feature coding," *IEEE Geoscience and Remote Sensing Letters,* vol. 19, pp. 1–5, 2022.
47. G. Q. Wu, S. W. Chen, Y. Z. Li, and X. S. Wang, "Null-pol response pattern in polarimetric rotation domain: Characterization and application," *IEEE Geoscience and Remote Sensing Letters,* vol. 19, pp. 1–5, 2022.
48. H. L. Li, M. D. Li, and S. W. Chen, "Man-made targets characterization with polarimetric correlation pattern interpretation tool," in *IEEE International Geoscience and Remote Sensing Symposium*, Brussels, Belgium, 2021, pp. 3057–3060.
49. H. L. Li, C. L. Yang, and S. W. Chen, "Characterization of complex corner reflectors in polarimetric rotation domain," in *CIE International Conference on Radar*, Haikou, China, 2021, pp. 590–593.
50. G. Q. Wu, S. W. Chen, Y. Z. Li, and X. S. Wang, "Manmade target scattering structure null-pol recognition and application," in *CIE International Conference on Radar*, Haikou, China, 2021, pp. 619–622.
51. H. L. Li, S. W. Chen, and X. S. Wang, "Study on characterization of sea corner reflectors in polarimetric rotation domain," *Systems Engineering and Electronics,* vol. 44, no. 7, pp. 2065–2073, 2022.
52. X. C. Cui, H. L. Li, Y. W. Fu, S. W. Chen, and Y. Su, "Scattering structure recognition of space target in polarimetric rotation domain," *Journal of Electronics and Information Technology,* vol. 45, no. 6, pp. 2105–2114, 2022.
53. H. L. Li and S. W. Chen, "Corner reflector jamming identification with polarimetric rotation domain features," in *The 17th Chinese National Symposium on Radio Propagation*, Yan'an, China, 2022, pp. 201–204.
54. H. L. Li, G. Y. Wang, H. Zhang, and S. W. Chen, "Compact polarimetric radar seeker adaptive corner reflector identification method in the polarization rotation domain," in *2022 Missile Penetration Technology Conference*, Beijing, China, 2022, pp. 1–9.
55. X. C. Cui, H. L. Li, and S. W. Chen, "Space target scattering structure recognition based on polarimetric rotation domain," in *The 17th Chinese National Symposium on Radio Propagation*, Yan'an, China, 2022, pp. 48–52.
56. S. W. Chen, X. S. Wang, and S. P. Xiao, "Urban damage level mapping based on co-polarization coherence pattern using multitemporal polarimetric SAR data," *IEEE Journal of Selected Topics in Applied Earth Observations and Remote Sensing,* vol. 11, no. 8, pp. 2657–2667, 2018.
57. S. W. Chen, X. S. Wang, S. P. Xiao, and Y. Su, "Urban damage mapping using fully polarimetric SAR data," in *Progress in Electromagnetics Research Symposium*, Toyama, Japan, 2018, pp. 2239–2244.
58. S. W. Chen, "PolSAR urban damage level estimation based on scattering mechanism interpretation," in *The 3rd Advanced Symposium on Imaging Radar Earth Observation*, Changsha, China, 2016, p. 90.
59. S. R. Cloude, *Polarisation Application in Remote Sensing*. New York: Oxford University Press, 2009.
60. D. L. Schuler, J. S. Lee, and T. L. Ainsworth, "Compensation of terrain azimuthal slope effects in geophysical parameter studies using polarimetric SAR data," *Remote Sensing of Environment,* vol. 69, no. 2, pp. 139–155, 1999.
61. J. S. Lee, D. L. Schuler, and T. L. Ainsworth, "Polarimetric SAR data compensation for terrain azimuth slope variation," *IEEE Transactions on Geoscience and Remote Sensing,* vol. 38, no. 5, pp. 2153–2163, 2000.
62. J. R. Huynen, "Phenomenological theory of radar targets," Ph.D. dissertation, Technical University of Delft, The Netherlands, 1970.
63. F. Xu and Y. Q. Jin, "Deorientation theory of polarimetric scattering targets and application to terrain surface classification," *IEEE Transactions on Geoscience and Remote Sensing,* vol. 43, no. 10, pp. 2351–2364, 2005.

64. W. T. An, Y. Cui, and J. Yang, "Three-component model-based decomposition for polarimetric SAR data," *IEEE Transactions on Geoscience and Remote Sensing,* vol. 48, no. 6, pp. 2732–2739, 2010.
65. Y. Yamaguchi, A. Sato, W. M. Boerner, R. Sato, and H. Yamada, "Four-component scattering power decomposition with rotation of coherency matrix," *IEEE Transactions on Geoscience and Remote Sensing,* vol. 49, no. 6, pp. 2251–2258, 2011.
66. Y. Yamaguchi, T. Moriyama, M. Ishido, and H. Yamada, "Four-component scattering model for polarimetric SAR image decomposition," *IEEE Transactions on Geoscience and Remote Sensing,* vol. 43, no. 8, pp. 1699–1706, 2005.
67. J. R. Huynen, *Phenomenological Theory of Radar Target.* Amsterdam: Elsevier, 1978.
68. J. R. Huynen, "The Stokes matrix parameters and their interpretation in terms of physical target properties," in *Proceedings of the SPIE,* York, 1990[0], 15–18 April, pp. 195–207.
69. A. Freeman and S. L. Durden, "A three-component scattering model for polarimetric SAR data," *IEEE Transactions on Geoscience and Remote Sensing,* vol. 36, no. 3, pp. 963–973, 1998.
70. T. L. Ainsworth, D. L. Schuler, and J. S. Lee, "Polarimetric SAR characterization of man-made structures in urban areas using normalized circular-pol correlation coefficients," *Remote Sensing of Environment,* vol. 112, no. 6, pp. 2876–2885, 2008.
71. Y. Yamaguchi, Y. Yamamoto, H. Yamada, J. Yang, and W. M. Boerner, "Classification of terrain by implementing the correlation coefficient in the circular polarization basis using X-band POLSAR data," *IEICE Transactions on Communications,* vol. E91B, no. 1, pp. 297–301, 2008.
72. S. W. Chen, X. S. Wang, and M. Sato, "PolInSAR complex coherence estimation based on covariance matrix similarity test," *IEEE Transactions on Geoscience and Remote Sensing,* vol. 50, no. 11, pp. 4699–4710, 2012.
73. R. Touzi, A. Lopes, J. Bruniquel, and P. W. Vachon, "Coherence estimation for SAR imagery," *IEEE Transactions on Geoscience and Remote Sensing,* vol. 37, no. 1, pp. 135–149, 1999.
74. J. M. Lopez Sanchez, S. R. Cloude, and J. David Ballester-Berman, "Rice phenology monitoring by means of SAR polarimetry at X-band," *IEEE Transactions on Geoscience and Remote Sensing,* vol. 50, no. 7, pp. 2695–2709, 2012.
75. X. Blaes, L. Vanhalle, and P. Defourny, "Efficiency of crop identification based on optical and SAR image time series," *Remote Sensing of Environment,* vol. 96, no. 3–4, pp. 352–365, 2005.
76. X. Blaes, F. Holecz, H. J. C. Van Leeuwen, and P. Defourny, "Regional crop monitoring and discrimination based on simulated ENVISAT ASAR wide swath mode images," *International Journal of Remote Sensing,* vol. 28, no. 1–2, pp. 371–393, 2007.
77. H. McNairn, T. J. Jackson, G. Wiseman, S. Belair, A. Berg, P. Bullock, et al., "The soil moisture active passive validation experiment 2012 (SMAPVEX12): Prelaunch calibration and validation of the SMAP soil moisture algorithms," *IEEE Transactions on Geoscience and Remote Sensing,* vol. 53, no. 5, pp. 2784–2801, 2015.
78. V. A. Toponogov, *Differential Geometry of Curves and Surfaces.* Boston: Birkhäuser, 2006.

3 Polarimetric Rotation Domain Roll-Invariant Features

3.1 INTRODUCTION

Polarimetric radar can acquire full-polarization information and is one of the mainstreams in microwave remote sensing [1, 2]. In order to effectively characterize and utilize the target polarization effect, in the past almost 80 years of developments of radar polarimetry [1, 3–5], radar polarimetry scholars have proposed many polarimetric information characterization methods [1, 6–10], including the Sinclair matrix [3] that connects the Jones vector of the transmitting and receiving electromagnetic waves, the Kennaugh and Mueller matrix [11–13] that characterizes the Stokes vector, and the Graves power matrix [10] that characterizes received energy. In addition, the Pauli scattering vector and the Lexicographic scattering vector can be obtained from the Sinclair matrix. Meanwhile, the polarimetric coherency/covariance matrix can be further constructed [1]. In order to study and understand the target polarimetric effect and its scattering mechanism, Huynen proposed the concept of radar target phenomenology [14], which inspired the development of polarimetric target decomposition theory. Polarimetric target decomposition is an effective method to interpret polarimetric radar data and target scattering mechanisms. Polarimetric target decomposition mainly includes coherent polarization target decomposition and incoherent polarization target decomposition [15, 16].

The target scattering diversity effect brings challenges to polarimetric radar data interpretation and application development. The polarimetric rotation domain roll-invariant features (abbreviated as polarimetric roll-invariant features), which are independent of target attitude rotation around the radar line of sight, have been emphasized and studied by plenty of researchers. In 1965, Bickel and Huynen independently proposed a set of polarimetric roll-invariant features [14, 17–19]. After that, with the developments of polarimetric scattering mechanism interpretation theory, a number of polarimetric roll-invariant features have been reported [1, 4, 20–26]. Meanwhile, polarimetric roll-invariant features have also achieved many successful applications such as target classification [1, 4, 20–26], target detection [27–30], structure recognition [31–33], disaster evaluation [34, 35], and so on. In recent years, with the recent developments of radar polarimetry such as the polarimetric rotation domain interpretation tool, series of polarimetric roll-invariant features have been further reported [36–39]. The main development of polarimetric roll-invariant

DOI: 10.1201/9781003461296-3

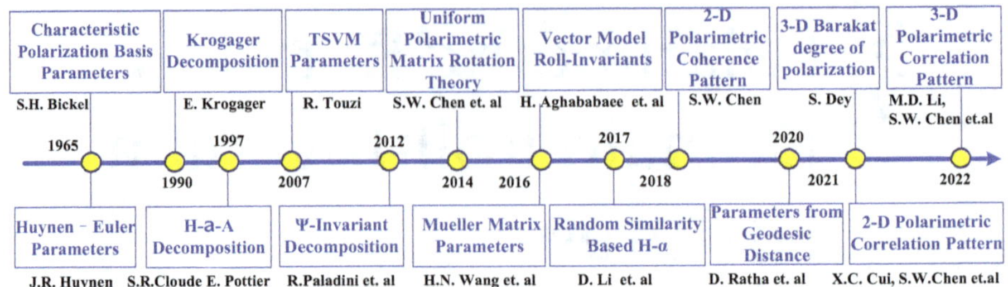

FIGURE 3.1.1 The main development of polarimetric roll-invariant features. TSVM is an abbreviation of the target scattering vector modeling.

features is summarized in Figure 3.1.1. Note that the dates are based on the corresponding journal publications.

From this introduction, it is clear that polarimetric roll-invariant features are essentially valuable in the radar polarimetry field. However, it is also clear that these features are scattered in a number of sources across about six decades. In this vein, this chapter will provide a comprehensive summary of these polarimetric roll-invariant features, which can be convenient for current studies and further developments [40, 41].

Based on their original derivation fashions, these polarimetric roll-invariant features are partitioned into seven groups: derived from the Sinclair matrix, derived from the Graves matrix, derived from the polarimetric coherency and covariance matrix, derived from eigenvalue-eigenvector-based decompositions, derived from two-dimension polarimetric coherence/correlation patterns, derived from the three-dimension polarimetric correlation pattern, and derived from the Kennaugh matrix.

Meanwhile, their expressions, inter-relationships, and physical interpretations are further examined and presented.

3.2 POLARIMETRIC ROLL-INVARIANT FEATURES

The typical polarimetric roll-invariant features are investigated and summarized mainly in terms of horizontal and vertical polarization basis $(\mathrm{H,V})$ and circular polarization basis $(\mathrm{L,R})$. Without additional instructions, for horizontal and vertical polarization basis $(\mathrm{H,V})$, the subscript $(\mathrm{H,V})$ is omitted for polarimetric matrix representations of $\mathbf{S}$, $\mathbf{T}$, $\mathbf{C}$, and $\mathbf{K}$.

3.2.1 Derived from the Sinclair Matrix

The Sinclair matrix is the basic representation of scattering information. From the literature, polarimetric roll-invariant features are mainly derived from the mathematical properties of the Sinclair matrix, the Huynen–Euler parameters [14, 18, 19], the characteristic polarization parameters [17], and the Krogager decomposition [42].

It is clear that there are two fundamental polarimetric roll-invariant features of the total backscattering power *SPAN* and the reciprocity indicator *Reciprocity*:

$$SPAN = |S_{HH}|^2 + |S_{HV}|^2 + |S_{VH}|^2 + |S_{VV}|^2 \tag{3.2.1}$$

$$Reciprocity = S_{HV} - S_{VH} \tag{3.2.2}$$

The Sinclair matrix **S** can be expressed with eigenvalues and eigenvectors:

$$\mathbf{S} = \mathbf{U}_0^{H} \boldsymbol{\Lambda} \mathbf{U}_0 = \mathbf{U}_0^{H} \begin{bmatrix} \lambda_1^{S} & 0 \\ 0 & \lambda_2^{S} \end{bmatrix} \mathbf{U}_0 \tag{3.2.3}$$

where $\mathbf{U}_0$ is the unitary matrix, and $\boldsymbol{\Lambda}$ is a diagonal matrix of λ_1^{S} and λ_2^{S}, which are eigenvalues of the Sinclair matrix **S**. Without any loss of generality, $|\lambda_1^{S}| > |\lambda_2^{S}|$ is assumed.

Thus, $\mathbf{S}(\theta)$ can be expressed in the polarimetric rotation domain:

$$\mathbf{S}(\theta) = \left[\mathbf{U}_0 \mathbf{R}_2(\theta)\right]^{T} \boldsymbol{\Lambda} \left[\mathbf{U}_0 \mathbf{R}_2(\theta)\right], \theta \in [-\pi, \pi) \tag{3.2.4}$$

Note that $\mathbf{U}_0 \mathbf{R}_2(\theta)$ is also the unitary matrix, and thus the eigenvalues of $\mathbf{S}(\theta)$ are the same as those of **S**, which means λ_1^{S} and λ_2^{S} are polarimetric roll-invariant features.

In this vein, the matrix eigenvalues $\lambda_i(\mathbf{S})$, matrix trace $\mathrm{Tr}(\mathbf{S})$, matrix determination $\mathrm{Det}(\mathbf{S})$, matrix two norm $\|\mathbf{S}\|_2$, and matrix Frobenius norm $\|\mathbf{S}\|_F$ are polarimetric roll-invariant features:

$$\lambda_i\left[\mathbf{S}(\theta)\right] = \lambda_i(\mathbf{S}) = \lambda_i^{S} \tag{3.2.5}$$

$$\mathrm{Tr}\left[\mathbf{S}(\theta)\right] = \mathrm{Tr}(\mathbf{S}) \tag{3.2.6}$$

$$\mathrm{Det}\left[\mathbf{S}(\theta)\right] = \mathrm{Det}(\mathbf{S}) \tag{3.2.7}$$

$$\|\mathbf{S}(\theta)\|_2 = \|\mathbf{S}\|_2 = \sqrt{\max\left[\lambda_i\left(\mathbf{S}^{H}\mathbf{S}\right)\right]} \tag{3.2.8}$$

$$\|\mathbf{S}\|_F = \sqrt{\mathrm{Tr}\left(\mathbf{S}^{H}\mathbf{S}\right)} = \sqrt{SPAN} \tag{3.2.9}$$

where $\lambda_i(\cdot), \mathrm{Tr}(\cdot), \mathrm{Det}(\cdot), \|\cdot\|_2$, and $\|\cdot\|_F$ are matrix eigenvalues, matrix trace, matrix determination, matrix two norm, and matrix Frobenius norm, respectively.

Furthermore, polarimetric roll-invariant features λ_i^{S}, $\mathrm{Tr}(\mathbf{S})$, and $\mathrm{Det}(\mathbf{S})$ can be expressed using elements of the Sinclair matrix **S**:

$$\lambda_i^{S} = \frac{(S_{HH} + S_{VV}) \pm \sqrt{(S_{HH} - S_{VV})^2 + 4 S_{HV} S_{VH}}}{2} \tag{3.2.10}$$

$$\mathrm{Tr}(\mathbf{S}) = \lambda_1^{\mathrm{S}} + \lambda_2^{\mathrm{S}} = S_{\mathrm{HH}} + S_{\mathrm{VV}} \tag{3.2.11}$$

$$\mathrm{Det}(\mathbf{S}) = \left|\lambda_1^{\mathrm{S}}\lambda_2^{\mathrm{S}}\right| = S_{\mathrm{HH}}S_{\mathrm{VV}} - S_{\mathrm{HV}}S_{\mathrm{VH}} \tag{3.2.12}$$

The reciprocity feature can also be deduced with the matrix trace operation:

$$Reciprocity = S_{\mathrm{HV}} - S_{\mathrm{VH}} = \mathrm{Tr}\left(\begin{bmatrix} S_{\mathrm{HH}} & S_{\mathrm{HV}} \\ S_{\mathrm{VH}} & S_{\mathrm{VV}} \end{bmatrix}\begin{bmatrix} 0 & -1 \\ 1 & 0 \end{bmatrix}\right) \tag{3.2.13}$$

Except for the basic matrix operations, the Sinclair matrix can also be represented with the framework of Huynen–Euler parameters [14, 18, 19]:

$$\mathbf{S}{=}me^{j\phi}\mathbf{U}_2^{*}\left(\theta,\varepsilon,\alpha-v\right)\begin{bmatrix} 1 & 0 \\ 0 & \tan^2\gamma \end{bmatrix}\mathbf{U}_2\left(\theta,\varepsilon,\alpha-v\right) \tag{3.2.14}$$

$$\mathbf{U}_2\left(\theta,\varepsilon,\alpha-v\right) = \begin{bmatrix} \cos\theta & -\sin\theta \\ \sin\theta & \cos\theta \end{bmatrix}\begin{bmatrix} \cos\varepsilon & \mathrm{j}\sin\theta \\ \mathrm{j}\sin\theta & \cos\varepsilon \end{bmatrix}\begin{bmatrix} e^{\mathrm{j}(\alpha-v)} & 0 \\ 0 & e^{-\mathrm{j}(\alpha-v)} \end{bmatrix} \tag{3.2.15}$$

where $m$is the maximum polarization, ϕ is the absolute phase of the Sinclair matrix, γ is the characteristic angle, θ is the polarization orientation angle of the eigen-basis of the object relative to the radar coordinate system, ε is the ellipticity angle of the eigen-basis of the object, v is the skip angle, and α is the balance angle of S_{HH} and S_{VV}.

Therefore, Huynen–Euler parameters m, v, and γ can be expressed using the matrix eigenvalues derived from the Sinclair matrix:

$$m{=}\sqrt{\max\left[\lambda_i\left(\mathbf{S}^{\mathrm{H}}\mathbf{S}\right)\right]} = \lambda_1^{\mathrm{S}} \tag{3.2.16}$$

$$v = -\frac{1}{4}\arg\left(\frac{\lambda_2^{\mathrm{S}}}{\lambda_1^{\mathrm{S}}}\right) \tag{3.2.17}$$

$$\gamma = \tan^{-1}\sqrt{\frac{\lambda_2^{\mathrm{S}}}{\lambda_1^{\mathrm{S}}}} \tag{3.2.18}$$

On the other hand, with the characteristic polarization basis, the Sinclair matrix is [17]

$$\tilde{\mathbf{S}} = \mathbf{Q}^{\mathrm{T}}\mathbf{S}\mathbf{Q} \tag{3.2.19}$$

where $\mathbf{Q}$ is the unitary matrix.

On the other hand, the polarimetric roll-invariant features under the characteristic polarization basis have also been reported [17]. They are the depolarization D, the ellipticity of the eigenpolarization α_{d}, and the relative orientation between the antenna and the axis of the eigenpolarization ellipse θ_{d}. Their definitions are as follows:

$$D = 1 - \frac{\left|S_{\mathrm{HH}} + S_{\mathrm{VV}}\right|^2}{2\left(\left|S_{\mathrm{HH}}\right|^2 + \left|S_{\mathrm{HV}}\right|^2 + \left|S_{\mathrm{VH}}\right|^2 + \left|S_{\mathrm{VV}}\right|^2\right)} \tag{3.2.20}$$

$$\alpha_{\mathrm{d}} = \frac{1}{2}\tan^{-1}\frac{\mathrm{j}2\tilde{S}_{12}'}{\tilde{S}_1} \tag{3.2.21}$$

$$\theta_{\mathrm{d}} = \frac{1}{2}\tan^{-1}\frac{2\,\mathrm{Re}\left(\tilde{S}_1^*\tilde{S}_{12}\right)}{\mathrm{Re}\left(\tilde{S}_1^*\tilde{S}_2\right)} \tag{3.2.22}$$

where $\tilde{S}_1$, $\tilde{S}_2$, $\tilde{S}_{12}$, and $\tilde{S}_{12}'$ are:

$$\begin{cases} \tilde{S}_1 = S_{\mathrm{HH}} + S_{\mathrm{VV}} \\ \tilde{S}_2 = S_{\mathrm{HH}} - S_{\mathrm{VV}} \\ \tilde{S}_{12} = S_{\mathrm{HV}} \\ \tilde{S}_{12}' = \tilde{S}_{12}\cos\left(2\theta_{\mathrm{d}}\right) - \frac{1}{2}\tilde{S}_2\sin\left(2\theta_{\mathrm{d}}\right) \end{cases} \tag{3.2.23}$$

Moreover, polarimetric roll-invariant features can be obtained from the Krogager decomposition [42], which decomposes a Sinclair matrix **S** into three coherent scattering components in terms of the sphere, diplane, and helix scattering mechanism:

$$\begin{aligned} \mathbf{S} &= \mathrm{e}^{\mathrm{j}\phi}\left\{\mathrm{e}^{\mathrm{j}\phi_{\mathrm{s}}}k_{\mathrm{S}}\mathbf{S}_{\mathrm{sphere}} + k_{\mathrm{D}}\mathbf{S}_{\mathrm{diplane}}(\theta) + k_{\mathrm{H}}\mathbf{S}_{\mathrm{helix}}\right\} \\ &= \mathrm{e}^{\mathrm{j}\phi}\left\{\mathrm{e}^{\mathrm{j}\phi_{\mathrm{s}}}k_{\mathrm{S}}\begin{bmatrix}1 & 0\\ 0 & 1\end{bmatrix} + k_{\mathrm{D}}\begin{bmatrix}\cos 2\theta & \sin 2\theta \\ \sin 2\theta & -\cos 2\theta\end{bmatrix} + k_{\mathrm{H}}\mathrm{e}^{\mp\mathrm{j}2\theta}\begin{bmatrix}1 & \pm\mathrm{j} \\ \pm\mathrm{j} & -1\end{bmatrix}\right\} \end{aligned} \tag{3.2.24}$$

where $k_{\mathrm{S}}, k_{\mathrm{D}}$, and k_{H} correspond to the sphere, diplane, and helix scattering components; θ is the polarization orientation angle; ϕ is the absolute phase; and ϕ_s represents the displacement of the sphere scattering relative to the diplane scattering within a resolution cell.

The three scattering component coefficients $k_{\mathrm{S}}, k_{\mathrm{D}}$, and k_{H} are usually derived under the circular polarization basis (L,R):

$$\begin{aligned} \mathbf{S}_{(\mathrm{L},\mathrm{R})} &= \begin{bmatrix} S_{\mathrm{LL}} & S_{\mathrm{LR}} \\ S_{\mathrm{RL}} & S_{\mathrm{RR}} \end{bmatrix} \\ &= \mathrm{e}^{\mathrm{j}\phi}\left\{\mathrm{e}^{\mathrm{j}\phi_{\mathrm{s}}}k_{\mathrm{S}}\begin{bmatrix}0 & \mathrm{j}\\ \mathrm{j} & 0\end{bmatrix} + k_{\mathrm{D}}\begin{bmatrix}\mathrm{e}^{\mathrm{j}2\theta} & 0 \\ 0 & -\mathrm{e}^{-\mathrm{j}2\theta}\end{bmatrix} + k_{\mathrm{H}}\begin{bmatrix}\mathrm{e}^{\mathrm{j}2\theta} & 0 \\ 0 & 0\end{bmatrix}\right\} \end{aligned} \tag{3.2.25}$$

where S_{LL}, S_{RL}, S_{LR}, and S_{RR} are the elements of a Sinclair matrix under the circular polarization basis (L,R).

According to (3.2.25), the sphere scattering coefficient k_S can be derived:

$$k_S = |S_{RL}| \tag{3.2.26}$$

By examining the relative amplitudes of $|S_{RR}|$ and $|S_{LL}|$, the diplane and helix scattering coefficients of k_D and k_H are available:

$$|S_{RR}| \geq |S_{LL}| \Rightarrow \begin{cases} k_D = |S_{LL}| \\ k_H = |S_{RR}| - |S_{LL}| \Leftarrow \text{Left Helix} \end{cases} \tag{3.2.27}$$

$$|S_{RR}| < |S_{LL}| \Rightarrow \begin{cases} k_D = |S_{RR}| \\ k_H = |S_{LL}| - |S_{RR}| \Leftarrow \text{Right Helix} \end{cases} \tag{3.2.28}$$

Furthermore, the diplane and helix scattering coefficients of k_D and k_H can be deduced as

$$k_D = \min\left(|S_{LL}|, |S_{RR}|\right) \tag{3.2.29}$$

$$k_H = \left\| S_{RR}| - |S_{LL} \right\| \tag{3.2.30}$$

It can be verified that k_S, k_D, and k_H are all polarimetric roll-invariant features.

The expressions and physical meanings of these polarimetric roll-invariant features derived from the Sinclair matrix are summarized in Table 3.2.1.

3.2.2 Derived from the Graves Matrix

As for the Graves matrix, similar to the Sinclair matrix, the matrix operations, including the matrix trace $\mathrm{Tr}(\cdot)$, matrix determination $\mathrm{Det}(\cdot)$, eigenvalue calculation $\lambda_i(\cdot)$, matrix two norm $\|\cdot\|_2$, and matrix Frobenius norm $\|\cdot\|_F$, are also polarimetric roll-invariant features. Without any loss of generality, the eigenvalues of $\mathbf{G}$ are λ_1^{G} and λ_2^{G}, while $|\lambda_1^{\mathrm{G}}| > |\lambda_2^{\mathrm{G}}|$ is assumed:

$$\lambda_i(\mathbf{G}) = \lambda_i^{\mathrm{G}} \tag{3.2.31}$$

$$\mathrm{Tr}(\mathbf{G}) = SPAN = \lambda_1^{\mathrm{G}} + \lambda_2^{\mathrm{G}} \tag{3.2.32}$$

$$\mathrm{Det}(\mathbf{G}) = \mathrm{Det}(\mathbf{S}^{\mathrm{H}}\mathbf{S}) = \mathrm{Det}(\mathbf{S}^{\mathrm{H}})\mathrm{Det}(\mathbf{S}) = \lambda_1^{\mathrm{G}}\lambda_2^{\mathrm{G}} \tag{3.2.33}$$

$$\|\mathbf{G}\|_2 = \sqrt{\max\left[\lambda_i\left(\mathbf{G}^{\mathrm{H}}\mathbf{G}\right)\right]} = |\lambda_1^{\mathrm{G}}| \tag{3.2.34}$$

$$\|\mathbf{G}\|_F = \sqrt{\left(\lambda_1^{\mathrm{G}}\right)^2 + \left(\lambda_2^{\mathrm{G}}\right)^2} \tag{3.2.35}$$

TABLE 3.2.1
Polarimetric Roll-Invariant Features from the Sinclair Matrix

Feature	Expression	Physical Meaning
SPAN	$\lvert S_{HH}\rvert^2+\lvert S_{HV}\rvert^2+\lvert S_{VH}\rvert^2+\lvert S_{VV}\rvert^2$	The target's ability to intercept and reflect the incident waves.
Reciprocity	$S_{HV}-S_{VH}$	The reciprocity of the scattering mechanism.
$\lambda_i(\mathbf{S})$	$\frac{(S_{HH}+S_{VV})\pm\sqrt{(S_{HH}-S_{VV})^2+4S_{HV}S_{VH}}}{2}$	Eigenvalues of the Sinclair matrix.
$\mathrm{Tr}(\mathbf{S})$	$\lambda_1^S+\lambda_2^S=S_{HH}+S_{VV}$	First element of Pauli vector, which represents the odd-bounce scattering.
$\mathrm{Det}(\mathbf{S})$	$\lvert\lambda_1^S\lambda_2^S\rvert=S_{HH}S_{VV}-S_{HV}S_{VH}$	Representation of the thickness of the target to a certain extent. When the horizontal direction of the target is significantly greater than the vertical one, $\mathrm{Det}(\mathbf{S})>1$, and vice versa.
$\lVert\mathbf{S}\rVert_2$	$\sqrt{\max\left[\lambda_i\left(\mathbf{S}^H\mathbf{S}\right)\right]}$	Characterization of the maximum reflection power density of a target to the incident wave.
$\lVert\mathbf{S}\rVert_F$	$\sqrt{\mathrm{Tr}\left(\mathbf{S}^H\mathbf{S}\right)}=\sqrt{SPAN}$	Similar to the *SPAN* feature.
m	λ_1^S	The maximum RCS.
ν	$-\frac{1}{4}\arg\left(\frac{\lambda_2^S}{\lambda_1^S}\right)$	Representation of the multiple scattering. For odd-bounce scattering, $\nu=0$, while $\nu=\pi/4$ for double-bounce scattering.
γ	$\tan^{-1}\sqrt{\frac{\lambda_2^S}{\lambda_1^S}}$	Related to target polarization sensitivity. For plane target, $\gamma=\pi/4$, while $\gamma=0$ for linear targets.
D	$1-\frac{\lvert S_{HH}+S_{VV}\rvert^2}{2\left(\lvert S_{HH}\rvert^2+\lvert S_{HV}\rvert^2+\lvert S_{VH}\rvert^2+\lvert S_{VV}\rvert^2\right)}$	Reflects the number of target scattering centers.

(Continued)

TABLE 3.2.1 ***(Continued)***

Polarimetric Roll-Invariant Features from the Sinclair Matrix

Feature	Expression	Physical Meaning
α_{d}	$\frac{1}{2}\tan^{-1}\frac{\mathrm{j}2\tilde{S}'_{12}}{\tilde{S}_1}$	Reflects the target symmetry.
θ_{d}	$\frac{1}{2}\tan^{-1}\frac{2\,\mathrm{Re}\left(\tilde{S}_1^*\tilde{S}_{12}\right)}{\mathrm{Re}\left(\tilde{S}_1^*\tilde{S}_2\right)}$	Reflects the relative orientation between the antenna and the axis of the eigenpolarization ellipse.
k_{S}	$\lvert S_{\mathrm{RL}}\rvert$	The coefficient of the sphere scattering component.
k_{D}	$\min\left(\lvert S_{\mathrm{LL}}\rvert,\lvert S_{\mathrm{RR}}\rvert\right)$	The coefficient of the diplane scattering component.
k_{H}	$\lVert S_{\mathrm{RR}}\rvert-\lvert S_{\mathrm{LL}}\rVert$	The coefficient of the helix scattering component.

$\lambda_i^{\mathbf{G}}$ can be expressed using elements of the Graves matrix $\mathbf{G}$:

$$\lambda_1^{\mathbf{G}} = \frac{(G_{11}+G_{22})+\sqrt{(G_{11}-G_{22})^2+4G_{12}G_{21}}}{2} \tag{3.2.36}$$

$$\lambda_2^{\mathbf{G}} = \frac{(G_{11}+G_{22})-\sqrt{(G_{11}-G_{22})^2+4G_{12}G_{21}}}{2} \tag{3.2.37}$$

The expressions and physical meanings of these derived polarimetric roll-invariant features from the Graves matrix are summarized in Table 3.2.2, where $\mathbf{G}_{11}$ is the $(1,1)$ element of the Graves matrix $\mathbf{G}$, and other terms are similarly defined. The relationships of these polarimetric roll-invariant features derived from the Sinclair matrix and the Graves matrix are displayed in Figure 3.2.1.

TABLE 3.2.2
Polarimetric Roll-Invariant Features from the Graves Matrix

Feature	Expression	Physical Meaning
$\lambda_i(\mathbf{G})$	$\frac{(G_{11}+G_{22})\pm\sqrt{(G_{11}-G_{22})^2+4G_{12}G_{21}}}{2}$	Characterization of the maximum reflection power density of target to incident wave.
$\mathrm{Tr}(\mathbf{G})$	$\lambda_1^{\mathrm{G}}+\lambda_2^{\mathrm{G}}=\lvert S_{\mathrm{HH}}\rvert^2+\lvert S_{\mathrm{HV}}\rvert^2+\lvert S_{\mathrm{VH}}\rvert^2+\lvert S_{\mathrm{VV}}\rvert^2$	The target's ability to intercept and reflect the incident waves.
$\mathrm{Det}(\mathbf{G})$	$\lambda_1^{\mathrm{G}}\lambda_2^{\mathrm{G}}=\lvert S_{\mathrm{HH}}\rvert^2\lvert S_{\mathrm{VV}}\rvert^2+\lvert S_{\mathrm{HV}}\rvert^2\lvert S_{\mathrm{VH}}\rvert^2 - S_{\mathrm{HH}}S_{\mathrm{VV}}S_{\mathrm{HV}}^*S_{\mathrm{VH}}^* - S_{\mathrm{HH}}^*S_{\mathrm{VV}}^*S_{\mathrm{HV}}S_{\mathrm{VH}}$	The determinant of Graves matrix.
$\lVert\mathbf{G}\rVert_2$	$\sqrt{\max\left[\lambda_i\left(\mathbf{G}^{\mathrm{H}}\mathbf{G}\right)\right]}=\lvert\lambda_1^{\mathrm{G}}\rvert$	Target's maximum scattering power.
$\lVert\mathbf{G}\rVert_{\mathrm{F}}$	$\sqrt{(\lambda_1^{\mathrm{G}})^2+(\lambda_2^{\mathrm{G}})^2}$	The Frobenius norm of the Graves matrix.

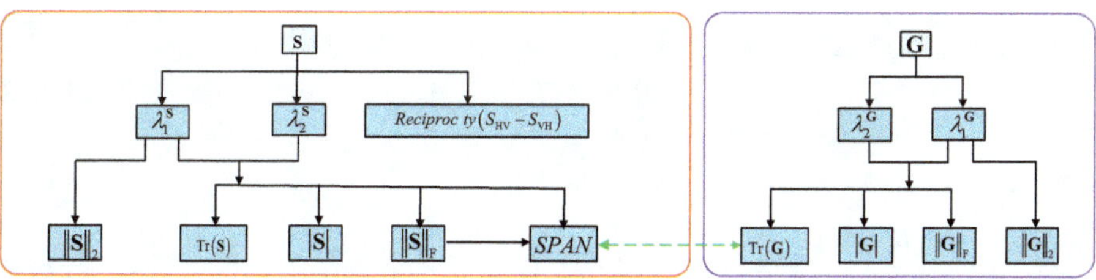

FIGURE 3.2.1 The relationships of the polarimetric roll-invariant features derived from the Sinclair matrix and the Graves matrix.

3.2.3 Derived from Polarimetric Coherency and Covariance Matrix

The secondary moment information of the Sinclair matrix is usually represented by the polarimetric coherency and covariance matrix. From the literature, polarimetric roll-invariant features from polarimetric coherency and covariance matrix are mainly derived through Huynen decomposition [14, 43, 44], uniform polarimetric matrix rotation theory [45], and the polarimetric covariance matrix with a circular polarization basis.

Huynen decomposition [14, 43, 44] is the pioneer approach in radar polarimetry. The core idea is to decompose a polarimetric matrix (e.g. Kennaugh matrix, polarimetric coherency matrix) into two components in terms of a single target and N targets. Taking the polarimetric coherency matrix as an example, the decomposition is

$$\mathbf{T}=\begin{bmatrix} 2\langle A_0\rangle & \langle C\rangle-\mathrm{j}\langle D\rangle & \langle H\rangle+\mathrm{j}\langle G\rangle \\ \langle C\rangle+\mathrm{j}\langle D\rangle & \langle B_0\rangle+\langle B\rangle & \langle E\rangle+\mathrm{j}\langle F\rangle \\ \langle H\rangle-\mathrm{j}\langle G\rangle & \langle E\rangle-\mathrm{j}\langle F\rangle & \langle B_0\rangle-\langle B\rangle \end{bmatrix}=\mathbf{T}_0+\mathbf{T}_{\mathrm{N}} \tag{3.2.38}$$

where $\mathbf{T}_0$ represents a single target and $\mathbf{T}_{\mathrm{N}}$ indicates N targets:

$$\mathbf{T}_0=\begin{bmatrix} 2\langle A_0\rangle & \langle C\rangle-\mathrm{j}\langle D\rangle & \langle H\rangle+\mathrm{j}\langle G\rangle \\ \langle C\rangle+\mathrm{j}\langle D\rangle & B_{0\mathrm{T}}+B_{\mathrm{T}} & E_{\mathrm{T}}+\mathrm{j}F_{\mathrm{T}} \\ \langle H\rangle-\mathrm{j}\langle G\rangle & E_{\mathrm{T}}-\mathrm{j}F_{\mathrm{T}} & B_{0\mathrm{T}}-B_{\mathrm{T}} \end{bmatrix} \tag{3.2.39}$$

$$\mathbf{T}_{\mathrm{N}}=\begin{bmatrix} 0 & 0 & 0 \\ 0 & B_{0\mathrm{N}}+B_{\mathrm{N}} & E_{\mathrm{N}}+jF_{\mathrm{N}} \\ 0 & E_{\mathrm{N}}-jF_{\mathrm{N}} & B_{0\mathrm{N}}-B_{\mathrm{N}} \end{bmatrix} \tag{3.2.40}$$

where A_0, B_0, B, $B_{0\mathrm{T}}$, B_{T}, $B_{0\mathrm{N}}$, B_{N}, C, D, E, E_{T}, E_{N},F, F_{T}, F_{N}, G, and H are the elements of $\mathbf{T},\mathbf{T}_0$ and $\mathbf{T}_{\mathrm{N}}$.

Among them, polarimetric roll-invariant features include$A_0, B_0, B_{0\mathrm{N}}, B_{0\mathrm{T}}, F, F_{\mathrm{T}}$, and F_{N}. Their definitions in terms of polarimetric coherency matrix elements are summarized in Table 3.2.3 and shown as follows

$$A_0=\frac{1}{2}T_{11} \tag{3.2.41}$$

$$B_0=\frac{1}{2}\left(T_{22}+T_{33}\right) \tag{3.2.42}$$

$$F=\mathrm{Im}\left[T_{23}\right] \tag{3.2.43}$$

$$B_{0\mathrm{N}}=\frac{1}{2}\left[T_{22}+T_{33}-\left(\left|T_{12}\right|^2+\left|T_{13}\right|^2\right)/T_{11}\right] \tag{3.2.44}$$

$$F_{\mathrm{N}}=\mathrm{Im}\left[T_{23}\right]-\mathrm{Im}\left[T_{12}^{*}T_{13}\right]/2T_{11} \tag{3.2.45}$$

TABLE 3.2.3
Polarimetric Roll-Invariant Features from the Huynen Decomposition

Feature	Expression	Physical Meaning
A_0	$\frac{1}{2}T_{11}$	The total scattering power from the regular, smooth, and convex parts of the scatterers.
B_0	$\frac{1}{2}(T_{22}+T_{33})$	The total scattering power from depolarized components such as irregular, rough, and nonconvex scatterers.
F	$\mathrm{Im}[T_{23}]$	The depolarization components caused by the scatterers' overall distortion.
B_{0N}	$\frac{1}{2}\left[T_{22}+T_{33}-\left(\lvert T_{12}\rvert^2+\lvert T_{13}\rvert^2\right)/T_{11}\right]$	The depolarized components' total scattering power coming from the irregular, rough and nonconvex N targets.
F_N	$\mathrm{Im}[T_{23}]-\mathrm{Im}\left[T_{12}^*T_{13}\right]/2T_{11}$	The depolarized components coming from the helicity of the asymmetric N targets.
B_{0T}	$B_0-B_{0N}=\left(\lvert T_{12}\rvert^2+\lvert T_{13}\rvert^2\right)/2T_{11}$	The depolarized components' total scattering power coming from the irregular, rough, and nonconvex single scatterers.
F_T	$F-F_N=\mathrm{Im}\left[T_{12}^*T_{13}\right]/T_{11}$	The depolarized components coming from the helicity of the asymmetric single targets.

$$B_{0T}=\left(\lvert T_{12}\rvert^2+\lvert T_{13}\rvert^2\right)/2T_{11} \tag{3.2.46}$$

$$F_T=\mathrm{Im}\left[T_{12}^*T_{13}\right]/T_{11} \tag{3.2.47}$$

where T_{ij} represents the (i,j) entry of $\mathbf{T}$.

In addition, according to the uniform polarimetric matrix rotation theory [45] introduced in Chapter 2, all the elements of a rotated polarimetric coherency matrix in the polarimetric rotation domain can be represented as a sinusoidal function:

$$f(\theta)=A\sin\left[\omega(\theta+\theta_0)\right]+B \tag{3.2.48}$$

where A is oscillation amplitude, B is oscillation center, ω is angular frequency, and θ_0 is the initial angle.

With the representation of (3.2.48), polarimetric roll-invariant features, including the oscillation amplitudes $A\_\mathrm{Re}[T_{12}]$, $A\_\mathrm{Im}[T_{12}]$, A_T_{22}, $A\_\lvert T_{12}\rvert^2$, and the oscillation centers B_T_{22}, $B\_\lvert T_{12}\rvert^2$, and $B\_\lvert T_{23}\rvert^2$, can be obtained. Their expressions in terms of polarimetric coherency matrix elements are summarized in Table 3.2.4 and shown as follows:

$$A\_\mathrm{Re}[T_{12}]=\mathrm{Re}^2[T_{12}]+\mathrm{Re}^2[T_{13}] \tag{3.2.49}$$

TABLE 3.2.4
Polarimetric Roll-Invariant Features from Uniform Polarimetric Matrix Rotation Theory

Feature	Expression	Physical Meaning
$A\_\text{Re}[T_{12}]$	$\text{Re}^2[T_{12}]+\text{Re}^2[T_{13}]$	Oscillation amplitude of $\text{Re}[T_{12}]$.
$A\_\text{Im}[T_{12}]$	$\text{Im}^2[T_{12}]+\text{Im}^2[T_{13}]$	Oscillation amplitude of $\text{Im}[T_{12}]$.
A_T_{22}	$\frac{1}{4}(T_{33}-T_{22})^2+\text{Re}^2[T_{23}]$	Oscillation amplitude of T_{22}.
$A\_\|T_{12}\|^2$	$\text{Re}^2[T_{12}T_{13}^*]+\frac{1}{4}(\|T_{13}\|^2-\|T_{12}\|^2)^2$	Oscillation amplitude of $\|T_{12}\|^2$.
B_T_{22}	$\frac{1}{2}(T_{22}+T_{33})$	Oscillation center of T_{22}.
$B\_\|T_{12}\|^2$	$\frac{1}{2}(\|T_{12}\|^2+\|T_{13}\|^2)$	Oscillation center of $\|T_{12}\|^2$.
$B\_\|T_{23}\|^2$	$\frac{1}{8}\{(T_{33}-T_{22})^2+4\text{Re}^2[T_{23}]\}+\text{Im}^2[T_{23}]$	Oscillation center of $\|T_{23}\|^2$.

$$A\_\text{Im}[T_{12}]=\text{Im}^2[T_{12}]+\text{Im}^2[T_{13}] \tag{3.2.50}$$

$$A\_T_{22}=\frac{1}{4}(T_{33}-T_{22})^2+\text{Re}^2[T_{23}] \tag{3.2.51}$$

$$A\_|T_{12}|^2=\text{Re}^2[T_{12}T_{13}^*]+\frac{1}{4}\left(|T_{13}|^2-|T_{12}|^2\right)^2 \tag{3.2.52}$$

$$B\_T_{22}=\frac{1}{2}(T_{22}+T_{33}) \tag{3.2.53}$$

$$B\_|T_{12}|^2=\frac{1}{2}\left(|T_{12}|^2+|T_{13}|^2\right) \tag{3.2.54}$$

$$B\_|T_{23}|^2=\frac{1}{8}\left\{(T_{33}-T_{22})^2+4\text{Re}^2[T_{23}]\right\}+\text{Im}^2[T_{23}] \tag{3.2.55}$$

where $A\_\text{Re}[T_{12}]$ indicates the oscillation amplitude of $\text{Re}[T_{12}]$. The other terms are similarly defined.

Furthermore, polarimetric roll-invariant features can also be directly derived from polarimetric covariance matrix $\mathbf{C}_{(\text{L,R})}$ with circular polarization basis (L,R):

$$\mathbf{C}_{(\text{L,R})}=\begin{bmatrix} \langle|S_{\text{LL}}|^2\rangle & \sqrt{2}\langle S_{\text{LL}}S_{\text{LR}}^*\rangle & \langle S_{\text{LL}}S_{\text{RR}}^*\rangle \\ \sqrt{2}\langle S_{\text{LL}}^*S_{\text{LR}}\rangle & \langle|S_{\text{LR}}|^2\rangle & \sqrt{2}\langle S_{\text{LR}}S_{\text{RR}}^*\rangle \\ \langle S_{\text{LL}}^*S_{\text{RR}}\rangle & \sqrt{2}\langle S_{\text{LR}}^*S_{\text{RR}}\rangle & \langle|S_{\text{RR}}|^2\rangle \end{bmatrix} \tag{3.2.56}$$

Polarimetric roll-invariant features include the power terms of $C_{11(\mathrm{L,R})}$, $C_{22(\mathrm{L,R})}$, and $C_{33(\mathrm{L,R})}$ and the correlation terms of $\left|C_{12(\mathrm{L,R})}\right|$, $\left|C_{13(\mathrm{L,R})}\right|$, and $\left|C_{23(\mathrm{L,R})}\right|$. Their expressions can also be represented by the polarimetric coherency matrix with horizontal and vertical polarization basis $(\mathrm{H,V})$, which are summarized in Table 3.2.5 and shown as follows:

$$C_{11(\mathrm{L,R})} = \frac{1}{2}\left(T_{22} + T_{33} - 2\,\mathrm{Im}\left[T_{23}\right]\right) \tag{3.2.57}$$

$$C_{22(\mathrm{L,R})} = \frac{1}{2}T_{11} \tag{3.2.58}$$

$$C_{33(\mathrm{L,R})} = \frac{1}{2}\left(T_{22} + T_{33} + 2\,\mathrm{Im}\left[T_{23}\right]\right) \tag{3.2.59}$$

$$\left|C_{12(\mathrm{L,R})}\right| = \frac{\sqrt{2}}{2}\left|T_{13} - \mathrm{j}T_{12}\right| = \sqrt{\frac{1}{2}\left(\left|T_{12}\right|^2 + \left|T_{13}\right|^2\right) + \mathrm{Im}\left[T_{12}T_{13}^*\right]} \tag{3.2.60}$$

$$\left|C_{13(\mathrm{L,R})}\right| = \frac{1}{2}\left|T_{33} - T_{22} - 2\mathrm{j}\,\mathrm{Re}\left[T_{23}\right]\right| = \frac{1}{2}\sqrt{\left(T_{33} - T_{22}\right)^2 + 4\,\mathrm{Re}^2\left[T_{23}\right]} \tag{3.2.61}$$

$$\left|C_{23(\mathrm{L,R})}\right| = \frac{\sqrt{2}}{2}\left|T_{13} + \mathrm{j}T_{12}\right| = \sqrt{\frac{1}{2}\left(\left|T_{12}\right|^2 + \left|T_{13}\right|^2\right) - \mathrm{Im}\left[T_{12}T_{13}^*\right]} \tag{3.2.62}$$

Through this analysis, the polarimetric roll-invariant features from the Huynen decomposition, uniform polarimetric matrix rotation theory, and polarimetric covariance

TABLE 3.2.5
Polarimetric Roll-Invariant Features from the Polarimetric Covariance Matrix with Circular Polarization Basis

Feature	Expression	Physical Meaning
$C_{11(\mathrm{L,R})}$	$\frac{1}{2}\left(T_{22} + T_{33} - 2\,\mathrm{Im}\left[T_{23}\right]\right)$	The LL co-polarization component.
$C_{22(\mathrm{L,R})}$	$\frac{1}{2}T_{11}$	The LR cross-polarization component.
$C_{33(\mathrm{L,R})}$	$\frac{1}{2}\left(T_{22} + T_{33} + 2\,\mathrm{Im}\left[T_{23}\right]\right)$	The RR co-polarization component.
$\left\|C_{12(\mathrm{L,R})}\right\|$	$\sqrt{\frac{1}{2}\left(\left\|T_{12}\right\|^2 + \left\|T_{13}\right\|^2\right) + \mathrm{Im}\left[T_{12}T_{13}^*\right]}$	The polarimetric correlation value between LL and LR.
$\left\|C_{13(\mathrm{L,R})}\right\|$	$\frac{1}{2}\sqrt{\left(T_{33} - T_{22}\right)^2 + 4\,\mathrm{Re}^2\left[T_{23}\right]}$	The polarimetric correlation value between LL and RR.
$\left\|C_{23(\mathrm{L,R})}\right\|$	$\sqrt{\frac{1}{2}\left(\left\|T_{12}\right\|^2 + \left\|T_{13}\right\|^2\right) - \mathrm{Im}\left[T_{12}T_{13}^*\right]}$	The polarimetric correlation value between LR and RR.

matrix with circular polarization basis can be mainly derived from eight independent elements or element combinations of polarimetric coherency matrix: T_{11}, $T_{22}+T_{33}$, $\mathrm{Im}[T_{23}]$, $\mathrm{Re}^2[T_{12}]+\mathrm{Re}^2[T_{13}]$, $\mathrm{Im}^2[T_{12}]+\mathrm{Im}^2[T_{13}]$, $\left(|T_{13}|^2-|T_{12}|^2\right)^2+4\mathrm{Re}^2\left[T_{12}T_{13}^*\right]$, $[T_{33}-T_{22}]^2+4\mathrm{Re}^2[T_{23}]$, and $\mathrm{Im}\left[T_{12}T_{13}^*\right]$. These common polarimetric roll-invariant features are summarized in Table 3.2.6. The relationships of these derived polarimetric roll-invariant features are shown in Figure 3.2.2, and their relationships with the Sinclair matrix are also given.

3.2.4 Derived from Eigenvalue-Eigenvector Based Decomposition

Except those features shown in Section 3.2.3, other polarimetric roll-invariant features from the polarimetric coherency matrix are mainly derived through matrix operations and eigenvalue-eigenvector-based decomposition methods, including Cloude-Pottier decomposition [23], Touzi decomposition [46], Paladini decomposition [47], and Aghababaee decomposition [48].

The polarimetric coherency matrix $\mathbf{T}$ can be expressed with eigenvalues and eigenvectors:

$$\mathbf{T}=\mathbf{U}^{\mathrm{H}}\Lambda\mathbf{U}=\mathbf{U}^{\mathrm{H}}\begin{bmatrix}\lambda_1^{\mathrm{T}} & & \\ & \lambda_2^{\mathrm{T}} & \\ & & \lambda_3^{\mathrm{T}}\end{bmatrix}\mathbf{U} \tag{3.2.63}$$

where $\mathbf{U}=[\mathbf{u}_1,\mathbf{u}_2,\mathbf{u}_3]$ is the unitary matrix and contains the eigenvectors $\mathbf{u}_1,\mathbf{u}_2,\mathbf{u}_3$. λ_1^{T}, λ_2^{T}, and λ_3^{T} are eigenvalues, which are polarimetric roll-invariant. Without any loss of generality, $\lambda_1^{\mathrm{T}}\geq\lambda_2^{\mathrm{T}}\geq\lambda_3^{\mathrm{T}}$ is assumed.

In this vein, the following equations can be derived and the roll-invariance characteristics can be verified:

$$\lambda_i\left[\mathbf{T}(\theta)\right]=\lambda_i(\mathbf{T})=\lambda_i^{\mathrm{T}} \tag{3.2.64}$$

$$\mathrm{Tr}\left[\mathbf{T}(\theta)\right]=\mathrm{Tr}(\mathbf{T})=T_{11}+T_{22}+T_{33} \tag{3.2.65}$$

$$\mathrm{Det}\left[\mathbf{T}(\theta)\right]=\mathrm{Det}(\mathbf{T})=\lambda_1^{\mathrm{T}}\lambda_2^{\mathrm{T}}\lambda_3^{\mathrm{T}} \tag{3.2.66}$$

$$\|\mathbf{T}(\theta)\|_2=\|\mathbf{T}\|_2=\sqrt{\max\left[\lambda_i\left(\mathbf{T}^{\mathrm{H}}\mathbf{T}\right)\right]} \tag{3.2.67}$$

$$\|\mathbf{T}\|_{\mathrm{F}}=\sqrt{\mathrm{Tr}\left(\mathbf{T}^{\mathrm{H}}\mathbf{T}\right)} \tag{3.2.68}$$

Cloude-Pottier decomposition [23] is a pioneer eigenvalue-eigenvector-based decomposition, which mainly derives polarimetric features by characterizing the eigenvectors $\mathbf{u}_i$:

$$\mathbf{u}_i=\left[\cos\alpha_i \quad \sin\alpha_i\cos\beta_i e^{j\delta_i} \quad \sin\alpha_i\sin\beta_i e^{j\gamma_i}\right]^{\mathrm{T}} \tag{3.2.69}$$

TABLE 3.2.6
Common Polarimetric Roll-Invariant Features from the Polarimetric Coherency Matrix

Feature	Expression	Physical Meaning
T_{11}	$\frac{1}{2}\lvert S_{\mathrm{HH}}+S_{\mathrm{VV}}\rvert^2$	Reflects the odd-bounce scattering.
$T_{22}+T_{33}$	$\frac{1}{2}\left(\lvert S_{\mathrm{HH}}-S_{\mathrm{VV}}\rvert^2+4\lvert S_{\mathrm{HV}}\rvert^2\right)$	Sum of the double-bounce scattering and the cross-polarization.
$\mathrm{Im}[T_{23}]$	$\mathrm{Im}\left[(S_{\mathrm{HH}}-S_{\mathrm{VV}})S_{\mathrm{HV}}^*\right]$	Reflects the helix scattering.
$\mathrm{Re}^2[T_{12}]+\mathrm{Re}^2[T_{13}]$	$\left(\lvert S_{\mathrm{HH}}\rvert^2-\lvert S_{\mathrm{VV}}\rvert^2\right)^2+4\left(\mathrm{Re}\left[S_{\mathrm{HH}}S_{\mathrm{HV}}^*\right]+\mathrm{Re}\left[S_{\mathrm{VV}}S_{\mathrm{HV}}^*\right]\right)^2$	–
$\mathrm{Im}^2[T_{12}]+\mathrm{Im}^2[T_{13}]$	$\left(S_{\mathrm{HH}}S_{\mathrm{VV}}^*-S_{\mathrm{HH}}^*S_{\mathrm{VV}}\right)^2+4\left(\mathrm{Im}\left[S_{\mathrm{HH}}S_{\mathrm{HV}}^*\right]+\mathrm{Im}\left[S_{\mathrm{VV}}S_{\mathrm{HV}}^*\right]\right)^2$	–
$\left(\lvert T_{13}\rvert^2-\lvert T_{12}\rvert^2\right)^2+4\,\mathrm{Re}^2\left[T_{12}T_{13}^*\right]$	$\frac{1}{16}\left[4S_{\mathrm{HV}}^2(S_{\mathrm{HH}}+S_{\mathrm{VV}})^2+\left(S_{\mathrm{HH}}^2-S_{\mathrm{VV}}^2\right)^2\right]\left[4\left(S_{\mathrm{HV}}^*\right)^2\left(S_{\mathrm{HH}}^*-S_{\mathrm{VV}}^*\right)^2+\left[\left(S_{\mathrm{HH}}^*\right)^2-\left(S_{\mathrm{VV}}^*\right)^2\right]^2\right]$	–
$[T_{33}-T_{22}]^2+4\,\mathrm{Re}^2[T_{23}]$	$\frac{1}{4}\left[4S_{\mathrm{HV}}^2+(S_{\mathrm{HH}}-S_{\mathrm{VV}})^2\right]\left[4\left(S_{\mathrm{HV}}^*\right)^2+\left(S_{\mathrm{HH}}^*-S_{\mathrm{VV}}^*\right)^2\right]$	–
$\mathrm{Im}\left[T_{12}T_{13}^*\right]$	$\frac{1}{4}\lvert S_{\mathrm{HH}}+S_{\mathrm{VV}}\rvert^2\left[(S_{\mathrm{HH}}-S_{\mathrm{VV}})^*S_{\mathrm{HV}}-(S_{\mathrm{HH}}-S_{\mathrm{VV}})S_{\mathrm{HV}}^*\right]$	–

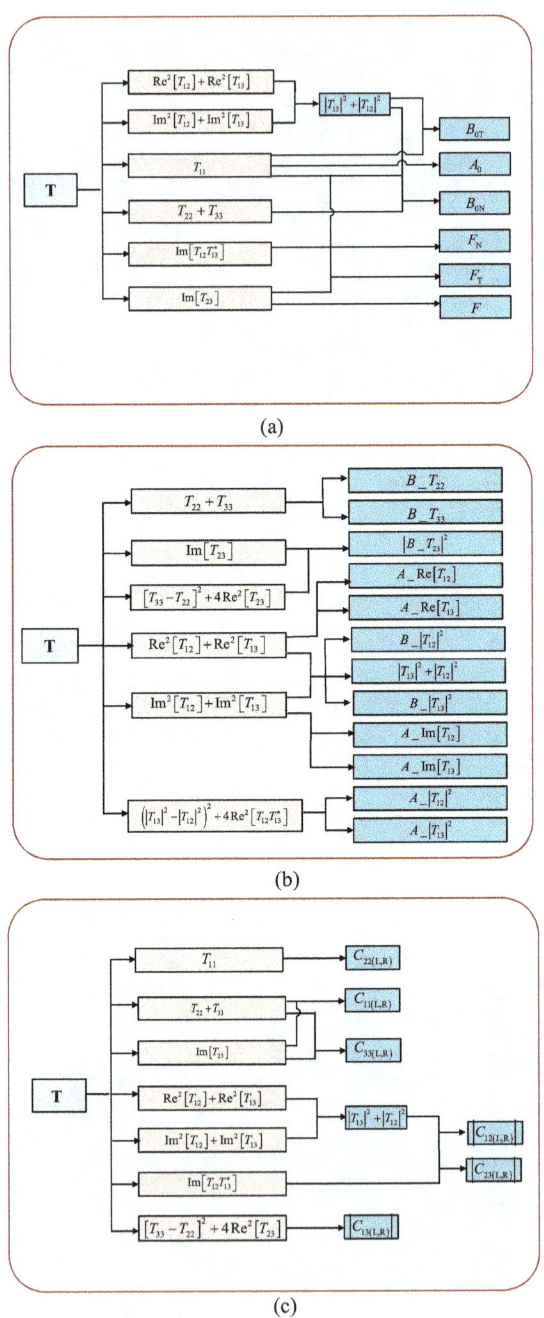

FIGURE 3.2.2 The relationships of polarimetric roll-invariant features derived from (a) Huynen decomposition, (b) uniform polarimetric matrix rotation theory, and (c) polarimetric covariance matrix with a circular polarization basis.

where β_i is the orientation angle, and α_i is a coefficient factor. The subscript i corresponds to the three eigenvectors, and $i = 1,2,3$.

The derived polarimetric roll-invariant features include the entropy H, the mean alpha angle $\bar{\alpha}$, and the anisotropy A:

$$H = -\sum_{i=1}^{3} P_i \log_3 P_i \tag{3.2.70}$$

$$\bar{\alpha} = \sum_{i=1}^{3} P_i \alpha_i \tag{3.2.71}$$

$$A = \frac{\lambda_2^{\mathrm{T}} - \lambda_3^{\mathrm{T}}}{\lambda_2^{\mathrm{T}} + \lambda_3^{\mathrm{T}}} \tag{3.2.72}$$

where $P_i = \dfrac{\lambda_i^{\mathrm{T}}}{\lambda_1^{\mathrm{T}} + \lambda_2^{\mathrm{T}} + \lambda_3^{\mathrm{T}}}$.

Meanwhile, other eigenvalue-eigenvector–based decomposition methods have been proposed to represent the eigenvectors with physically meaningful parameters. Among them, the polarimetric scattering vector model of (3.2.73) has been adopted for the development of Touzi decomposition [46], which includes coherent and incoherent target decomposition modes. For polarimetric incoherent decomposition, the derived polarimetric scattering vector $\mathbf{k}_{\mathrm{T}}^{\mathrm{SV}}$ is one of the eigenvectors of the polarimetric coherency matrix. For polarimetric coherent decomposition, the polarimetric scattering vector $\mathbf{k}_{\mathrm{T}}^{\mathrm{SV}}$ can be obtained by projecting the polarimetric scattering matrix onto the Pauli basis:

$$\mathbf{k}_{\mathbf{T}}^{\mathrm{SV}} = m_{\mathbf{T}} \exp^{\mathrm{j}\Phi_{\mathrm{s}}} \begin{bmatrix} 1 & 0 & 0 \\ 0 & \cos 2\theta & -\sin 2\theta \\ 0 & \sin 2\theta & \cos 2\theta \end{bmatrix} \begin{bmatrix} \cos\alpha_{\mathrm{s}} \cos 2\tau_{\mathrm{m}} \\ \sin\alpha_{\mathrm{s}} e^{\mathrm{j}\Phi_{\alpha_{\mathrm{s}}}} \\ -\mathrm{j}\cos\alpha_{\mathrm{s}} \sin 2\tau_{\mathrm{m}} \end{bmatrix} \tag{3.2.73}$$

where θ is the orientation angle, $m_{\mathrm{T}} = \|\mathbf{u}_1\|$ is the maximum amplitude, and τ_{m} stands for the helicity:

$$\tau_{\mathrm{m}} = \frac{1}{2}\tan^{-1}\left\{-\mathrm{Im}\left[v_3^\theta\right] / \mathrm{Re}\left[v_1^\theta\right]\right\} \tag{3.2.74}$$

where $v_j^\theta, j = 1,2,3$ are the elements of de-oriented eigenvectors $\mathbf{u}_i^\theta = \begin{bmatrix} v_1^\theta & v_2^\theta & v_3^\theta \end{bmatrix}^{\mathrm{T}}$. The de-orientation procedure is represented in (3.2.75) and (3.2.76). First, the eigenvector $\mathbf{u}_i$ is rotated with the orientation θ:

$$\mathbf{u}_i^\theta = \begin{bmatrix} 1 & 0 & 0 \\ 0 & \cos 2\theta & -\sin 2\theta \\ 0 & \sin 2\theta & \cos 2\theta \end{bmatrix} \mathbf{u}_i \tag{3.2.75}$$

Then, $\mathbf{u}_i^\theta$ is rotated with a rotation matrix of the helicity angle τ_{m} and transformed into $\mathbf{u}_i^{\theta,\tau} = \begin{bmatrix} v_1^{\theta,\tau} & v_2^{\theta,\tau} & v_3^{\theta,\tau} \end{bmatrix}^{\text{T}}$:

$$\mathbf{u}_i^{\theta,\tau} = \begin{bmatrix} \cos 2\tau_m & 0 & \text{j}\sin 2\tau_m \\ 0 & 1 & 0 \\ \text{j}\sin 2\tau_m & 0 & \cos 2\tau_m \end{bmatrix} \mathbf{u}_i^\theta \tag{3.2.76}$$

Meanwhile, α_{s} and $\Phi_{\alpha_{\text{s}}}$ are the magnitude and phase of symmetric scattering type, respectively:

$$\alpha_{\text{s}} = \cos^{-1}\left\{\text{Re}\left[v_1^{\theta,\tau}\right]\right\} \tag{3.2.77}$$

$$\Phi_{\alpha_{\text{s}}} = \tan^{-1}\left\{\text{Im}\left[v_2^\theta\right] / \text{Re}\left[v_2^\theta\right]\right\} \tag{3.2.78}$$

In this vein, polarimetric roll-invariant features $\left\{\tau_{\text{m}}, m_{\text{T}}, \alpha_{\text{s}}, \Phi_{\alpha_{\text{s}}}\right\}$ are available.

In Paladini et al. decomposition [47], a sufficient representation for the physical characteristics of the observed polarimetric data in terms of a minimum set of orientation invariant parameters has been developed with circular polarization basis:

$$\mathbf{k}_{\text{L}_{(\text{L,R})}} = \begin{bmatrix} S_{\text{LL}} \\ \sqrt{2}S_{\text{LR}} \\ S_{\text{RR}} \end{bmatrix} = \begin{bmatrix} \sin(\alpha_{\text{C}})\cos(\beta_{\text{C}})\text{e}^{\text{j}\left(-\frac{4}{3}\Upsilon - 2\psi\right)} \\ \cos(\alpha_{\text{C}})\text{e}^{\text{j}\frac{8}{3}\Upsilon} \\ -\sin(\alpha_{\text{C}})\cos(\beta_{\text{C}})\text{e}^{\text{j}\left(-\frac{4}{3}\Upsilon + 2\psi\right)} \end{bmatrix} \tag{3.2.79}$$

where $\mathbf{k}_{\text{L}_{(\text{L,R})}}$ is the polarimetric scattering vector with the circular polarization basis. In coherent decomposition, eigenvector $\mathbf{u}_i$ is applied to $\mathbf{k}_{\text{L}_{(\text{L,R})}}$. Generally, this representation is applied for the dominant eigenvector.

Therefore, a parameter set $\{\psi, \Upsilon, \alpha_{\text{C}}, \beta_{\text{C}}, Hel\}$ can be derived. ψ represents the orientation angle of eigenvectors:

$$\psi = \frac{1}{4}\arg\left(-S_{\text{RR}}S_{\text{LL}}^*\right) \tag{3.2.80}$$

Υ represents the phase difference between de-oriented co-polarization and cross-polarization scattering coefficients:

$$\begin{aligned} \Upsilon &= \frac{3}{8}\arg\left[\mathbf{u}_1(2)\right] \\ &= \frac{1}{4}\arg\left\{\left[\mathbf{k}_{\text{C}}^{\text{DeO}}(1)\right]^* \mathbf{k}_{\text{C}}^{\text{DeO}}(2)\right\} \end{aligned} \tag{3.2.81}$$

where $\mathbf{k}_{\mathrm{C}}^{\mathrm{DeO}} = \mathbf{R}(-\theta)\mathbf{k}_{\mathrm{L}_{(\mathrm{L,R})}}$ is the de-oriented eigenvector, and $\mathbf{R}(\theta)$ is the rotation matrix of the polarization orientation angle:

$$\mathbf{R}(\theta) = \begin{bmatrix} 1 & 0 & 0 \\ 0 & \cos 2\theta & -\sin 2\theta \\ 0 & \sin 2\theta & \cos 2\theta \end{bmatrix} \tag{3.2.82}$$

α_{C} is the main parameter for estimating the odd- and double-bounce nature of the target, which is similar to the parameter $\bar{\alpha}$ in Cloude-Pottier decomposition:

$$\begin{aligned} \alpha_{\mathrm{C}} &= \cos^{-1}\left(\left|\mathbf{u}_1(2)\right|\right) \\ &= \cos^{-1}\left(\sqrt{\frac{1}{2}\left|S_{\mathrm{HH}} + S_{\mathrm{VV}}\right|^2} \Big/ \left\|\mathbf{k}_{\mathrm{L}_{(\mathrm{L,R})}}\right\|\right) \end{aligned} \tag{3.2.83}$$

β_{C} is the degree of balance between the co-polarized coefficients of the circular polarization scattering vector and represents the scattering of left and right helices:

$$\beta_{\mathrm{C}} = \cos^{-1}\left(\left\|\mathbf{k}_{\mathrm{L}_{(\mathrm{L,R})}}(1)\right\| \Big/ \sin(\alpha_{\mathrm{C}})\left\|\mathbf{k}_{\mathrm{L}_{(\mathrm{L,R})}}\right\|\right) \tag{3.2.84}$$

Hel is the degree of asymmetry of the target vector:

$$Hel = \sin(\alpha_{\mathrm{C}})^2\left[\cos(\beta_{\mathrm{C}})^2 - \sin(\beta_{\mathrm{C}})^2\right] \tag{3.2.85}$$

For $Hel \to 1$, left helix scattering is indicated, while right helix scattering is identified for $Hel \to -1$. In this vein, the absolute value of *Hel* can represent the degree of target asymmetry.

Aghababaee et al. [48] proposed a polarimetric scattering vector model combined with Cameron decomposition. Characterizations of the minimum and maximum symmetric scattering components are considered in a unified model and projected onto the Pauli basis to derive a new polarimetric scattering vector $\hat{\mathbf{k}}$ for the representation of coherent target scattering. The polarization scattering vector $\hat{\mathbf{k}} = \begin{bmatrix} \hat{k}_1 & \hat{k}_2 & \hat{k}_3 \end{bmatrix}^{\mathrm{T}}$ is obtained by eigenvalue-eigenvector decomposition of the polarimetric coherency matrix $\mathbf{T}$. Furthermore, the polarimetric scattering matrix $\hat{\mathbf{S}}$ can be derived. Considering the maximum eigenvalue of the polarimetric coherency matrix $\mathbf{T}$, $\hat{\mathbf{k}}$ and $\hat{\mathbf{S}}$ are:

$$\hat{\mathbf{k}} = \mathbf{u}_1 = \sqrt{2}\begin{pmatrix} 1 & 0 & 0 \\ 0 & \cos 2\psi_{\max} & -\sin 2\psi_{\max} \\ 0 & \sin 2\psi_{\max} & \cos 2\psi_{\max} \end{pmatrix}\begin{pmatrix} \mu_1 \\ \mu_2 \\ \mu_3 \end{pmatrix} \tag{3.2.86}$$

$$\hat{\mathbf{S}} = \frac{\sqrt{2}}{2}\begin{bmatrix} \hat{k}_1 + \hat{k}_2 & \hat{k}_3 \\ \hat{k}_3 & \hat{k}_1 - \hat{k}_2 \end{bmatrix} \tag{3.2.87}$$

where $\psi_{\max}$ is the phase angle of the maximum symmetrical component.

Then, polarimetric roll-invariant features are derived as $\{\tau_{\mathrm{p}}, \alpha_{\mathrm{p}}, \varphi_{\mathrm{p}}, m_{\mathrm{p}}\}$. Among them, τ_{p} is the degree of asymmetry:

$$\tau_{\mathrm{p}} = \sin^{-1}\left(\left|\lambda_3^{\mathrm{T}}\right| \Big/ \sqrt{\left|\lambda_1^{\mathrm{T}}\right|^2 + \left|\lambda_2^{\mathrm{T}}\right|^2 + \left|\lambda_3^{\mathrm{T}}\right|^2}\right) \tag{3.2.88}$$

α_{p} represents the scattering type, and φ_{p} represents the phase difference of two eigenvectors:

$$\alpha_{\mathrm{p}} = \tan^{-1}\left|\frac{\mu_2}{\mu_1}\right| \tag{3.2.89}$$

$$\varphi_{\mathrm{p}} = \arg\left(\frac{\mu_2}{\mu_1}\right) \tag{3.2.90}$$

m_{p} is an estimation of the maximum amplitude return:

$$m_{\mathrm{p}} = \sqrt{\left|\mu_1\right|^2 + \left|\mu_2\right|^2 + \left|\mu_3\right|^2} \tag{3.2.91}$$

The expressions of polarimetric roll-invariant features derived from eigenvalue-eigenvector-based decomposition are listed in Table 3.2.7. Since Cloude and Pottier proposed the $H - \bar{\alpha} - A$ decomposition, researchers have been devoted to parameterizing the eigenvalue and eigenvector of the polarimetric coherency matrix. These polarimetric roll-invariant features are derived from different forms. However, the physical meanings of these features are similar. Among them, the scattering type parameter and the degree of helicity are the main parameters. As for scattering type parameters, they indicate the scattering type of radar targets, such as the mean alpha angle $\bar{\alpha}$ in Cloude-Pottier decomposition. There are three specific states for the mean alpha angle $\bar{\alpha}$:

1) $\bar{\alpha} \to 0$, the scattering mechanism relates to single-bounce scattering induced by a rough surface.
2) $\bar{\alpha} \to \pi/4$, the scattering mechanism corresponds to volume scattering.
3) $\bar{\alpha} \to \pi/2$, the scattering mechanism is due to double-bounce scattering.

Similarly, α_{s} in Touzi decomposition, α_1 in Paladini decomposition, and α_{p} in Aghababaee decomposition show similar physical meaning. As for the degree of helicity, the concept of a helix mechanism or helicity is adopted in the polarimetric decomposition framework. Helicity is related to the target symmetry such as τ_{m} in Touzi decomposition, *Hel* in Paladini decomposition, and τ_{p} in Aghababaee decomposition.

TABLE 3.2.7
Polarimetric Roll-Invariant Features Derived from Eigenvalue-Eigenvector Based Decomposition

Feature	Expression	Physical Meaning
$\lambda_i(\mathbf{T})$	–	Eigenvalues of polarimetric coherency matrix.
$\mathrm{Tr}(\mathbf{T})$	$T_{11}+T_{22}+T_{33}$	Target's ability to intercept and reflect the incident waves.
$\mathrm{Det}(\mathbf{T})$	$\lambda_1^{\mathrm{T}}\lambda_2^{\mathrm{T}}\lambda_3^{\mathrm{T}}$	The determinant of polarimetric coherency matrix.
$\Vert\mathbf{T}\Vert_2$	$\sqrt{\max\left[\lambda_i\left(\mathbf{T}^{\mathrm{H}}\mathbf{T}\right)\right]}$	The two-norm of polarimetric coherency matrix.
$\Vert\mathbf{T}\Vert_{\mathrm{F}}$	$\sqrt{\mathrm{Tr}\left(\mathbf{T}^{\mathrm{H}}\mathbf{T}\right)}$	Similar to *SPAN* feature.
H	$-\sum_{i=1}^{3}P_i\log_3 P_i$	Statistical disorder of the polarization scattering types.
$\bar{\alpha}$	$\sum_{i=1}^{3}P_i\alpha_i$	Characterization of the target scattering types.
A	$\left(\lambda_2^{\mathrm{T}}-\lambda_3^{\mathrm{T}}\right)/\left(\lambda_2^{\mathrm{T}}+\lambda_3^{\mathrm{T}}\right)$	The ratio of polarimetric coherency matrix's last two eigenvalues.
τ_{m}	$\frac{1}{2}\tan^{-1}\left\{-\mathrm{Im}\left[v_3^{\theta}\right]/\mathrm{Re}\left[v_1^{\theta}\right]\right\}$	The helicity of a target.
m_{T}	$\Vert\mathbf{u}_1\Vert$	The intensity of the eigenvalue.
α_{s}	$\cos^{-1}\left\{\mathrm{Re}\left[v_1^{\theta,\tau}\right]\right\}$	The scattering type.
$\Phi_{\alpha_{\mathrm{s}}}$	$\tan^{-1}\left\{\mathrm{Im}\left[v_2^{\theta}\right]/\mathrm{Re}\left[v_2^{\theta}\right]\right\}$	The phase of the scattering type parameter.
ψ	$\frac{1}{4}\arg\left(-S_{\mathrm{RR}}S_{\mathrm{LL}}^{*}\right)$	The rotation of the eigenvectors.
Υ	$\frac{1}{4}\arg\left\{\left[\mathbf{k}_{\mathrm{C}}^{\mathrm{DeO}}(1)\right]^{*}\mathbf{k}_{\mathrm{C}}^{\mathrm{DeO}}(2)\right\}$	The phase difference of co-polarization and cross-polarization's scattering factors.
α_{C}	$\cos^{-1}\left(\sqrt{\frac{1}{2}\lvert S_{\mathrm{HH}}+S_{\mathrm{VV}}\rvert^2}\Big/\left\Vert\mathbf{k}_{\mathrm{L(L,R)}}\right\Vert\right)$	The scattering type.
β_{C}	$\cos^{-1}\left(\left\Vert\mathbf{k}_{\mathrm{L(L,R)}}(1)\right\Vert/\sin(\alpha_{\mathrm{C}})\left\Vert\mathbf{k}_{\mathrm{L(L,R)}}\right\Vert\right)$	Degree of balance between the co-polarized coefficients of the circular polarization scattering vector; represents the scattering of left and right helices.
Hel	$\sin^2(\alpha_{\mathrm{C}})\left[\cos^2(\beta_{\mathrm{C}})-\sin^2(\beta_{\mathrm{C}})\right]$	The helicity of the target.
τ_{p}	$\sin^{-1}\left(\lvert\lambda_3^{\mathrm{T}}\rvert\Big/\sqrt{\lvert\lambda_1^{\mathrm{T}}\rvert^2+\lvert\lambda_2^{\mathrm{T}}\rvert^2+\lvert\lambda_3^{\mathrm{T}}\rvert^2}\right)$	The helicity of the target.
α_{p}	$\tan^{-1}\lvert\mu_2/\mu_1\rvert$	The scattering type.
φ_{p}	$\arg(\mu_2/\mu_1)$	The phase angle of the scattering type parameter.
m_{p}	$\sqrt{\lvert\mu_1\rvert^2+\lvert\mu_2\rvert^2+\lvert\mu_3\rvert^2}$	The matrix two-norm of eigenvector.

3.2.5 Derived from Two-Dimension Polarimetric Coherence/Correlation Patterns

The two-dimension polarimetric coherence/correlation pattern interpretation tools introduced in Chapter 2 have the capability to represent the rich information from target scattering diversity [29, 49]. Due to the derivation procedure of polarimetric coherence/correlation patterns, the majority of polarimetric coherence/correlation features are roll-invariant.

The core idea of two-dimension polarimetric coherence patterns is to extend the traditional polarimetric coherence at a given rotation state ($\theta = 0$) to the rotation domain ($\theta \in [-\pi, \pi)$) for hidden information exploration. For two polarization channels s_1 and s_2, the definition of the polarimetric coherence pattern [49] is

$$\left|\gamma_{1-2}(\theta)\right| = \frac{\left|\left\langle s_1(\theta) s_2^*(\theta)\right\rangle\right|}{\sqrt{\left\langle\left|s_1(\theta)\right|^2\right\rangle\left\langle\left|s_2(\theta)\right|^2\right\rangle}}, \qquad \theta \in [-\pi, \pi) \tag{3.2.92}$$

The two-dimension polarimetric correlation pattern can be similarly defined [29]:

$$\left|\hat{\gamma}_{1-2}(\theta)\right| = \left|\left\langle s_1(\theta) s_2^*(\theta)\right\rangle\right|, \qquad \theta \in [-\pi, \pi) \tag{3.2.93}$$

According to Table 2.4.1, ten polarimetric features can be derived for each two-dimension polarimetric coherence pattern. Among them, there are seven polarimetric roll-invariant features, as follows:

1) Coherence Degree $\left|\gamma_{1-2}(\theta)\right|_{\text{mean}}$ is

$$\left|\gamma_{1-2}(\theta)\right|_{\text{mean}} = \frac{1}{2\pi}\int_0^{2\pi}\left|\gamma_{1-2}(\theta)\right| \mathrm{d}\theta \tag{3.2.94}$$

2) Coherence Fluctuation $\left|\gamma_{1-2}(\theta)\right|_{\text{std}}$ is

$$\left|\gamma_{1-2}(\theta)\right|_{\text{std}} = \sqrt{\frac{1}{2\pi}\int_0^{2\pi}\left(\left|\gamma_{1-2}(\theta)\right| - \left|\gamma_{1-2}(\theta)\right|_{\text{mean}}\right)^2 \mathrm{d}\theta} \tag{3.2.95}$$

3) Maximum Coherence $\left|\gamma_{1-2}(\theta)\right|_{\max}$ is

$$\left|\gamma_{1-2}(\theta)\right|_{\max} = \max\left(\left|\gamma_{1-2}(\theta)\right|\right) \tag{3.2.96}$$

4) Minimum Coherence $\left|\gamma_{1-2}(\theta)\right|_{\min}$ is

$$\left|\gamma_{1-2}(\theta)\right|_{\min} = \min\left(\left|\gamma_{1-2}(\theta)\right|\right) \tag{3.2.97}$$

5) Coherence Contrast $\left|\gamma_{1-2}(\theta)\right|_{\text{contrast}}$ is

$$\left|\gamma_{1-2}(\theta)\right|_{\text{contrast}} = \left|\gamma_{1-2}(\theta)\right|_{\max} - \left|\gamma_{1-2}(\theta)\right|_{\min} \tag{3.2.98}$$

6) Coherence Anisotropy $\left|\gamma_{1-2}(\theta)\right|_{\text{A}}$ is

$$\left|\gamma_{1-2}(\theta)\right|_{\text{A}} = \frac{\left|\gamma_{1-2}(\theta)\right|_{\max} - \left|\gamma_{1-2}(\theta)\right|_{\min}}{\left|\gamma_{1-2}(\theta)\right|_{\max} + \left|\gamma_{1-2}(\theta)\right|_{\min}} \tag{3.2.99}$$

7) Coherence Beamwidth $\left|\gamma_{1-2}(\theta)\right|_{\text{bw0.95}}$ is

$$\left|\gamma_{1-2}(\theta)\right|_{\text{bw0.95}} = \left\{\theta \middle| \left|\gamma_{1-2}(\theta)\right|_{\max} \geq \left|\gamma_{1-2}(\theta)\right| \geq 0.95 \times \left|\gamma_{1-2}(\theta)\right|_{\max}\right\} \tag{3.2.100}$$

In addition, according to Table 2.5.1, each two-dimension polarimetric correlation pattern can derive ten polarimetric features. Among them, there are also seven polarimetric roll-invariant features:

1) Correlation Degree $\left|\hat{\gamma}_{1-2}(\theta)\right|_{\text{mean}}$ is

$$\left|\hat{\gamma}_{1-2}(\theta)\right|_{\text{mean}} = \frac{1}{2\pi}\int_{0}^{2\pi}\left|\hat{\gamma}_{1-2}(\theta)\right|\mathrm{d}\theta \tag{3.2.101}$$

2) Correlation Fluctuation $\left|\hat{\gamma}_{1-2}(\theta)\right|_{\text{std}}$ is

$$\left|\hat{\gamma}_{1-2}(\theta)\right|_{\text{std}} = \sqrt{\frac{1}{2\pi}\int_{0}^{2\pi}\left(\left|\hat{\gamma}_{1-2}(\theta)\right| - \left|\hat{\gamma}_{1-2}(\theta)\right|_{\text{mean}}\right)^{2}\mathrm{d}\theta} \tag{3.2.102}$$

3) Maximum Correlation $\left|\hat{\gamma}_{1-2}(\theta)\right|_{\max}$ is

$$\left|\hat{\gamma}_{1-2}(\theta)\right|_{\max} = \max\left(\left|\hat{\gamma}_{1-2}(\theta)\right|\right) \tag{3.2.103}$$

4) Minimum Correlation $\left|\hat{\gamma}_{1-2}(\theta)\right|_{\min}$ is

$$\left|\hat{\gamma}_{1-2}(\theta)\right|_{\min} = \min\left(\left|\hat{\gamma}_{1-2}(\theta)\right|\right) \tag{3.2.104}$$

5) Correlation Contrast $\left|\hat{\gamma}_{1-2}(\theta)\right|_{\text{contrast}}$ is

$$\left|\hat{\gamma}_{1-2}(\theta)\right|_{\text{contrast}} = \left|\hat{\gamma}_{1-2}(\theta)\right|_{\max} - \left|\hat{\gamma}_{1-2}(\theta)\right|_{\min} \tag{3.2.105}$$

6) Correlation Anisotropy $\left|\hat{\gamma}_{1-2}(\theta)\right|_{\mathrm{A}}$ is

$$\left|\hat{\gamma}_{1-2}(\theta)\right|_{\mathrm{A}} = \frac{\left|\hat{\gamma}_{1-2}(\theta)\right|_{\max} - \left|\hat{\gamma}_{1-2}(\theta)\right|_{\min}}{\left|\hat{\gamma}_{1-2}(\theta)\right|_{\max} + \left|\hat{\gamma}_{1-2}(\theta)\right|_{\min}} \tag{3.2.106}$$

7) Correlation Beamwidth $\left|\hat{\gamma}_{1-2}(\theta)\right|_{\mathrm{bw}0.95}$ is

$$\left|\hat{\gamma}_{1-2}(\theta)\right|_{\mathrm{bw}0.95} = \left\{\theta \,\middle|\, \left|\hat{\gamma}_{1-2}(\theta)\right|_{\max} \geq \left|\hat{\gamma}_{1-2}(\theta)\right| \geq 0.95 \times \left|\hat{\gamma}_{1-2}(\theta)\right|_{\max}\right\} \tag{3.2.107}$$

The expression and physical meanings of polarimetric roll-invariant features derived from two-dimension polarimetric coherence and correlation patterns are listed in Table 3.2.8.

TABLE 3.2.8
Polarimetric Roll-Invariant Features Derived from Two-Dimension Polarimetric Coherence and Correlation Patterns

Feature	Expression	Physical Meaning
$\left|\gamma_{1-2}(\theta)\right|_{\mathrm{mean}}$	$\frac{1}{2\pi}\int_0^{2\pi}\left|\gamma_{1-2}(\theta)\right| d\theta$	The general coherence level and a measure for targets' average decorrelation effect in the polarimetric rotation domain.
$\left|\gamma_{1-2}(\theta)\right|_{\mathrm{std}}$	$\sqrt{\frac{1}{2\pi}\int_0^{2\pi}\left(\left|\gamma_{1-2}(\theta)\right| - \left|\gamma_{1-2}(\theta)\right|_{\mathrm{mean}}\right)^2 d\theta}$	The coherence fluctuation and a measure for target scattering orientation diversity. For polarimetric roll-invariant scatterers, coherence fluctuation $\left|\gamma_{1-2}(\theta)\right|_{\mathrm{std}}$ will shrink to 0.
$\left|\gamma_{1-2}(\theta)\right|_{\max}$	$\max\left(\left|\gamma_{1-2}(\theta)\right|\right)$	The upper limit of polarimetric coherence of two polarization channels via orientation adjustment.
$\left|\gamma_{1-2}(\theta)\right|_{\min}$	$\min\left(\left|\gamma_{1-2}(\theta)\right|\right)$	The lower limit of polarimetric coherence of two polarization channels via orientation adjustment.
$\left|\gamma_{1-2}(\theta)\right|_{\mathrm{contrast}}$	$\left|\gamma_{1-2}(\theta)\right|_{\max} - \left|\gamma_{1-2}(\theta)\right|_{\min}$	The coherence contrast in the polarimetric rotation domain.
$\left|\gamma_{1-2}(\theta)\right|_{\mathrm{A}}$	$\frac{\left|\gamma_{1-2}(\theta)\right|_{\max} - \left|\gamma_{1-2}(\theta)\right|_{\min}}{\left|\gamma_{1-2}(\theta)\right|_{\max} + \left|\gamma_{1-2}(\theta)\right|_{\min}}$	The relative coherence contrast in the polarimetric rotation domain.
$\left|\gamma_{1-2}(\theta)\right|_{\mathrm{bw}0.95}$	$\left\{\theta \,\middle|\, \left|\gamma_{1-2}(\theta)\right|_{\max} \geq \left|\gamma_{1-2}(\theta)\right| \geq 0.95 \times \left|\gamma_{1-2}(\theta)\right|_{\max}\right\}$	The measure of the orientation dependency effect. The smaller the value is, the greater the decorrelation effect and orientation dependency are.

(Continued)

TABLE 3.2.8 *(Continued)*
Polarimetric Roll-Invariant Features Derived from Two-Dimension Polarimetric Coherence and Correlation Patterns

Feature	Expression	Physical Meaning
$\|\hat{\gamma}_{1-2}(\theta)\|_{\text{mean}}$	$\frac{1}{2\pi}\int_0^{2\pi}\|\hat{\gamma}_{1-2}(\theta)\|d\theta$	The general correlation level and a measure for targets' average decorrelation effect in the polarimetric rotation domain.
$\|\hat{\gamma}_{1-2}(\theta)\|_{\text{std}}$	$\sqrt{\frac{1}{2\pi}\int_0^{2\pi}\left(\|\hat{\gamma}_{1-2}(\theta)\|-\|\hat{\gamma}_{1-2}(\theta)\|_{\text{mean}}\right)^2 d\theta}$	The correlation fluctuation and a measure for target scattering orientation diversity. For polarimetric roll-invariant scatterers, correlation fluctuation $\|\hat{\gamma}_{1-2}(\theta)\|_{\text{std}}$ will shrink to 0.
$\|\hat{\gamma}_{1-2}(\theta)\|_{\text{max}}$	$\max\left(\|\hat{\gamma}_{1-2}(\theta)\|\right)$	The upper limit of polarimetric correlation of two polarization channels via orientation adjustment.
$\|\hat{\gamma}_{1-2}(\theta)\|_{\text{min}}$	$\min\left(\|\hat{\gamma}_{1-2}(\theta)\|\right)$	The lower limit of polarimetric correlation of two polarization channels via orientation adjustment.
$\|\hat{\gamma}_{1-2}(\theta)\|_{\text{contrast}}$	$\|\hat{\gamma}_{1-2}(\theta)\|_{\text{max}}-\|\hat{\gamma}_{1-2}(\theta)\|_{\text{min}}$	The correlation contrast in the polarimetric rotation domain.
$\|\hat{\gamma}_{1-2}(\theta)\|_{\text{A}}$	$\frac{\|\hat{\gamma}_{1-2}(\theta)\|_{\text{max}}-\|\hat{\gamma}_{1-2}(\theta)\|_{\text{min}}}{\|\hat{\gamma}_{1-2}(\theta)\|_{\text{max}}+\|\hat{\gamma}_{1-2}(\theta)\|_{\text{min}}}$	The relative correlation contrast in the polarimetric rotation domain.
$\|\hat{\gamma}_{1-2}(\theta)\|_{\text{bw0.95}}$	$\left\{\theta \,\middle\|\, \|\hat{\gamma}_{1-2}(\theta)\|_{\text{max}}\geq\|\hat{\gamma}_{1-2}(\theta)\|\geq 0.95\times\|\hat{\gamma}_{1-2}(\theta)\|_{\text{max}}\right\}$	The measure of the orientation dependency effect. The smaller the value is, the greater the decorrelation effect and orientation dependency are.

3.2.6 Derived from Three-Dimension Polarimetric Correlation Patterns

By extending the two-dimension polarimetric correlation pattern along the polarimetric ellipticity angle dimension, a three-dimension polarimetric correlation pattern can be obtained, which can explore and utilize more complete target scattering diversity [36], which is introduced in Chapter 2. For two polarization channels s_1 and s_2, the three-dimension polarimetric correlation pattern $|\hat{\gamma}_{1-2}(\theta,\tau)|$ is:

$$|\hat{\gamma}_{1-2}(\theta,\tau)| = \left|\langle s_1(\theta,\tau)s_2^*(\theta,\tau)\rangle\right|, \qquad \theta\in[-\pi,\pi), \quad \tau\in[-\pi/4,\pi/4) \tag{3.2.108}$$

According to Section 2.6.4, three types of polarimetric features can be derived from each three-dimension polarimetric correlation pattern, including global features, local features, and mutual features. Among the local features, only the three-dimension polarimetric correlation pattern features obtained from the polarimetric ellipticity angle profile may be polarimetric roll-invariant features. Therefore, for a given polarimetric ellipticity angle $\tau = \alpha$, for each three-dimension polarimetric correlation pattern, 12 polarimetric roll-invariant features, including 5 global features, 6 local features, and 1 mutual feature, can be derived.

1) Global Features:

1) Maximum Correlation $\left|\hat{\gamma}_{1-2}(\theta,\tau)\right|_{\text{all-max}}$ is

$$\left|\hat{\gamma}_{1-2}(\theta,\tau)\right|_{\text{all-max}} = \max\left(\left|\hat{\gamma}_{1-2}(\theta,\tau)\right|\right) \tag{3.2.109}$$

2) Minimum Correlation $\left|\hat{\gamma}_{1-2}(\theta,\tau)\right|_{\text{all-min}}$ is

$$\left|\hat{\gamma}_{1-2}(\theta,\tau)\right|_{\text{all-min}} = \min\left(\left|\hat{\gamma}_{1-2}(\theta,\tau)\right|\right) \tag{3.2.110}$$

3) Maximum Gaussian Curvature $\left|\hat{\gamma}_{1-2}(\theta,\tau)\right|_{\text{MGC}}$ is

$$\left|\hat{\gamma}_{1-2}(\theta,\tau)\right|_{\text{MGC}} = \max\left(\left|\text{GC}(\theta,\tau)\right|\right) \tag{3.2.111}$$

4) Maximum Mean Curvature $\left|\hat{\gamma}_{1-2}(\theta,\tau)\right|_{\text{MMC}}$ is

$$\left|\hat{\gamma}_{1-2}(\theta,\tau)\right|_{\text{MMC}} = \max\left(\left|\text{MC}(\theta,\tau)\right|\right) \tag{3.2.112}$$

5) Maximum Principal Curvature $\left|\hat{\gamma}_{1-2}(\theta,\tau)\right|_{\text{MPC}}$ is

$$\left|\hat{\gamma}_{1-2}(\theta,\tau)\right|_{\text{MPC}} = \max\left[\max\left(\left|\kappa_1(\theta,\tau)\right|\right), \max\left(\left|\kappa_2(\theta,\tau)\right|\right)\right] \tag{3.2.113}$$

where the definitions of GC, MC, and κ_1, κ_2 can be found in Section 2.6.4.

2) Local Features:

1) Slice Maximum Correlation $\left|\hat{\gamma}_{1-2}(\theta,\tau)\right|_{\text{mean}}^{\tau=\alpha}$ is

$$\left|\hat{\gamma}_{1-2}(\theta,\tau)\right|_{\text{mean}}^{\tau=\alpha} = \frac{1}{2\pi}\int_0^{2\pi}\left|\hat{\gamma}_{1-2}(\theta,\tau)\right|^{\tau=\alpha} \mathrm{d}\theta \tag{3.2.114}$$

2) Slice Correlation Degree $\left|\hat{\gamma}_{1-2}(\theta,\tau)\right|_{\text{std}}^{\tau=\alpha}$ is

$$\left|\hat{\gamma}_{1-2}(\theta,\tau)\right|_{\text{std}}^{\tau=\alpha} = \sqrt{\frac{1}{2\pi}\int_0^{2\pi}\left(\left|\hat{\gamma}_{1-2}(\theta,\tau)\right|^{\tau=\alpha} - \left|\hat{\gamma}_{1-2}(\theta,\tau)\right|_{\text{mean}}^{\tau=\alpha}\right)^2 \mathrm{d}\theta} \tag{3.2.115}$$

3) Slice Maximum Correlation $\left|\hat{\gamma}_{1-2}(\theta,\tau)\right|_{\text{max}}^{\tau=\alpha}$ is

$$\left|\hat{\gamma}_{1-2}(\theta,\tau)\right|_{\text{max}}^{\tau=\alpha} = \max\left(\left|\hat{\gamma}_{1-2}(\theta,\tau)\right|^{\tau=\alpha}\right) \tag{3.2.116}$$

4) Slice Minimum Correlation $\left|\hat{\gamma}_{1-2}(\theta,\tau)\right|_{\min}^{\tau=\alpha}$ is

$$\left|\hat{\gamma}_{1-2}(\theta,\tau)\right|_{\min}^{\tau=\alpha} = \min\left(\left|\hat{\gamma}_{1-2}(\theta,\tau)\right|^{\tau=\alpha}\right) \tag{3.2.117}$$

5) Slice Correlation Contrast $\left|\hat{\gamma}_{1-2}(\theta,\tau)\right|_{\text{contrast}}^{\tau=\alpha}$ is

$$\left|\hat{\gamma}_{1-2}(\theta,\tau)\right|_{\text{contrast}}^{\tau=\alpha} = \left|\hat{\gamma}_{1-2}(\theta,\tau)\right|_{\max}^{\tau=\alpha} - \left|\hat{\gamma}_{1-2}(\theta,\tau)\right|_{\min}^{\tau=\alpha} \tag{3.2.118}$$

6) Slice Maximum Curve Curvature $\left|\hat{\gamma}_{1-2}(\theta,\tau)\right|_{\text{MCC}}^{\tau=\alpha}$ is

$$\left|\hat{\gamma}_{1-2}(\theta,\tau)\right|_{\text{MCC}}^{\tau=\alpha} = \max\left(\left|\text{CC}(\theta)\right|\right) \tag{3.2.119}$$

where α is the polarimetric ellipticity angle and $\alpha \in \{0^\circ, 45^\circ\}$ is usually adopted. The definition of CC can be found in Section 2.6.4.

3) Mutual Features:

1) Correlation Difference $\left|\hat{\gamma}_{1-2}(\theta,\tau)\right|_{\text{diff}}$ is

$$\left|\hat{\gamma}_{1-2}(\theta,\tau)\right|_{\text{diff}} = \left|\left|\hat{\gamma}_{1-2}(\theta,\tau)\right|_{\max}^{\tau=\alpha_1} - \left|\hat{\gamma}_{1-2}(\theta,\tau)\right|_{\max}^{\tau=\alpha_2}\right| \tag{3.2.120}$$

where α_1 and α_2 are two different polarimetric ellipticity angles. $\alpha_1 = 0^\circ$ and $\alpha_2 = 45^\circ$ are usually adopted. The expressions and the physical meanings are shown in Table 3.2.9.

3.2.7 Derived from the Kennaugh Matrix

The Kennaugh matrix is a 4 × 4 real matrix to represent the backscattering of targets. The polarimetric roll-invariant features of the Kennaugh matrix are mainly derived through geodesic distance [39] and the non-conventional three-dimension Barakat degree of polarization [37, 38].

With the reciprocity condition, the Kennaugh matrix can be represented in terms of the elements of the polarimetric coherency matrix:

$$\mathbf{K} = \begin{bmatrix} \frac{T_{11}+T_{22}+T_{33}}{2} & \text{Re}[T_{12}] & \text{Re}[T_{13}] & \text{Im}[T_{23}] \\ \text{Re}[T_{12}] & \frac{T_{11}+T_{22}-T_{33}}{2} & \text{Re}[T_{23}] & \text{Im}[T_{13}] \\ \text{Re}[T_{13}] & \text{Re}[T_{23}] & \frac{T_{11}-T_{22}+T_{33}}{2} & -\text{Im}[T_{12}] \\ \text{Im}[T_{23}] & \text{Im}[T_{13}] & -\text{Im}[T_{12}] & \frac{-T_{11}+T_{22}+T_{33}}{2} \end{bmatrix} \tag{3.2.121}$$

TABLE 3.2.9
Polarimetric Roll-Invariant Features Derived from Three-Dimension Polarimetric Correlation Patterns

Feature	Expression	Physical Meaning
$\lvert\hat{\gamma}_{1-2}(\theta,\tau)\rvert_{\text{all-max}}$	$\max\left(\lvert\hat{\gamma}_{1-2}(\theta,\tau)\rvert\right)$	The upper limit of two polarimetric channels' correlation value.
$\lvert\hat{\gamma}_{1-2}(\theta,\tau)\rvert_{\text{all-min}}$	$\min\left(\lvert\hat{\gamma}_{1-2}(\theta,\tau)\rvert\right)$	The lower limit of two polarimetric channels' correlation value.
$\lvert\hat{\gamma}_{1-2}(\theta,\tau)\rvert_{\text{MGC}}$	$\max\left(\lvert\text{GC}(\theta,\tau)\rvert\right)$	The maximum bending degree of the three-dimension polarimetric correlation pattern.
$\lvert\hat{\gamma}_{1-2}(\theta,\tau)\rvert_{\text{MMC}}$	$\max\left(\lvert\text{MC}(\theta,\tau)\rvert\right)$	The maximum mean bending degree of the three-dimension polarimetric correlation pattern.
$\lvert\hat{\gamma}_{1-2}(\theta,\tau)\rvert_{\text{MPC}}$	$\max\left[\max\left(\lvert\kappa_1(\theta,\tau)\rvert\right),\max\left(\lvert\kappa_2(\theta,\tau)\rvert\right)\right]$	The maximum primary bending degree of the three-dimension polarimetric correlation pattern.
$\lvert\hat{\gamma}_{1-2}(\theta,\tau)\rvert_{\text{mean}}^{\tau=\alpha}$	$\frac{1}{2\pi}\int_0^{2\pi}\lvert\hat{\gamma}_{1-2}(\theta,\tau)\rvert^{\tau=\alpha}d\theta$	The mean correlation at the $\tau=\alpha$ plane.
$\lvert\hat{\gamma}_{1-2}(\theta,\tau)\rvert_{\text{std}}^{\tau=\alpha}$	$\sqrt{\frac{1}{2\pi}\int_0^{2\pi}\left(\lvert\hat{\gamma}_{1-2}(\theta,\tau)\rvert^{\tau=\alpha}-\lvert\hat{\gamma}_{1-2}(\theta,\tau)\rvert_{\text{mean}}^{\tau=\alpha}\right)^2d\theta}$	The standard deviation of the correlation value at the $\tau=\alpha$ plane.
$\lvert\hat{\gamma}_{1-2}(\theta,\tau)\rvert_{\text{max}}^{\tau=\alpha}$	$\max\left(\lvert\hat{\gamma}_{1-2}(\theta,\tau)\rvert^{\tau=\alpha}\right)$	The maximum correlation value at the $\tau=\alpha$ plane.
$\lvert\hat{\gamma}_{1-2}(\theta,\tau)\rvert_{\text{min}}^{\tau=\alpha}$	$\min\left(\lvert\hat{\gamma}_{1-2}(\theta,\tau)\rvert^{\tau=\alpha}\right)$	The minimum correlation value at the $\tau=\alpha$ plane.
$\lvert\hat{\gamma}_{1-2}(\theta,\tau)\rvert_{\text{contrast}}^{\tau=\alpha}$	$\lvert\hat{\gamma}_{1-2}(\theta,\tau)\rvert_{\text{max}}^{\tau=\alpha}-\lvert\hat{\gamma}_{1-2}(\theta,\tau)\rvert_{\text{min}}^{\tau=\alpha}$	The correlation contrast at the $\tau=\alpha$ plane.
$\lvert\hat{\gamma}_{1-2}(\theta,\tau)\rvert_{\text{MCC}}^{\tau=\alpha}$	$\max\left(\lvert\text{CC}(\theta)\rvert\right)$	The maximum change degree at the $\tau=\alpha$ plane.
$\lvert\hat{\gamma}_{1-2}(\theta,\tau)\rvert_{\text{diff}}$	$\left\lvert\lvert\hat{\gamma}_{1-2}(\theta,\tau)\rvert_{\text{max}}^{\tau=\alpha_1}-\lvert\hat{\gamma}_{1-2}(\theta,\tau)\rvert_{\text{max}}^{\tau=\alpha_2}\right\rvert$	The difference of the maximum correlation values of the $\tau=\alpha_1$ and $\tau=\alpha_2$ planes.

Ratha et al. proposed a scattering power factorization framework, and polarimetric roll-invariant features were determined using a geodesic distance [39]. The geodesic distance between pairs of 4 × 4 Kennaugh matrices is used to derive the similarity between two targets. The geodesic distance between two Kennaugh matrices $\mathbf{K}_1$ and $\mathbf{K}_2$ is defined:

$$\text{GD}(\mathbf{K}_1,\mathbf{K}_2)=\frac{2}{\pi}\cos^{-1}\frac{\text{Tr}\left(\mathbf{K}_1^{\text{T}}\mathbf{K}_2\right)}{\sqrt{\text{Tr}\left(\mathbf{K}_1^{\text{T}}\mathbf{K}_1\right)}\sqrt{\text{Tr}\left(\mathbf{K}_2^{\text{T}}\mathbf{K}_2\right)}} \tag{3.2.122}$$

The proposed framework, along with the geodesic distance, is used to obtain polarimetric roll-invariant features, including scattering type parameter α_{GD}, helicity parameter τ_{GD}, and purity parameter P_{GD}. Among them, the scattering type parameter α_{GD} is derived through the dissimilarity with respect to the trihedral scattering:

$$\alpha_{\mathrm{GD}} = \frac{\pi}{2}\mathrm{GD}\left(\mathbf{K}, \mathbf{K}_{\text{Trihedral}}\right) \tag{3.2.123}$$

where $\mathbf{K}_{\text{Trihedral}}$ is the Kennaugh matrix of a canonical trihedral:

$$\mathbf{K}_{\text{Trihedral}} = \begin{bmatrix} 1 & 0 & 0 & 0 \\ 0 & 1 & 0 & 0 \\ 0 & 0 & 1 & 0 \\ 0 & 0 & 0 & -1 \end{bmatrix} \tag{3.2.124}$$

The helicity parameter τ_{GD} is obtained by replacing the geodesic distance with the mean of the distances from the left and right helices in the definition of similarity:

$$\tau_{\mathrm{GD}} = \frac{\pi}{4}\left[1 - \sqrt{\mathrm{GD}\left(\mathbf{K}, \mathbf{K}_{\text{Left-Helix}}\right)\mathrm{GD}\left(\mathbf{K}, \mathbf{K}_{\text{Right-Helix}}\right)}\right] \tag{3.2.125}$$

where $\mathbf{K}_{\text{Left-Helix}}$ and $\mathbf{K}_{\text{Right-Helix}}$ are the Kennaugh matrix of the left and right helices:

$$\mathbf{K}_{\text{Left-Helix}} = \begin{bmatrix} 1 & 0 & 0 & -1 \\ 0 & 0 & 0 & 0 \\ 0 & 0 & 0 & 0 \\ -1 & 0 & 0 & 1 \end{bmatrix}, \mathbf{K}_{\text{Right-Helix}} = \begin{bmatrix} 1 & 0 & 0 & 1 \\ 0 & 0 & 0 & 0 \\ 0 & 0 & 0 & 0 \\ 1 & 0 & 0 & 1 \end{bmatrix} \tag{3.2.126}$$

A depolarization index is defined:

$$P_{\mathrm{GD}} = \left[\frac{3}{2}\mathrm{GD}\left(\mathbf{K}, \mathbf{K}_{\mathrm{ID}}\right)\right]^2 \tag{3.2.127}$$

where $\mathbf{K}_{\mathrm{ID}}$ is the ideal depolarization scatterer.

$$\mathbf{K}_{\mathrm{ID}} = \begin{bmatrix} 1 & 0 & 0 & 0 \\ 0 & 0 & 0 & 0 \\ 0 & 0 & 0 & 0 \\ 0 & 0 & 0 & 0 \end{bmatrix} \tag{3.2.128}$$

Other polarimetric roll-invariant features are derived from the non-conventional three-dimension Barakat degree of polarization [37, 38], including the

TABLE 3.2.10
Polarimetric Roll-Invariant Features Derived from Kennaugh Matrix

Feature	Expression	Physical Meaning
α_{GD}	$\frac{\pi}{2}\text{GD}\left(\mathbf{K},\mathbf{K}_{\text{Trihedral}}\right)$	The scattering type.
τ_{GD}	$\frac{\pi}{4}\left[1-\sqrt{\text{GD}\left(\mathbf{K},\mathbf{K}_{\text{Left-Helix}}\right)\text{GD}\left(\mathbf{K},\mathbf{K}_{\text{Right-Helix}}\right)}\right]$	The helicity of a target.
P_{GD}	$\left[\frac{3}{2}\text{GD}\left(\mathbf{K},\mathbf{K}_{\text{ID}}\right)\right]^2$	The depolarization of a target. $P_{\text{GD}}=0$ corresponds to the ideal depolarizer, and $P_{\text{GD}}=1$ corresponds to non-depolarizing targets.
m_{FP}	$m_{\text{FP}}=\sqrt{1-\frac{27\lvert\mathbf{T}\rvert}{\left[\text{Tr}(\mathbf{T})\right]^3}}$	The 3D Barakat degree of polarization.
θ_{FP}	$\tan^{-1}\frac{4m_{\text{FP}}K_{11}K_{44}}{K_{44}^2-\left(1+4m_{\text{FP}}^2\right)K_{11}^2}$	The scattering type.
τ_{FP}	$\tan^{-1}\frac{\lvert K_{14}\rvert}{K_{11}}$	The helicity of a target.

three-dimension Barakat degree of polarization m_{FP}, the scattering-type parameter θ_{FP}, and the target scattering asymmetry τ_{FP}. The three-dimension Barakat degree of polarimetric m_FP can be derived as:

$$m_{\text{FP}}=\sqrt{1-\frac{27\lvert\mathbf{T}\rvert}{\left[\text{Tr}(\mathbf{T})\right]^3}} \tag{3.2.129}$$

The scattering type parameter θ_{FP} can be derived as:

$$\theta_{\text{FP}}=\tan^{-1}\frac{4m_{\text{FP}}K_{11}K_{44}}{K_{44}^2-(1+4m_{\text{FP}}^2)K_{11}^2},\ \theta_{\text{FP}}\in\left[-\frac{\pi}{4},\frac{\pi}{4}\right] \tag{3.2.130}$$

A scattering asymmetry parameter τ_{FP} can be derived as:

$$\tau_{\text{FP}}=\tan^{-1}\frac{\lvert K_{14}\rvert}{K_{11}},\ \tau_{\text{FP}}\in\left[0,\frac{\pi}{4}\right] \tag{3.2.131}$$

where K_{11},K_{14},K_{44} are elements of Kennaugh matrix. The expressions and the physical meanings are shown in Table 3.2.10.

3.3 DEMONSTRATIONS AND COMPARISONS

Polarimetric roll-invariant features have been successfully applied in many fields. PolSAR ship detection is adopted as the application scenario. In general, features with higher target-to-clutter ratio (TCR) are expected to achieve better target detection performance. Therefore, this section focuses on the TCR index examinations of these polarimetric roll-invariant features, which is a basis for the PolSAR ship detection application in Chapter 5.

3.3.1 PolSAR Data Description

In this study, two spaceborne PolSAR datasets from GaoFen-3 and Radarsat-2 are used. The data information, including imaging area, acquisition date, data size, and resolution, is shown in Table 3.3.1. The Pauli RGB images are shown in Figures 3.3.1–3.3.2.

3.3.2 TCR Experiments

Five ship regions of interest (ROIs) and five sea clutter ROIs with the size of 51×51 are randomly selected from each PolSAR data. These ship and sea clutter ROIs are indicated by green and yellow boxes shown in Figures 3.3.1–3.3.2, respectively. The TCR index is

$$\text{TCR}=10\log_{10}\left\{\frac{\text{ship}_{\text{mean}}^{\text{All}}}{\text{sea}_{\text{mean}}^{\text{All}}}\right\} \tag{3.3.1}$$

TABLE 3.3.1
PolSAR Data Information

Data	Imaging Area	Acquisition Date	Size (pixels)	Resolution (range × azimuth, m)
GaoFen-3 data	Hong Kong, China	15 March 2017	3450 × 2150	8 × 8
Radarsat-2 data	Hong Kong, China	16 December 2008	500 × 300	4.82 × 4.73

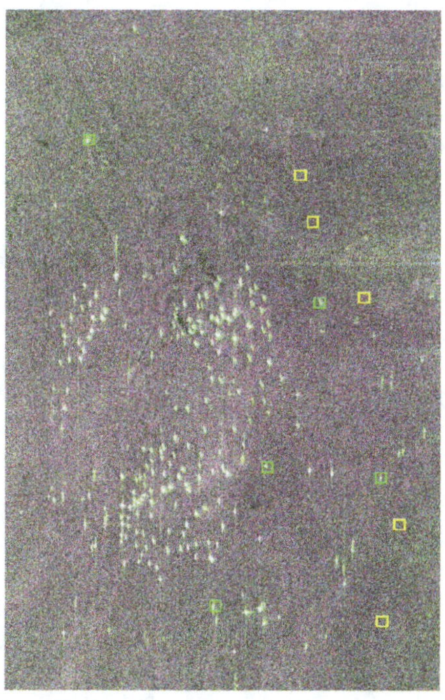

FIGURE 3.3.1 Pauli image of GaoFen-3 data.

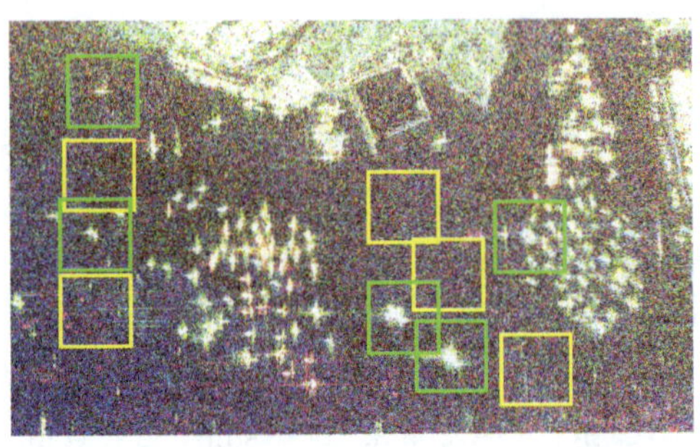

FIGURE 3.3.2 Pauli image of Radarsat-2 data.

where $\text{ship}_{\text{mean}}^{\text{All}}$ and $\text{sea}_{\text{mean}}^{\text{All}}$ indicate the averaged feature values within each ROI for ship and sea pixels, respectively.

This section focuses on the analysis of the representative polarimetric roll-invariant features summarized in Tables 3.2.1–3.2.10. Considering that the Chapter 5 will investigate the characteristics of the three-dimension polarimetric correlation pattern features in detail, this section will omit them. Therefore, in this section, 71 independent polarimetric roll-invariant features are considered for demonstration. In detail, there are 12 independent polarimetric roll-invariant features, $SPAN$, $\lambda_i(\mathbf{S})$, $\|\mathbf{S}\|_2$, $\|\mathbf{S}\|_F$, ν, γ, D, θ_d, α_d, k_S, k_D, and k_H, derived from the Sinclair matrix; 3 independent polarimetric roll-invariant features, $\text{Tr}(\mathbf{G})$, $\|\mathbf{G}\|_2$, and $\|\mathbf{G}\|_F$, derived from the Graves matrix; 8 independent polarimetric roll-invariant features, T_{11}, $T_{22}+T_{33}$, $\text{Im}(T_{23})$, $\text{Re}^2[T_{12}]+\text{Re}^2[T_{13}]$, $\text{Im}^2[T_{12}]+\text{Im}^2[T_{13}]$, $\left(|T_{13}|^2-|T_{12}|^2\right)^2+4\,\text{Re}^2\left[T_{12}T_{13}^*\right]$, $[T_{33}-T_{22}]^2+4\,\text{Re}^2[T_{23}]$, and $\text{Im}\left[T_{12}T_{13}^*\right]$, derived from the polarimetric coherency matrix; 18 independent polarimetric roll-invariant features, $\|\mathbf{T}\|_2$, $\|\mathbf{T}\|_F$, H, $\bar{\alpha}$, A, τ_m, m_T, α_s, Φ_{α_s}, α_C, β_C, Υ, ψ, Hel, τ_p, α_p, φ_p, and m_p, derived from the eigenvalue-eigenvector-based decomposition; 6 independent polarimetric roll-invariant features, α_{GD}, τ_{GD}, P_{GD}, m_{FP}, θ_{FP}, and τ_{FP}, derived from the Kennaugh matrix; and 24 independent polarimetric roll-invariant features, $\left|\hat{\gamma}_{\text{HH-VV}}(\theta)\right|_{\text{std}}$, $\left|\hat{\gamma}_{\text{HH-VV}}(\theta)\right|_{\text{constrast}}$, $\left|\hat{\gamma}_{\text{HH-VV}}(\theta)\right|_{\text{A}}$, $\left|\hat{\gamma}_{\text{HH-VV}}(\theta)\right|_{\text{mean}}$, $\left|\hat{\gamma}_{\text{HH-VV}}(\theta)\right|_{\text{max}}$, $\left|\hat{\gamma}_{\text{HH-VV}}(\theta)\right|_{\text{min}}$, $\left|\hat{\gamma}_{\text{HH-HV}}(\theta)\right|_{\text{std}}$, $\left|\hat{\gamma}_{\text{HH-HV}}(\theta)\right|_{\text{constrast}}$, $\left|\hat{\gamma}_{\text{HH-HV}}(\theta)\right|_{\text{A}}$, $\left|\hat{\gamma}_{\text{HH-HV}}(\theta)\right|_{\text{mean}}$, $\left|\hat{\gamma}_{\text{HH-HV}}(\theta)\right|_{\text{max}}$, $\left|\hat{\gamma}_{\text{HH-HV}}(\theta)\right|_{\text{min}}$, $\left|\hat{\gamma}_{\text{(HH+VV)-(HH-VV)}}(\theta)\right|_{\text{std}}$, $\left|\hat{\gamma}_{\text{(HH+VV)-(HH-VV)}}(\theta)\right|_{\text{std}}$, $\left|\hat{\gamma}_{\text{(HH+VV)-(HH-VV)}}(\theta)\right|_{\text{A}}$, $\left|\hat{\gamma}_{\text{(HH+VV)-(HH-VV)}}(\theta)\right|_{\text{mean}}$, $\left|\hat{\gamma}_{\text{(HH+VV)-(HH-VV)}}(\theta)\right|_{\text{max}}$, $\left|\hat{\gamma}_{\text{(HH+VV)-(HH-VV)}}(\theta)\right|_{\text{min}}$, $\left|\hat{\gamma}_{\text{(HH-VV)-HV}}(\theta)\right|_{\text{std}}$, $\left|\hat{\gamma}_{\text{(HH-VV)-HV}}(\theta)\right|_{\text{constrast}}$, $\left|\hat{\gamma}_{\text{(HH-VV)-HV}}(\theta)\right|_{\text{A}}$, $\left|\hat{\gamma}_{\text{(HH-VV)-HV}}(\theta)\right|_{\text{mean}}$, $\left|\hat{\gamma}_{\text{(HH-VV)-HV}}(\theta)\right|_{\text{max}}$, and $\left|\hat{\gamma}_{\text{(HH-VV)-HV}}(\theta)\right|_{\text{min}}$, derived from the two-dimension polarimetric correlation pattern.

For GaoFen-3 and Radarsat-2 PolSAR datasets, the TCR comparison results of these polarimetric roll-invariant features are shown in Figures 3.3.3 (a–b), respectively. It can be seen that for these datasets, the polarimetric roll-invariant features derived from the Graves matrix, the polarimetric coherency matrix, and the two-dimension polarimetric correlation patterns generally have higher TCR values. At the same time, the polarimetric roll-invariant feature with the highest TCR value from the two datasets comes from the

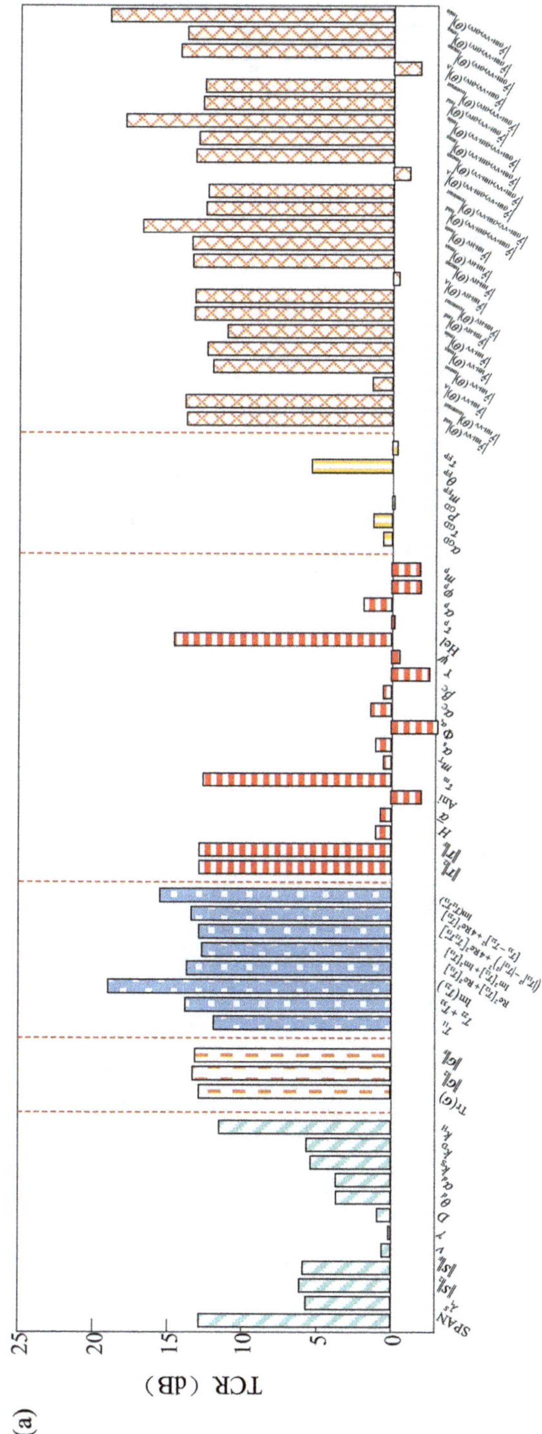

FIGURE 3.3.3 TCR results of polarimetric roll-invariant features. (a) GaoFen-3 data, (b) Radarsat-2 data.

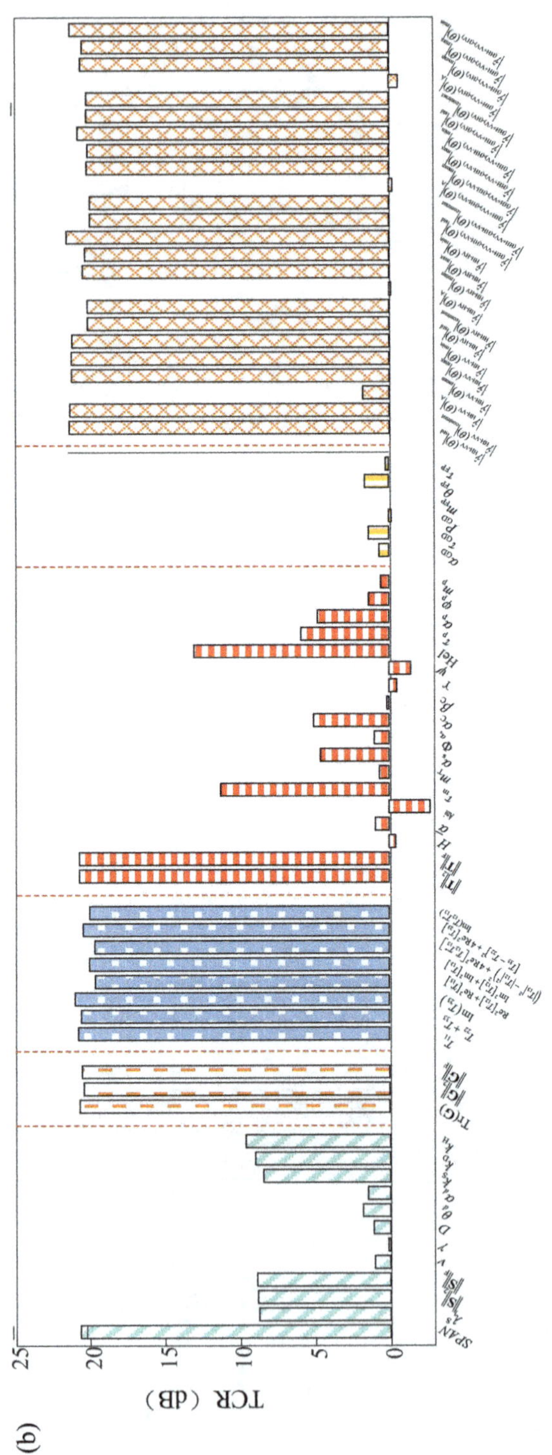

FIGURE 3.3.3 (Continued).

two-dimension polarimetric correlation pattern interpretation tool. For GaoFen-3 data, the TCR value of the $\left|\hat{\gamma}_{\text{(HH+VV)-HV}}(\theta)\right|_{\min}$ feature is the highest, 19.72 dB. For Radarsat-2 data, the TCR value of the $\left|\hat{\gamma}_{\text{HH-HV}}(\theta)\right|_{\min}$ feature is the highest, and it reaches 21.48 dB.

3.4 CONCLUSION

Polarimetric roll-invariant features are important in the field of radar polarimetry. This chapter systematically summarizes the polarimetric roll-invariant features that have been publicly reported in radar polarimetry and divides them into seven categories according to their derivation fashion. On this basis, the analytical expressions, internal relationships, and physical meanings of these polarimetric roll-invariant features are discussed and summarized. Finally, based on the contrast enhancement between ship and sea areas, performance comparisons and validations of these polarimetric roll-invariant features are carried out with two spaceborne PolSAR datasets, which will be further studied in Chapter 5.

REFERENCES

1. J. S. Lee and E. Pottier, *Polarimetric Radar Imaging: From Basics to Applications*. Boca Raton, FL: CRC Press, 2009.
2. X. S. Wang and S. W. Chen, "Polarimetric synthetic aperture radar interpretation and recognition: Advances and perspectives," *Journal of Radars,* vol. 9, no. 2, pp. 259–276, 2020.
3. G. Sinclair, "The transmission and reception of elliptically polarized waves," *Proceedings of the IRE,* vol. 38, no. 2, pp. 148–151, 1950.
4. S. R. Cloude, *Polarisation: Application in Remote Sensing*. New York: Oxford University Press, 2009.
5. S. W. Chen, X. S. Wang, S. P. Xiao, and M. Sato, *Target Scattering Mechanism in Polarimetric Synthetic Aperture Radar-Interpretation and Application*. Singapore: Springer, 2018.
6. W. M. Boerner, *Historical Development of Radar Polarimetry, Incentives for this Workshop, and Overview of Contributions to these Proceedings*. Dordrecht: Springer, 1992.
7. W. M. Boerner, H. Mott, E. Lüneburg, C. Livingston, B. Brisco, R. J. Brown, et al., *Polarimetry in Radar Remote Sensing: Basic and Applied Concepts (Chapter 5)*. New York: John Willy & Sons, 1998.
8. W. M. Boerner, "Recent advances in extra-wide-band polarimetry, interferometry and polarimetric interferometry in synthetic aperture remote sensing and its applications," *IEE Proceedings-Radar Sonar and Navigation,* vol. 150, no. 3, pp. 113–124, 2003.
9. A. Guissard, "Mueller and Kennaugh matrices in radar polarimetry," *IEEE Transactions on Geoscience and Remote Sensing,* vol. 32, no. 3, pp. 590–597, 1994.
10. C. D. Graves, "Radar polarization power scattering matrix," *Proceedings of the IRE,* vol. 44, no. 2, pp. 248–252, 1956.
11. E. M. Kennaugh, "Polarization properties of radar reflection," Ph.D. dissertation, Ohio State University, USA, 1952.
12. A. B. Kostinski and W. M. Boerner, "On foundations of radar polarimetry," *IEEE Transactions on Antennas and Propagation,* vol. 34, no. 12, pp. 1395–1404, 1986.
13. W. M. Boerner, W. L. Yan, A. Q. Xi, and Y. Yamaguchi, "On the basic principles of radar polarimetry—the target characteristic polarization state theory of Kennaugh, Huynen's polarization fork concept, and its extension to the partially polarized case," *Proceedings of the IEEE,* vol. 79, no. 10, pp. 1538–1550, 1991.
14. J. R. Huynen, "Phenomenological theory of radar targets," Ph.D. dissertation, Technical University of Delft, The Netherlands, 1970.

15. S. W. Chen, Y. Z. Li, X. S. Wang, S. P. Xiao, and M. Sato, "Modeling and interpretation of scattering mechanisms in polarimetric synthetic aperture radar: Advances and perspectives," *IEEE Signal Processing Magazine,* vol. 31, no. 4, pp. 79–89, 2014.
16. S. R. Cloude and E. Pottier, "A review of target decomposition theorems in radar polarimetry," *IEEE Transactions on Geoscience and Remote Sensing,* vol. 34, no. 2, pp. 498–518, 1996.
17. S. H. Bickel, "Some invariant properties of the polarization scattering matrix," *Proceedings of the IEEE,* vol. 53, no. 8, pp. 1070–1072, 1965.
18. J. R. Huynen, "Measurement of the target scattering matrix," *Proceedings of the IEEE,* vol. 53, no. 8, pp. 936–946, 1965.
19. V. Karnychev, V. A. Khlusov, L. P. Ligthart, and G. Sharygin, "Algorithms for estimating the complete group of polarization invariants of the scattering matrix (SM) based on measuring all SM elements," *IEEE Transactions on Geoscience and Remote Sensing,* vol. 42, no. 3, pp. 529–539, 2004.
20. C. S. Tao, S. W. Chen, Y. Z. Li, and S. P. Xiao, "Polarimetric SAR terrain classification using polarimetric features derived from rotation domain," *Journal of Radars,* vol. 6, no. 5, pp. 524–532, 2017.
21. C. S. Tao, S. W. Chen, Y. Z. Li, and S. P. Xiao, "PolSAR land cover classification based on roll-invariant and selected hidden polarimetric features in the rotation domain," *Remote Sensing,* vol. 9, no. 7, p. 660, 2017.
22. S. W. Chen and C. S. Tao, "PolSAR image classification using polarimetric-feature–driven deep convolutional neural network," *IEEE Geoscience and Remote Sensing Letters,* vol. 15, no. 4, pp. 627–631, 2018.
23. S. R. Cloude and E. Pottier, "An entropy based classification scheme for land applications of polarimetric SAR," *IEEE Transactions on Geoscience and Remote Sensing,* vol. 35, no. 1, pp. 68–78, 1997.
24. L. F. Famil, E. Pottier, and J. S. Lee, "Unsupervised classification of multifrequency and fully polarimetric SAR images based on the H/A/Alpha-Wishart classifier," *IEEE Transactions on Geoscience and Remote Sensing,* vol. 39, no. 11, pp. 2332–2342, 2001.
25. J. S. Lee, M. R. Grunes, T. L. Ainsworth, D. Li-Jen, D. L. Schuler, and S. R. Cloude, "Unsupervised classification using polarimetric decomposition and the complex Wishart classifier," *IEEE Transactions on Geoscience and Remote Sensing,* vol. 37, no. 5, pp. 2249–2258, 1999.
26. Y. Yamaguchi, Y. Yamamoto, H. Yamada, J. Yang, and W. M. Boerner, "Classification of terrain by implementing the correlation coefficient in the circular polarization basis using X-band POLSAR data," *IEICE Transactions on Communications,* vol. 9, no. 1, pp. 297–301, 2008.
27. T. L. Ainsworth, D. L. Schuler, and J. S. Lee, "Polarimetric SAR characterization of man-made structures in urban areas using normalized circular-pol correlation coefficients," *Remote Sensing of Environment,* vol. 112, no. 6, pp. 2876–2885, 2008.
28. X. C. Cui, Y. Su, and S. W. Chen, "Polarimetric SAR ship detection based on polarimetric rotation domain features and superpixel technique," *Journal of Radars,* vol. 10, no. 1, pp. 35–48, 2021.
29. X. C. Cui, C. S. Tao, Y. Su, and S. W. Chen, "PolSAR ship detection based on polarimetric correlation pattern," *IEEE Geoscience and Remote Sensing Letters,* vol. 18, no. 3, pp. 471–475, 2021.
30. H. L. Li, X. C. Cui, and S. W. Chen, "PolSAR ship detection with optimal polarimetric rotation domain features and SVM," *Remote Sensing,* vol. 13, no. 19, p. 3932, 2021.
31. G. Q. Wu, S. W. Chen, Y. Li, and X. S. Wang, "Null-pol response pattern in polarimetric rotation domain: Characterization and application," *IEEE Geoscience and Remote Sensing Letters*, pp. 1–5, 2021.
32. H. L. Li, M. D. Li, X. C. Cui, and S. W. Chen, "Man-made target structure recognition with polarimetric correlation pattern and roll-invariant feature coding," *IEEE Geoscience and Remote Sensing Letters,* vol. 19, pp. 1–5, 2022.

33. R. Paladini, M. Martorella, and F. Berizzi, "Classification of man-made targets via invariant coherency-matrix eigenvector decomposition of polarimetric SAR/ISAR images," *IEEE Transactions on Geoscience and Remote Sensing,* vol. 49, no. 8, pp. 3022–3034, 2011.
34. S. W. Chen, X. S. Wang, and S. P. Xiao, "Urban damage level mapping based on co-polarization coherence pattern using multi-temporal polarimetric SAR data," *IEEE Journal of Selected Topics in Applied Earth Observations and Remote Sensing,* vol. 11, no. 8, pp. 2657–2667, 2018.
35. M. Sato, S. Chen, and M. Satake, "Polarimetric SAR analysis of tsunami damage following the March 11, 2011 East Japan earthquake," *Proceedings of the IEEE,* vol. 100, no. 10, pp. 2861–2875, 2012.
36. M. D. Li, S. P. Xiao, and S. W. Chen, "Three-dimension polarimetric correlation pattern interpretation tool and its application," *IEEE Transactions on Geoscience and Remote Sensing,* vol. 60, pp. 1–16, 2022.
37. S. Dey, A. Bhattacharya, A. C. Frery, C. López-Martínez, and Y. S. Rao, "A model-free four component scattering power decomposition for polarimetric SAR data," *IEEE Journal of Selected Topics in Applied Earth Observations and Remote Sensing,* vol. 14, pp. 3887–3902, 2021.
38. S. Dey, A. Bhattacharya, D. Ratha, D. Mandal, and A. C. Frery, "Target characterization and scattering power decomposition for full and compact polarimetric SAR data," *IEEE Transactions on Geoscience and Remote Sensing,* vol. 59, no. 5, pp. 3981–3998, 2020.
39. D. Ratha, E. Pottier, A. Bhattacharya, and A. C. Frery, "A PolSAR scattering power factorization framework and novel roll-invariant parameter-based unsupervised classification scheme using a geodesic distance," *IEEE Transactions on Geoscience and Remote Sensing*, vol. 58, no. 5, pp. 3509–3525, 2020.
40. S. W. Chen, G. Q. Wu, D. H. Dai, X. S. Wang, and S. P. Xiao, "Roll-invariant features in radar polarimetry: A survey," in *IEEE International Geoscience and Remote Sensing Symposium*, Yokohama, Japan, 2019, 28 July–02 August, pp. 5015–5018.
41. S. W. Chen, M. D. Li, X. C. Cui, and H. L. Li, "Polarimetric roll-invariant features and application for PolSAR ship detection," *IEEE Geoscience and Remote Sensing Magazine*, vol. 12, no. 1, pp. 36–66, 2024.
42. E. Krogager, "New decomposition of the radar target scattering matrix," *Electronics Letters,* vol. 26, no. 18, pp. 1525–1527, 1990.
43. J. R. Huynen, "The Stokes matrix parameters and their interpretation in terms of physical target properties," in *Proceedings of the SPIE*, York, 1990, 15–18 April.
44. E. Pottier, "Dr. J. R. Huynen's main contributions in the development of polarimetric radar techniques and howthe 'radar targets phenomenological concept' becomes a theory," *Proceedings of the SPIE*, vol. 1748, pp. 72–85, 1993.
45. S. W. Chen, X. S. Wang, and M. Sato, "Uniform polarimetric matrix rotation theory and its applications," *IEEE Transactions on Geoscience and Remote Sensing,* vol. 52, no. 8, pp. 4756–4770, 2014.
46. R. Touzi, "Target scattering decomposition in terms of roll-invariant target parameters," *IEEE Transactions on Geoscience and Remote Sensing,* vol. 45, no. 1, pp. 73–84, 2007.
47. R. Paladini, L. F. Famil, E. Pottier, M. Martorella, F. Berizzi, and E. Dalle Mese, "Lossless and sufficient Ψ-invariant decomposition of random reciprocal target," *IEEE Transactions on Geoscience and Remote Sensing,* vol. 50, no. 9, pp. 3487–3501, 2012.
48. H. Aghababaee and M. R. Sahebi, "Incoherent target scattering decomposition of polarimetric SAR data based on vector model roll-invariant parameters," *IEEE Transactions on Geoscience and Remote Sensing,* vol. 54, no. 8, pp. 4392–4401, 2016.
49. S. W. Chen, "Polarimetric coherence pattern: A visualization and characterization tool for PolSAR data investigation," *IEEE Transactions on Geoscience and Remote Sensing,* vol. 56, no. 1, pp. 286–297, 2018.

4 Polarimetric Rotation Domain Land Cover Classification

4.1 INTRODUCTION

PolSAR, which can obtain fully polarimetric information, has become more and more important in microwave remote sensing [1–6]. Plenty of successful applications have been achieved with PolSAR data. Among them, land cover classification is an important application of PolSAR and has been widely studied [7–15]. Generally, there are two main steps for PolSAR land cover classification. The first is feature extraction and representation, while the second is classifier design and optimization. For feature extraction and representation, scattering mechanism modeling and interpretation techniques mainly with polarimetric target decomposition are commonly adopted [1, 4, 5, 16–19]. The derived polarimetric roll-invariant features of polarimetric entropy H, mean alpha angle $\bar{\alpha}$, and polarimetric anisotropy Ani from eigenvalue-eigenvector-based decomposition have been utilized for unsupervised and supervised classification [8]. Meanwhile, model-based decompositions and the derived polarimetric features for each basic scattering mechanism have also been explored for PolSAR land cover classification [1, 5, 9]. In addition, texture features, contexture features, and statistic features have also been involved to enhance PolSAR land cover classification performance. For classifier designing and optimization, various classifiers such as the Wishart classifier [7, 20], maximum likelihood classifier [11], support vector machine (SVM) classifier [12], and decision tree (DT) classifier [12], neural network classifier [13, 14] have been adopted for PolSAR land cover classification. Especially with novel polarimetric features and advanced classifiers, PolSAR land cover classification performances have been enhanced.

More recently, deep convolutional neural network (CNN) models [21], as a kind of deep learning technique, have achieved great successes in many image processing domains and have also been widely applied for SAR image target detection and classification fields. Compared with conventional classifiers, deep CNN classifiers have great potential to further enhance PolSAR classification performance. Meanwhile, as a kind of end-to-end technique, deep CNN classifiers have the advantage of automatically extracting useful features and producing final classification results. The point is that a well-trained deep CNN model needs a huge amount of various labeled samples. However, such large-scale ground-truth datasets are not always available or are very costly for the PolSAR field. Besides, PolSAR images are essentially different from optical images. PolSAR data acquisition and imaging include electromagnetic

DOI: 10.1201/9781003461296-4

wave radiation and propagation, target modulation and backscattering, polarimetric and radiometric calibration, imaging processing, and so on [16]. In this vein, direct implementation of a deep CNN classifier pre-trained by optical images may not be suitable for PolSAR images. As a result, how to properly combine the target electromagnetic scattering interpretation knowledge and deep CNN models is an important scientific problem.

In addition, target scattering diversity makes PolSAR land cover classification difficult [4, 22–24]. Polarimetric roll-invariant features, which are independent of the target scattering diversity, are preferred for PolSAR land cover classification. However, target scattering diversity also contains rich information. It is expected to further improve PolSAR land cover classification performance with polarimetric rotation domain interpretation theory via mining the target scattering diversity information and selecting highly discriminable polarimetric rotation domain features. In this vein, based on uniform polarimetric matrix rotation theory [22] and the two-dimension polarimetric coherence pattern interpretation tool [25] introduced in Chapter 2, this chapter mainly presents the polarimetric rotation domain feature selection scheme and the classifier designs, respectively, with classical machine learning and deep learning classifiers for PolSAR land cover classification.

4.2 POLARIMETRIC ROTATION DOMAIN CLASSIFICATION WITH CLASSICAL MACHINE LEARNING

Land cover classification is an important application for PolSAR data investigation. The polarimetric roll-invariant features of the total backscattering power $SPAN$, polarimetric entropy H, mean alpha angle $\bar{\alpha}$, and polarimetric anisotropy Ani have been popularly adopted for PolSAR land cover classification [1, 8, 12, 20]. Normally, these traditional polarimetric roll-invariant features cannot fully represent the scattering properties of various land covers, especially the target scattering diversity. In this vein, the selection of suitable polarimetric rotation domain features and the further combination with representative polarimetric roll-invariant features have the capability to boost the PolSAR land cover classification accuracy with classical machine learning classifiers.

4.2.1 POLARIMETRIC FEATURE SELECTION SCHEME

In Chapter 2, polarimetric rotation domain interpretation theory and a series of new polarimetric features are introduced. This section focuses on 11 independent polarimetric features derived from uniform polarimetric matrix rotation theory (shown in Table 2.3.1) and 40 polarimetric features derived from four independent two-dimension polarimetric coherence patterns (shown in Table 2.4.1). For these 51 polarimetric rotation domain features, a feature selection scheme is established by comprehensively utilizing the within-class distance, class separation distance, and class center distance criteria. The polarimetric feature selection flowchart is shown in Figure 4.2.1. It includes five main steps, as follows.

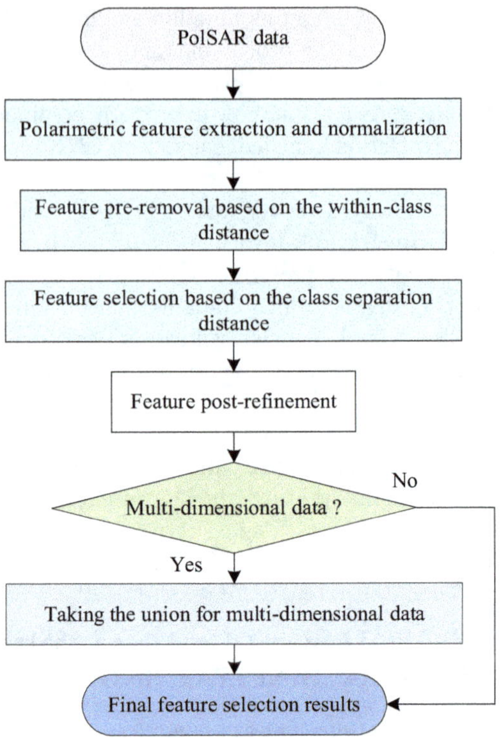

FIGURE 4.2.1 Flowchart of the polarimetric feature selection scheme.

1) Polarimetric feature extraction and normalization. Based on the speckle-filtered PolSAR data, independent polarimetric rotation domain features are extracted and normalized to the range of $[0,1]$. The total normalized feature set is $\mathbf{F}^{\text{all}} = \{f_i, i = 1,\ldots,I\}$, where I is the number of polarimetric rotation domain features.
2) Feature pre-removal based on the within-class distance. The within-class distance is a measure of the dispersal degree of samples within the same class. From the ground-truth map, there are X known land covers C_x, $x = 1,\ldots,X$. For feature f_i, the samples from each land cover C_x can be represented as $\mathbf{C}_x^i = \left\{f_i^{(x,k)}, k = 1,2,\ldots,N_x\right\}$, where $f_i^{(x,k)}$ is the feature value of the k th sample of land cover C_x and N_x is the total sample number of land cover C_x. The within-class distance of land cover C_x in terms of feature f_i is $d_{\text{Inner}}\left(\mathbf{C}_x^i\right)$:

$$d_{\text{Inner}}\left(\mathbf{C}_x^i\right) = \sqrt{\frac{1}{N_x}\sum_{k=1}^{N_x}\left(f_i^{(x,k)} - c_x^i\right)^2} \tag{4.2.1}$$

where the within-class distance of land cover C_x is $d_{\text{Inner}}\left(\mathbf{C}_x^i\right)$, $i = 1,\ldots,I$ represents polarimetric feature class, and c_x is the center of land cover C_x. With feature f_i, the center of land cover C_x is defined as

$$c_x^i = \frac{1}{N_x} \sum_{k=1}^{N_x} f_i^{(x,k)} \tag{4.2.2}$$

For each land cover, M polarimetric features with the largest within-class distance are selected. $M = 3$ is adopted in this section. For all land covers, the polarimetric features with large within-class distances can form a set $\mathbf{F}^{\text{Inner}}$. For a certain land cover, polarimetric features in $\mathbf{F}^{\text{Inner}}$ usually have significant differences for different samples within a class, which is not conducive for land cover classification. Therefore, $\mathbf{F}^{\text{Inner}}$ can be removed from the polarimetric feature set $\mathbf{F}^{\text{All}}$ and the pre-screened polarimetric feature set $\mathbf{F}^{\text{Pre}}$ can be obtained:

$$\mathbf{F}^{\text{Pre}} = \mathbf{F}^{\text{All}} - \mathbf{F}^{\text{Inner}} \tag{4.2.3}$$

where $\mathbf{F}^{\text{Pre}} = \left\{ \tilde{f}_j, j = 1,\ldots,J \right\}$, J is the feature number of $\mathbf{F}^{\text{Pre}}$, and $J < I$.

3) Feature selection based on the class separation distance. The class separation distance is defined as the summation of the distance between two classes and the distance between two class centers. For all classes, there are Y land cover pairs demoted as P_y, $y = 1,\ldots,Y$. For each land cover pair P_y, the containing two classes are denoted as C_{y_1} and C_{y_2}, where y_1 and y_2 are corresponding classes labels, respectively. For polarimetric feature $\tilde{f}_j$ from the pre-screened polarimetric feature set $\mathbf{F}^{\text{Pre}}$, the samples from land covers C_{y_1} and C_{y_2} are respectively represented as $\mathbf{C}_{y_1}^j = \left\{ \tilde{f}_j^{(y_1,l)}, l = 1,2,\ldots,N_{y_1} \right\}$ and $\mathbf{C}_{y_2}^j = \left\{ \tilde{f}_j^{(y_2,m)}, m = 1,2,\ldots,N_{y_2} \right\}$. For land cover pair P_y and polarimetric feature $\tilde{f}_j$, the distance $d_{\text{s}}\left(\text{C}_{y_1}^j, \text{C}_{y_2}^j\right)$ between two classes and the distance $d_{\text{c}}\left(c_{y_1}^j, c_{y_2}^j\right)$ between two class centers are as follows

$$d_{\text{s}}\left(\text{C}_{y_1}^j, \text{C}_{y_2}^j\right) = \sqrt{\frac{1}{N_{y_1} N_{y_2}} \sum_{l=1}^{N_{y_1}} \sum_{m=1}^{N_{y_2}} \left(\tilde{f}_j^{(y_1,l)} - \tilde{f}_j^{(y_2,m)} \right)^2} \tag{4.2.4}$$

$$d_c\left(c_{y_1}^j, c_{y_2}^j\right) = \left| \frac{1}{N_{y_1}} \sum_{l=1}^{N_{y_1}} \tilde{f}_j^{(y_1,l)} - \frac{1}{N_{y_2}} \sum_{m=1}^{N_{y_2}} \tilde{f}_j^{(y_2,m)} \right| \tag{4.2.5}$$

For land cover pair P_y and polarimetric feature $\tilde{f}_j$, both $d_{\text{s}}\left(\text{C}_{y_1}^j, \text{C}_{y_2}^j\right)$ and $d_{\text{c}}\left(c_{y_1}^j, c_{y_2}^j\right)$ are used to measure the separability of land cover pair. In this vein, the class separation distance $d_{\text{Between}}\left(\text{P}_y^j\right)$ of land cover pair P_y is obtained as

$$d_{\text{Between}}\left(\text{P}_y^j\right) = d_{\text{s}}\left(\text{C}_{y_1}^j, \text{C}_{y_2}^j\right) + d_{\text{c}}\left(c_{y_1}^j, c_{y_2}^j\right) \tag{4.2.6}$$

where the superscript j represents the class of polarimetric features in the pre-screened polarimetric feature set $\mathbf{F}^{\text{Pre}}$, and $j = 1,\ldots,J$.

For land cover pair P_y, the polarimetric feature with the largest class separation distance is selected as the initially selected polarimetric feature f_y^{Ini}:

$$f_y^{\mathrm{Ini}} = \arg\max_{\tilde{f}_j \in \mathbf{F}^{\mathrm{Pre}}} \left\{ d_{\mathrm{Inter}}\left(\mathrm{P}_y^j\right) \right\} \tag{4.2.7}$$

Traversing all land cover pairs, the initially selected polarimetric feature set $\mathbf{F}^{\mathrm{Ini}}$ can be obtained:

$$\mathbf{F}^{\mathrm{Ini}} = \left\{ f_1^{\mathrm{Ini}}, f_2^{\mathrm{Ini}}, \cdots f_Y^{\mathrm{Ini}} \right\} \tag{4.2.8}$$

Note that the same polarimetric features in the initially selected polarimetric feature set $\mathbf{F}^{\mathrm{Ini}}$ can be merged at first. Then, Y is the number of different polarimetric features in the initially selected polarimetric feature set $\mathbf{F}^{\mathrm{Ini}}$.

4) Feature post-refinement. After the preliminary selection, feature post-refinement is implemented to further determine polarimetric features with relatively high discriminant performance in $\mathbf{F}^{\mathrm{Ini}}$ according to the class separation distances. For each land cover pair, the polarimetric features which lead to the maximum class separation distances are recorded and counted. Polarimetric features with selected frequency higher than a predefined threshold r are collected. In this section, $r = 3$ is adopted. Then, the refined polarimetric feature set $\mathbf{F}^{\mathrm{Refined}}$ can be formed.
5) Final selected features. For single PolSAR data, the refined polarimetric feature set $\mathbf{F}^{\mathrm{Refined}}$ will be the final polarimetric feature set $\mathbf{F}^{\mathrm{Final}}$. For multiple PolSAR data (e.g. multi-temporal, multi-frequency), the union of all the refined polarimetric feature sets will be the final polarimetric feature set $\mathbf{F}^{\mathrm{Final}}$.

4.2.2 Classification Methodology Development

The core idea of the polarimetric rotation domain land cover classification method [12, 24] is to combine the traditional polarimetric roll-invariant features and the selected polarimetric rotation domain features as the input of a classifier to realize PolSAR land cover classification. This section adopts the traditional SVM and DT classifiers [26, 27]. The flowchart of the polarimetric rotation domain classification method is shown in Figure 4.2.2. First, speckle filtering is conducted for the input PolSAR data. The SimiTest speckle filtering method [28] is adopted in this section. Then, the classic polarimetric roll-invariant features, total backscattering power $SPAN$, polarimetric entropy H, mean alpha angle $\bar{\alpha}$, and polarimetric anisotropy Ani, are extracted. Meanwhile, the polarimetric rotation domain features are extracted using the uniform polarimetric matrix rotation theory [22] and the polarimetric coherence pattern interpretation tool [25]. The selected polarimetric rotation domain feature set can be obtained with the polarimetric feature selection method introduced in Section 4.2.1. Each of these polarimetric roll-invariant features and

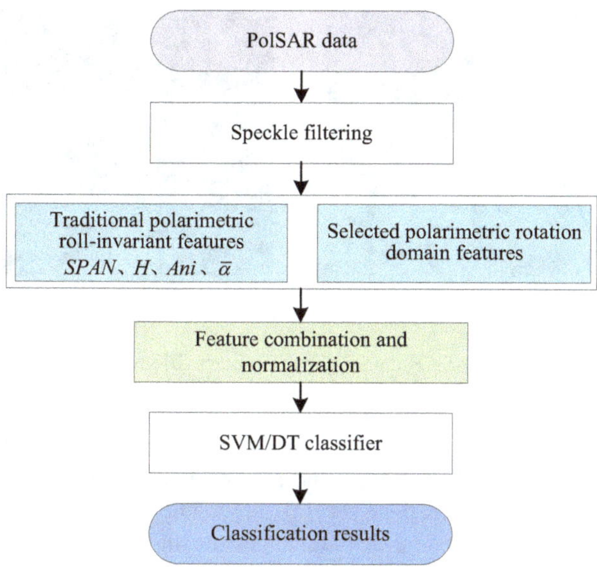

FIGURE 4.2.2 Flowchart of polarimetric rotation domain classification method.

polarimetric rotation domain features is normalized to the range of $[0,1]$. These normalized polarimetric features will be the classifier inputs. Finally, with the ground-truth data, both the SVM and DT classifiers can be trained and validated. Thereafter, the final PolSAR land cover classification results can be obtained.

4.2.3 Classification Comparison Studies

4.2.3.1 Data Description

In this section, experimental studies are carried out with L-band PolSAR datasets collected by the NASA/JPL AIRSAR and UAVSAR systems. All these PolSAR datasets are speckle-filtered by the SimiTest approach [28]. The AIRSAR PolSAR data mainly covers the Flevoland, the Netherlands. The range and azimuth resolutions are 6.6 m and 12.1 m, respectively. This study area contains various land covers, including water, rapeseed, grasses, bare soil, potatoes, beets, wheat, lucerne, forest, peas, and stembeans. The PolSAR Pauli image and the ground-truth map are shown in Figure 4.2.3. The UAVSAR collected multi-temporal PolSAR datasets over Manitoba, Canada [29]. Four multi-temporal datasets are adopted in this comparison. The acquisition dates are 17 June, 22 June, 5 July, and 17 July in 2012. The corresponding days of the year (DoYs) are 169, 174, 187, and 199. The range and azimuth resolutions are 5 m and 7 m, respectively. Pauli images of these four multi-temporal UAVSAR datasets are shown in Figure 4.2.4 (a)–(d). This study area mainly contains land covers of oats, rapeseed, wheat, corn, soybeans, forage crops, and broadleaf. The

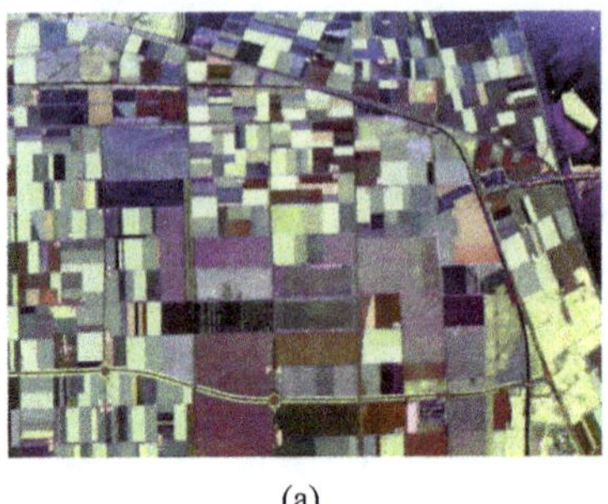

(a)

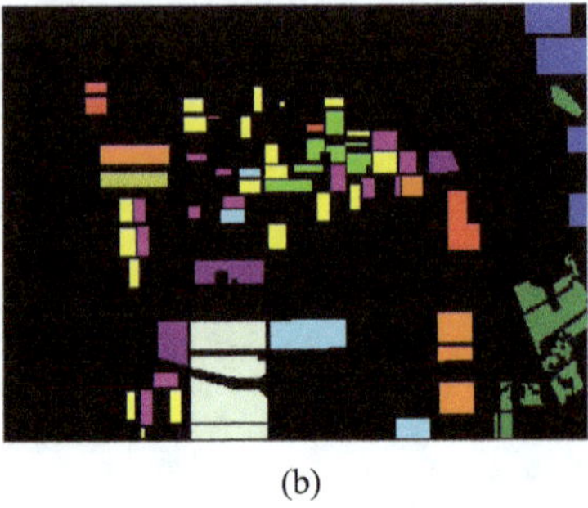

(b)

FIGURE 4.2.3 AIRSAR PolSAR data. (a) Pauli image, (b) ground-truth map.

ground-truth map is shown in Figure 4.2.4 (e). For these land covers, the main growth and maturation period is during June and July each year. The maximum temporal baseline of these four PolSAR datasets is 31 days. Therefore, some crops with rapid growth changes will exhibit significant changes during this period, and their corresponding polarimetric scattering responses will also change accordingly, which can be observed from the Pauli images shown in Figures 4.2.4 (a)–(d). Such scattering mechanism variations in terms of temporal baselines bring challenges to multi-temporal PolSAR land cover classification.

4.2.3.2 Selected Polarimetric Features

For each land cover class, 1000 samples are randomly selected for feature selection. For the AIRSAR data, according to the ground-truth map, there are 11 known types of land covers, which correspond to 55 land cover pairs. Meanwhile, for each temporal UAVSAR data, there are 7 known types of land covers, which can generate 21 land cover pairs. The initially selected feature sets $\mathbf{F}^{\text{Ini}}$ for different PolSAR datasets are shown in Table 4.2.1. The numbers in brackets indicate the total selected frequencies of each feature. For example, the polarimetric rotation domain feature $\left|\gamma_{\text{HH-VV}}(\theta)\right|_{\text{min}}$ can maximize the class separation distances of 14 land cover pairs for the AIRSAR data. In this vein, its selected frequency is 14.

With the initially selected polarimetric feature set summarized in Table 4.2.1 and the threshold $r=3$, the refined polarimetric feature set $\mathbf{F}^{\text{Refined}}$ can be obtained. In this vein, the polarimetric features with selected frequency at or over 3 among the initially selected polarimetric feature set $\mathbf{F}^{\text{Ini}}$ will be collected into the refined polarimetric feature set $\mathbf{F}^{\text{Refined}}$. For the AIRSAR PolSAR data, $\mathbf{F}^{\text{Refined}}$ becomes the final selection feature set $\mathbf{F}^{\text{Final}}$, which mainly includes the minimum coherence $\left|\gamma_{\text{HH-VV}}(\theta)\right|_{\text{min}}$, maximum coherence $\left|\gamma_{(\text{HH-VV})\text{-HV}}(\theta)\right|_{\text{max}}$, polarimetric null angle $\theta_{\text{null}}\_\text{Re}\left[T_{12}(\theta)\right]$, polarimetric null angle $\theta_{\text{null}}\_\text{Im}\left[T_{12}(\theta)\right]$, original coherence $\left|\gamma_{(\text{HH+VV})\text{-HV}}(\theta)\right|_{\text{org}}$, and coherence contrast $\left|\gamma_{(\text{HH-VV})\text{-HV}}(\theta)\right|_{\text{contrast}}$. Among them, the minimum coherence $\left|\gamma_{\text{HH-VV}}(\theta)\right|_{\text{min}}$, maximum coherence $\left|\gamma_{(\text{HH-VV})\text{-HV}}(\theta)\right|_{\text{max}}$, and coherence contrast $\left|\gamma_{(\text{HH-VV})\text{-HV}}(\theta)\right|_{\text{contrast}}$ are polarimetric roll-invariant features.

FIGURE 4.2.4 Multi-temporal UAVSAR PolSAR data. (a)–(d) Pauli images of DoYs 169, 174, 187, and 199, respectively, (e) ground-truth map.

TABLE 4.2.1
Initially Selected Feature Sets for Different PolSAR Datasets

PolSAR Data		Initially Selected Feature Set $\mathbf{F}^{\mathrm{Ini}}$ and Selected Frequency
AIRSAR		$\lvert\gamma_{\mathrm{HH-VV}}(\theta)\rvert_{\mathrm{min}}$ (14), $\lvert\gamma_{\mathrm{(HH-VV)-HV}}(\theta)\rvert_{\mathrm{max}}$ (11), $\theta_{\mathrm{null}}\_\mathrm{Re}[T_{12}(\theta)]$ (9), $\theta_{\mathrm{null}}\_\mathrm{Im}[T_{12}(\theta)]$ (6), $\lvert\gamma_{\mathrm{(HH+VV)-HV}}(\theta)\rvert_{\mathrm{org}}$ (3), $\lvert\gamma_{\mathrm{(HH-VV)-HV}}(\theta)\rvert_{\mathrm{contrast}}$ (3), $\lvert\gamma_{\mathrm{(HH+VV)-HV}}(\theta)\rvert_{\mathrm{max}}$ (2), $\lvert\gamma_{\mathrm{HH-HV}}(\theta)\rvert_{\mathrm{org}}$ (2), $\lvert\gamma_{\mathrm{HH-VV}}(\theta)\rvert_{\mathrm{org}}$ (2), $\lvert\gamma_{\mathrm{HH-VV}}(\theta)\rvert_{\mathrm{max}}$ (2), $\lvert\gamma_{\mathrm{HH-HV}}(\theta)\rvert_{\mathrm{max}}$ (1)
UAVSAR	DoY 169	$\theta_{\mathrm{null}}\_\mathrm{Im}[T_{12}(\theta)]$ (9), $\lvert\gamma_{\mathrm{(HH-VV)-HV}}(\theta)\rvert_{\mathrm{org}}$ (4), $\theta_{\mathrm{null}}\_\mathrm{Re}[T_{12}(\theta)]$ (2), $\lvert\gamma_{\mathrm{(HH-VV)-HV}}(\theta)\rvert_{\mathrm{max}}$ (2), $\lvert\gamma_{\mathrm{(HH+VV)-HV}}(\theta)\rvert_{\mathrm{mean}}$ (1), $\lvert\gamma_{\mathrm{(HH+VV)-HV}}(\theta)\rvert_{\mathrm{contrast}}$ (1), $\lvert\gamma_{\mathrm{(HH-VV)-HV}}(\theta)\rvert_{\mathrm{min}}$ (1), $\lvert\gamma_{\mathrm{HH-VV}}(\theta)\rvert_{\mathrm{org}}$ (1)
	DoY 174	$\theta_{\mathrm{null}}\_\mathrm{Im}[T_{12}(\theta)]$ (9), $\lvert\gamma_{\mathrm{(HH-VV)-HV}}(\theta)\rvert_{\mathrm{org}}$ (4), $\lvert\gamma_{\mathrm{(HH-VV)-HV}}(\theta)\rvert_{\mathrm{min}}$ (3), $\theta_{\mathrm{null}}\_\mathrm{Re}[T_{12}(\theta)]$ (1), $\lvert\gamma_{\mathrm{(HH+VV)-HV}}(\theta)\rvert_{\mathrm{mean}}$ (1), $\lvert\gamma_{\mathrm{(HH+VV)-HV}}(\theta)\rvert_{\mathrm{org}}$ (1), $\lvert\gamma_{\mathrm{HH-VV}}(\theta)\rvert_{\mathrm{org}}$ (1), $\lvert\gamma_{\mathrm{HH-VV}}(\theta)\rvert_{\mathrm{min}}$ (1)
	DoY 187	$\theta_{\mathrm{null}}\_\mathrm{Im}[T_{12}(\theta)]$ (7), $\theta_{\mathrm{null}}\_\mathrm{Re}[T_{12}(\theta)]$ (6), $\lvert\gamma_{\mathrm{HH-VV}}(\theta)\rvert_{\mathrm{min}}$ (3), $\lvert\gamma_{\mathrm{(HH+VV)-HV}}(\theta)\rvert_{\mathrm{org}}$ (3), $\lvert\gamma_{\mathrm{(HH+VV)-HV}}(\theta)\rvert_{\mathrm{contrast}}$ (1), $\lvert\gamma_{\mathrm{HH-VV}}(\theta)\rvert_{\mathrm{org}}$ (1)
	DoY 199	$\theta_{\mathrm{null}}\_\mathrm{Im}[T_{12}(\theta)]$ (6), $\theta_{\mathrm{null}}\_\mathrm{Re}[T_{12}(\theta)]$ (3), $\lvert\gamma_{\mathrm{(HH+VV)-HV}}(\theta)\rvert_{\mathrm{org}}$ (3), $\lvert\gamma_{\mathrm{HH-VV}}(\theta)\rvert_{\mathrm{min}}$ (3), $\lvert\gamma_{\mathrm{(HH-VV)-HV}}(\theta)\rvert_{\mathrm{org}}$ (2), $\lvert\gamma_{\mathrm{(HH-VV)-HV}}(\theta)\rvert_{\mathrm{max}}$ (1), $\lvert\gamma_{\mathrm{HH-HV}}(\theta)\rvert_{\mathrm{org}}$ (1), $\lvert\gamma_{\mathrm{HH-HV}}(\theta)\rvert_{\mathrm{max}}$ (1), $\lvert\gamma_{\mathrm{(HH+VV)-HV}}(\theta)\rvert_{\mathrm{max}}$ (1)

Similarly, for multi-temporal UAVSAR PolSAR datasets, with the initially selected polarimetric feature set shown in Table 4.2.1 and the threshold $r = 3$, the refined polarimetric feature set $\mathbf{F}^{\text{Refined}}$ for each temporal dataset can be obtained. Then, the union of these refined polarimetric feature sets becomes the final selection feature set $\mathbf{F}^{\text{Final}}$, which mainly includes the polarimetric null angle $\theta_{\text{null}}\_\text{Re}\left[T_{12}(\theta)\right]$, polarimetric null angle $\theta_{\text{null}}\_\text{Im}\left[T_{12}(\theta)\right]$, original coherence $\left|\gamma_{(\text{HH}-\text{VV})-\text{HV}}(\theta)\right|_{\text{org}}$, minimum coherence $\left|\gamma_{(\text{HH}-\text{VV})-\text{HV}}(\theta)\right|_{\text{min}}$, minimum coherence $\left|\gamma_{\text{HH}-\text{VV}}(\theta)\right|_{\text{min}}$, and original coherence $\left|\gamma_{(\text{HH}+\text{VV})-\text{HV}}(\theta)\right|_{\text{org}}$. Among them, the minimum coherence features $\left|\gamma_{(\text{HH}-\text{VV})-\text{HV}}(\theta)\right|_{\text{min}}$ and $\left|\gamma_{\text{HH}-\text{VV}}(\theta)\right|_{\text{min}}$ are polarimetric roll-invariant features.

Interestingly, for both single-temporal AIRSAR and multi-temporal UAVSAR PolSAR datasets, the final polarimetric feature sets only contain six polarimetric rotation domain features, which confirms that these selected polarimetric features have good representativeness and robustness. In addition, it is worth noting that the polarimetric null angle $\theta_{\text{null}}\_\text{Re}\left[T_{12}(\theta)\right]$, polarimetric null angle $\theta_{\text{null}}\_\text{Im}\left[T_{12}(\theta)\right]$, and minimum coherence $\left|\gamma_{\text{HH}-\text{VV}}(\theta)\right|_{\text{min}}$ are selected at the same time for these two kinds of PolSAR systems, which further verifies the good performance of these polarimetric rotation domain features. In particular, the polarimetric null angle features $\theta_{\text{null}}\_\text{Re}\left[T_{12}(\theta)\right]$ and $\theta_{\text{null}}\_\text{Im}\left[T_{12}(\theta)\right]$ achieve the highest selected frequency in the initially selected polarimetric feature set $\mathbf{F}^{\text{Ini}}$ for more land cover pairs and acquisition dates, which will be further studied in Section 4.3.

In order to investigate the performances of the selected polarimetric rotation domain features and their complementary advantages combined with the traditional polarimetric roll-invariant features, four traditional polarimetric roll-invariant features, including the total backscattering power $SPAN$, polarimetric entropy H, mean alpha angle $\bar{\alpha}$, and polarimetric anisotropy Ani, obtained from AIRSAR data are shown in Figures 4.2.5 (a)–(d), and the six selected polarimetric rotation domain features are shown in Figures 4.2.5 (e)–(j). In order to demonstrate the land cover discrimination capabilities of these polarimetric features, means and standard deviations of each polarimetric feature in terms of each known land cover are shown in Figures 4.2.6. With only the four polarimetric roll-invariant features H, Ani, $\bar{\alpha}$, and $SPAN$, it is observed that the grasses and wheat cannot be successfully discriminated. The separations between the grasses and lucerne, and the potatoes and forest are also limited. In comparison, with the polarimetric null angle $\theta_{\text{null}}\_\text{Re}\left[T_{12}(\theta)\right]$, the grasses and wheat can be clearly discriminated. Furthermore, with the minimum coherence $\left|\gamma_{\text{HH}-\text{VV}}(\theta)\right|_{\text{min}}$, better separations are achieved for the grasses and lucerne, and the potatoes and forest, accordingly. Other selected polarimetric rotation domain features are also able to enhance the separations for some land cover pairs. Thereby, the selected polarimetric rotation domain features exhibit the capability to further enhance the land cover separation and have the potential to boost land cover classification accuracy.

Similarly, the UAVSAR data of DoY 169 is also used as an example. The four polarimetric roll-invariant features and the six selected polarimetric rotation domain features are shown in Figure 4.2.7. Means and standard deviations of these polarimetric features for known land covers are shown in Figure 4.2.8. With only the

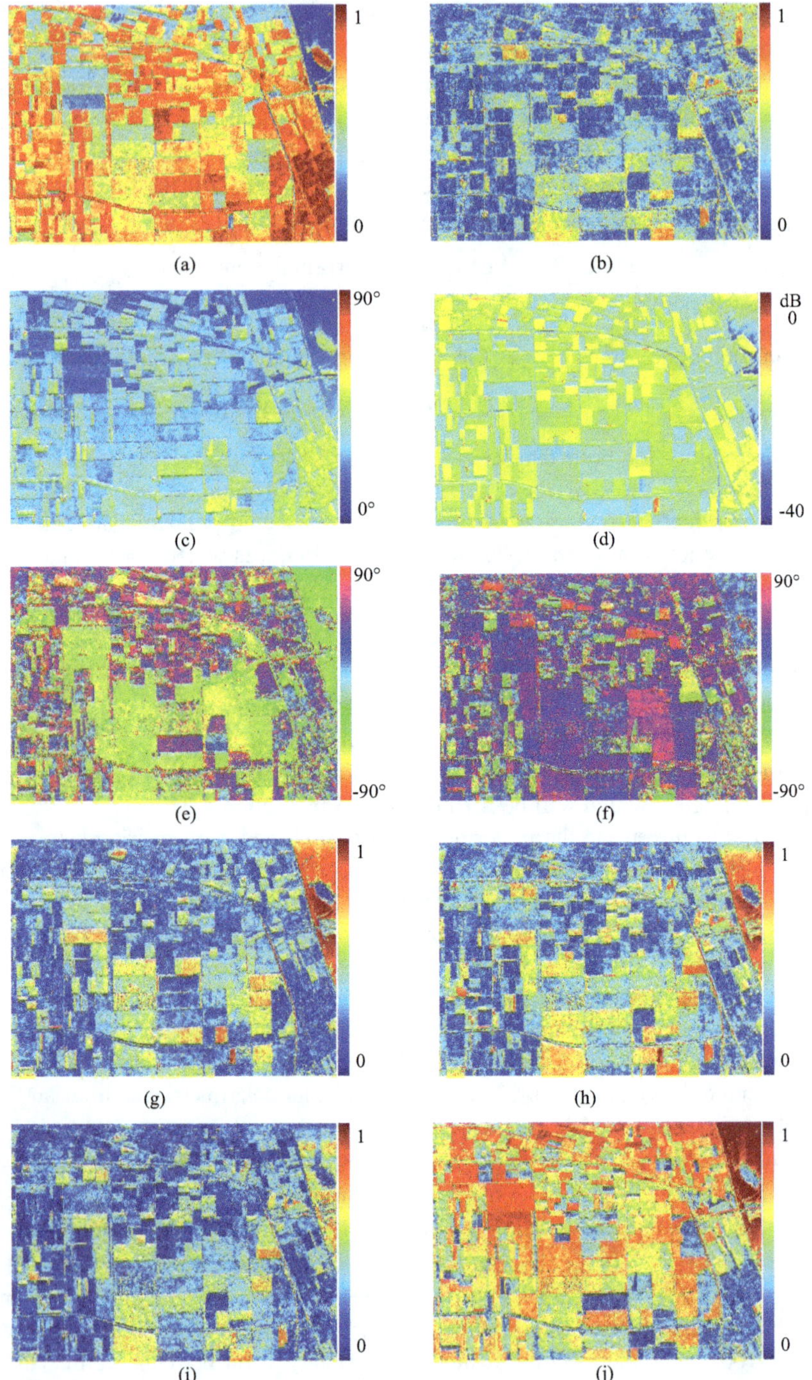

FIGURE 4.2.5 Traditional polarimetric roll-invariant features and selected polarimetric rotation domain features for AIRSAR data. (a) H, (b) Ani, (c) $\bar{\alpha}$, (d) $SPAN$, (e) $\theta_{\text{null}}\_\text{Re}\left[T_{12}(\theta)\right]$, (f) $\theta_{\text{null}}\_\text{Im}\left[T_{12}(\theta)\right]$, (g) $\left|\gamma_{(\text{HH+VV})-\text{HV}}(\theta)\right|_{\text{org}}$, (h) $\left|\gamma_{(\text{HH-VV})-\text{HV}}(\theta)\right|_{\text{max}}$, (i) $\left|\gamma_{(\text{HH-VV})-\text{HV}}(\theta)\right|_{\text{contrast}}$, (j) $\left|\gamma_{\text{HH-VV}}(\theta)\right|_{\text{min}}$.

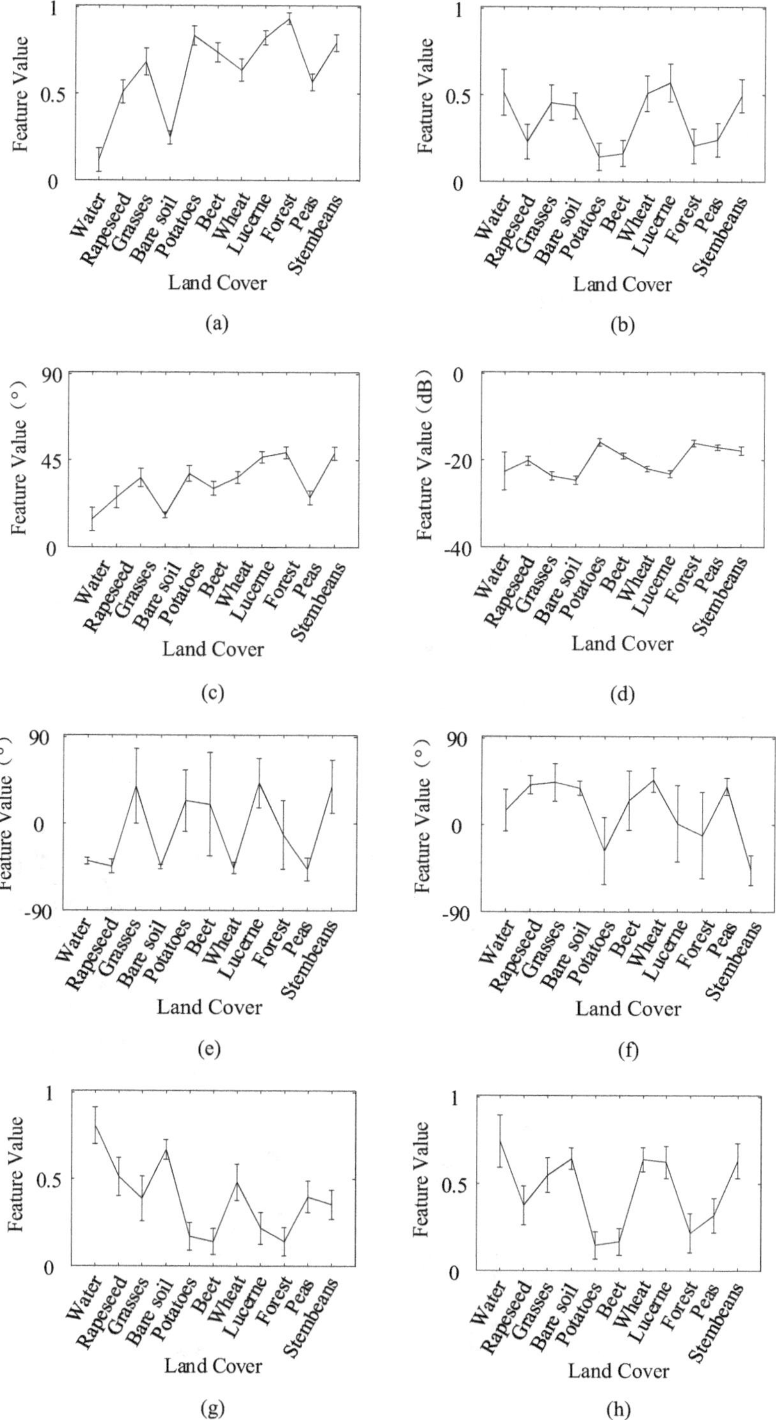

FIGURE 4.2.6 Polarimetric feature means and standard deviations comparison for AIRSAR data. (a) H, (b) Ani, (c) $\bar{\alpha}$, (d) $SPAN$, (e) $\theta_{\text{null}}\_\text{Re}\left[T_{12}(\theta)\right]$, (f) $\theta_{\text{null}}\_\text{Im}\left[T_{12}(\theta)\right]$, (g) $\left|\gamma_{(\text{HH+VV})-\text{HV}}(\theta)\right|_{\text{org}}$, (h) $\left|\gamma_{(\text{HH-VV})-\text{HV}}(\theta)\right|_{\max}$, (i) $\left|\gamma_{(\text{HH-VV})-\text{HV}}(\theta)\right|_{\text{contrast}}$, (j) $\left|\gamma_{\text{HH-VV}}(\theta)\right|_{\min}$.

(Continued)

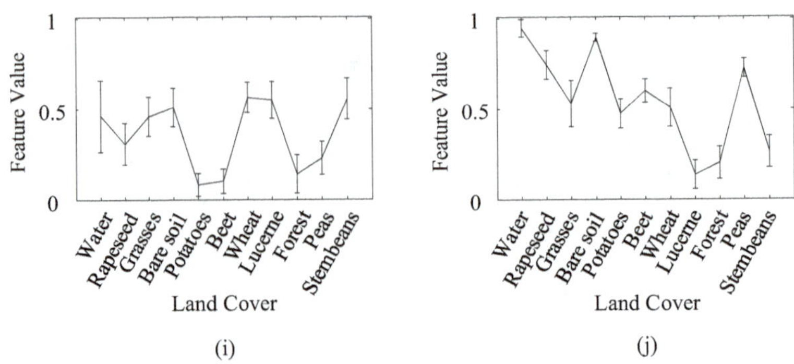

FIGURE 4.2.6 (Continued)

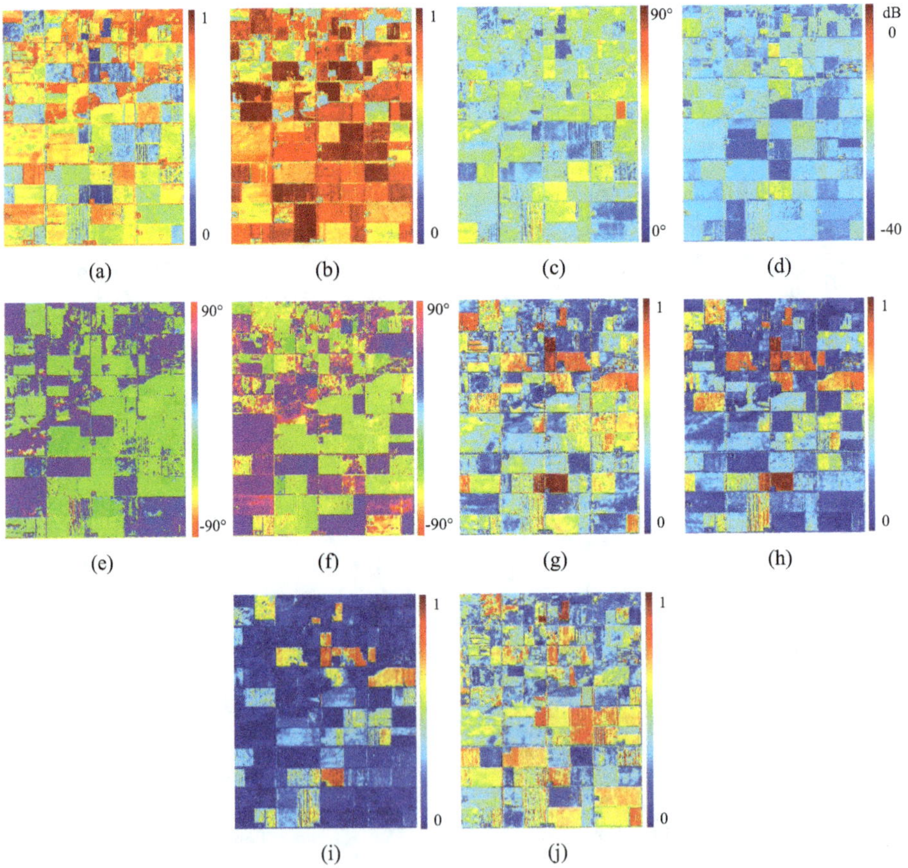

FIGURE 4.2.7 Traditional polarimetric roll-invariant features and selected polarimetric rotation domain features for UAVSAR data of DoY 169. (a) H, (b) Ani, (c) $\bar{\alpha}$, (d) $SPAN$, (e) $\theta_{\text{null}}\_\mathrm{Re}\left[T_{12}(\theta)\right]$, (f) $\theta_{\text{null}}\_\mathrm{Im}\left[T_{12}(\theta)\right]$, (g) $\left|\gamma_{(\text{HH+VV})-\text{HV}}(\theta)\right|_{\text{org}}$, (h) $\left|\gamma_{(\text{HH-VV})-\text{HV}}(\theta)\right|_{\max}$, (i) $\left|\gamma_{(\text{HH-VV})-\text{HV}}(\theta)\right|_{\min}$, (j) $\left|\gamma_{\text{HH-VV}}(\theta)\right|_{\min}$.

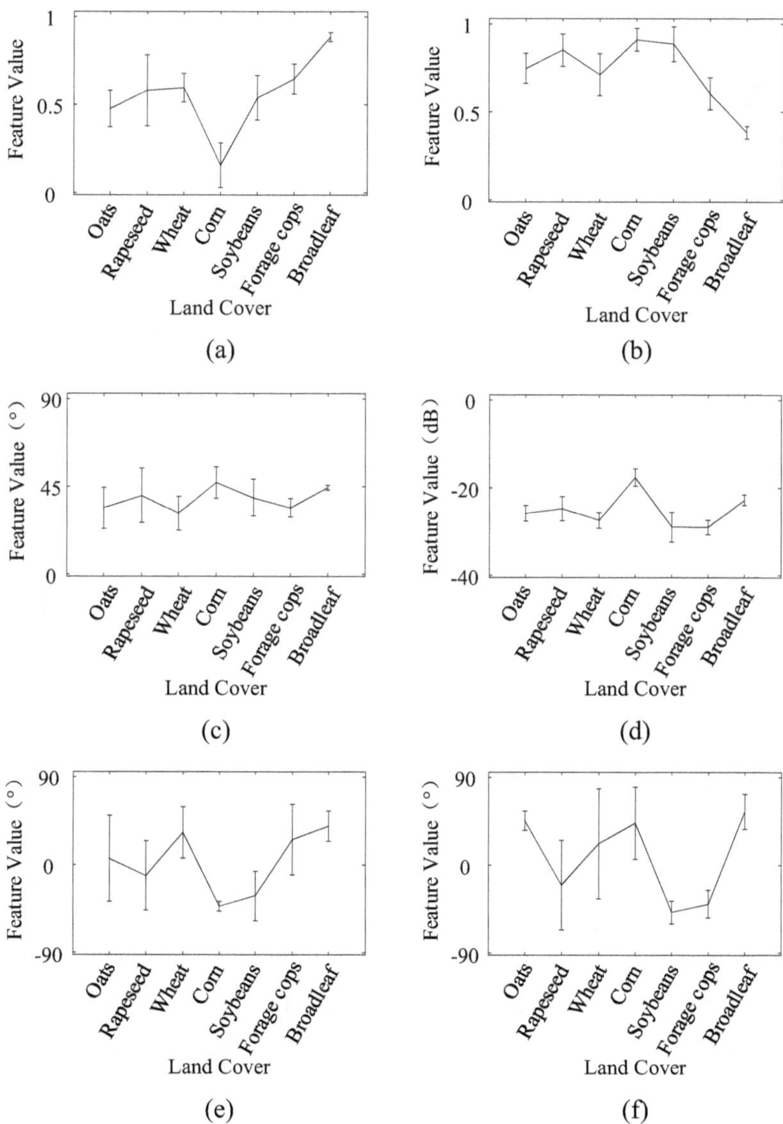

FIGURE 4.2.8 Polarimetric feature means and standard deviations comparison for UAVSAR data of DoY 169. (a) H, (b) Ani, (c) $\bar{\alpha}$, (d) $SPAN$, (e) $\theta_{\text{null}}\_\text{Re}\left[T_{12}(\theta)\right]$, (f) $\theta_{\text{null}}\_\text{Im}\left[T_{12}(\theta)\right]$, (g) $\left|\gamma_{(\text{HH+VV})-\text{HV}}(\theta)\right|_{\text{org}}$, (h) $\left|\gamma_{(\text{HH}-\text{VV})-\text{HV}}(\theta)\right|_{\max}$, (i) $\left|\gamma_{(\text{HH}-\text{VV})-\text{HV}}(\theta)\right|_{\min}$, (j) $\left|\gamma_{\text{HH}-\text{VV}}(\theta)\right|_{\min}$.

(Continued)

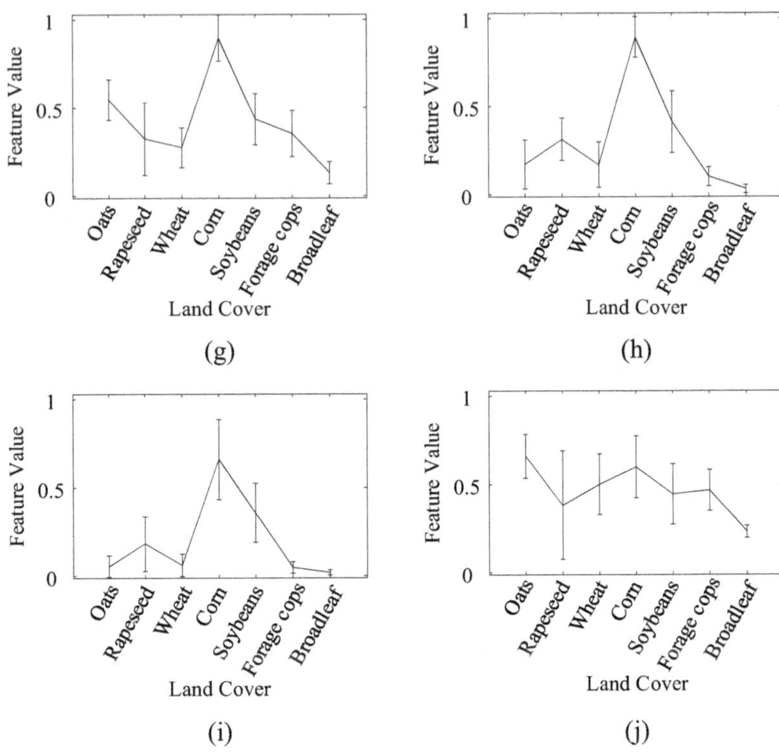

FIGURE 4.2.8 (Continued)

polarimetric roll-invariant features of H, Ani, $\bar{\alpha}$, and $SPAN$, it is clear that the oats and rapeseed, the oats and wheat, the oats and soybeans, and the wheat and soybeans may not be successfully separated. In comparison, they can be separated by the polarimetric rotation domain features of $\theta_{\text{null}}\_\text{Im}\left[T_{12}(\theta)\right]$, $\left|\gamma_{(\text{HH+VV})-\text{HV}}(\theta)\right|_{\text{org}}$, $\left|\gamma_{(\text{HH}-\text{VV})-\text{HV}}(\theta)\right|_{\text{min}}$, and $\theta_{\text{null}}\_\text{Re}\left[T_{12}(\theta)\right]$. This further verifies the necessity and importance of combining the polarimetric rotation domain features and the traditional polarimetric roll-invariant features.

4.2.3.3 Comparison with AIRSAR Data

In order to demonstrate the added benefits from polarimetric rotation domain features, the classification method introduced in Section 4.2.2 (noted as the polarimetric rotation domain classification) is compared with the conventional classification method, which only uses the polarimetric roll-invariant features of H, Ani, $\bar{\alpha}$, and $SPAN$. Both the SVM and DT classifiers are adopted for investigation. The training rate is set as 50%. The training rate is the proportion of the training samples to its ground-truth samples for each class. In this study, 50% of the samples are randomly selected for training the SVM and DT classifiers, while the remaining samples are used for validation.

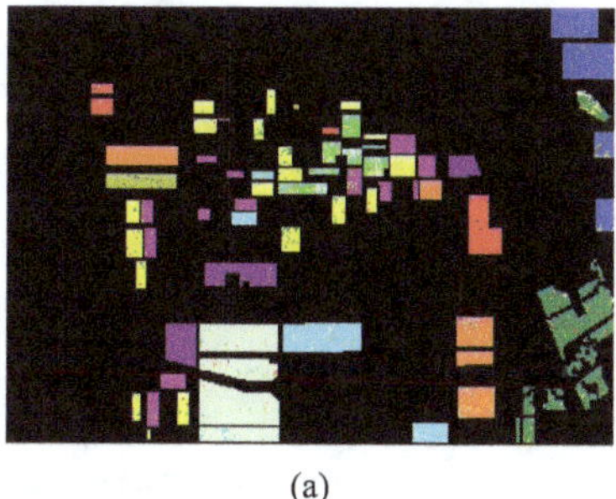

(a)

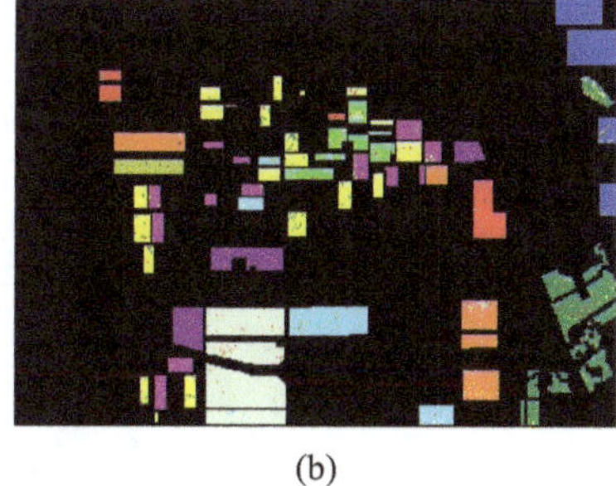

(b)

FIGURE 4.2.9 Classification results for AIRSAR data over 11 known land covers using the SVM classifier. (a) Conventional classification method, (b) polarimetric rotation domain classification method.

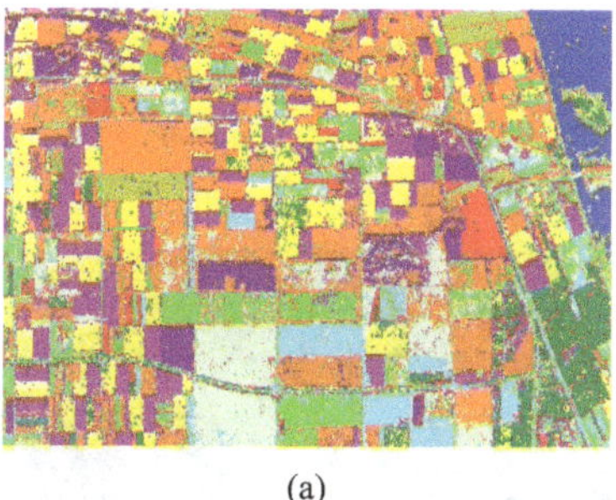

(a)

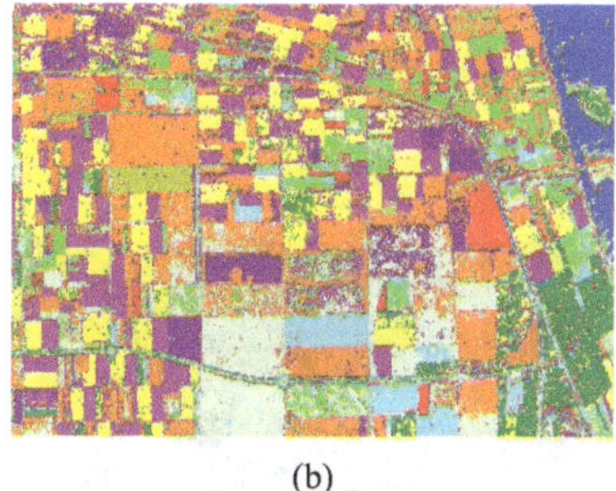

(b)

FIGURE 4.2.10 Classification results over the full-scene area of AIRSAR data using the SVM classifier. (a) Conventional classification method, (b) polarimetric rotation domain classification method.

First, comparison experiments are carried out with the SVM classifier. Classification results for AIRSAR data over 11 known land covers and the full-scene area are shown in Figures 4.2.9 and 4.2.10. The classification accuracies are listed in Table 4.2.2. It can be observed that the overall accuracy (OA) of the polarimetric rotation domain classification method is 95.37%, while that of the conventional classification method is 93.87%. Moreover, for 9 of these 11 land covers, the classification accuracies from the polarimetric rotation domain classification method are higher than those from the conventional classification method. Especially for the grasses, the classification accuracy enhancement is up by 14.35%, from 66.99% with the conventional method to 81.34% with the polarimetric rotation domain classification method. Overall, the performance of the polarimetric rotation domain classification method outperforms that of the conventional classification method.

In addition, with the DT classifier, the classification results over the 11 known land covers and the full-scene area are shown in Figures 4.2.11 and 4.2.12. The classification accuracies are listed in Table 4.2.3. The polarimetric rotation domain classification method still outperforms the conventional classification method. The OA of the polarimetric rotation domain classification method is 96.38%, which is higher than

TABLE 4.2.2
Classification Accuracies for AIRSAR Data Using the SVM Classifier (%)

Classes	Conventional Classification Method	Polarimetric Rotation Domain Classification Method
Stembeans	**98.07**	97.44
Peas	**97.85**	97.76
Forest	92.53	**94.08**
Lucerne	95.89	**96.33**
Wheat	96.12	**97.58**
Beet	94.89	**95.76**
Potatoes	92.81	**93.42**
Bare soil	95.84	**96.75**
Grasses	66.99	**81.34**
Rapeseed	94.89	**95.38**
Water	97.65	**98.39**
OA	93.87	**95.37**

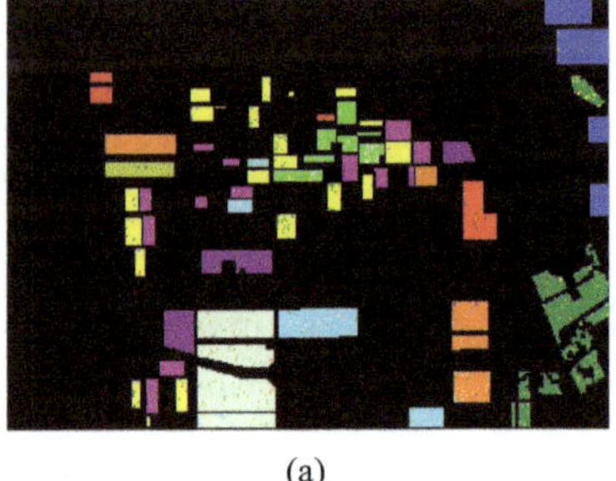
(a)

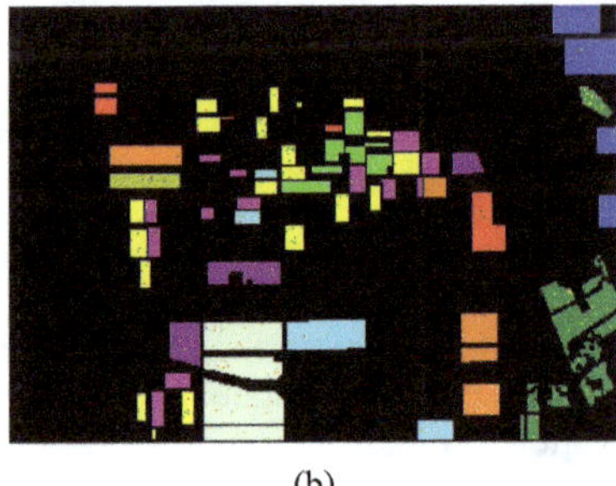
(b)

FIGURE 4.2.11 Classification results for AIRSAR data over 11 known land covers using the DT classifier. (a) Conventional classification method, (b) polarimetric rotation domain classification method.

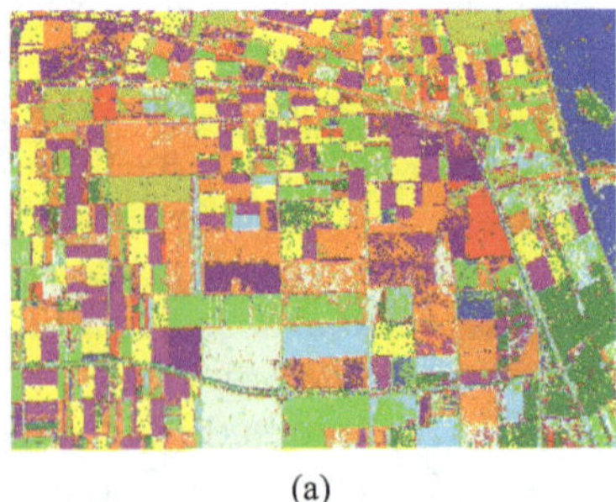
(a)

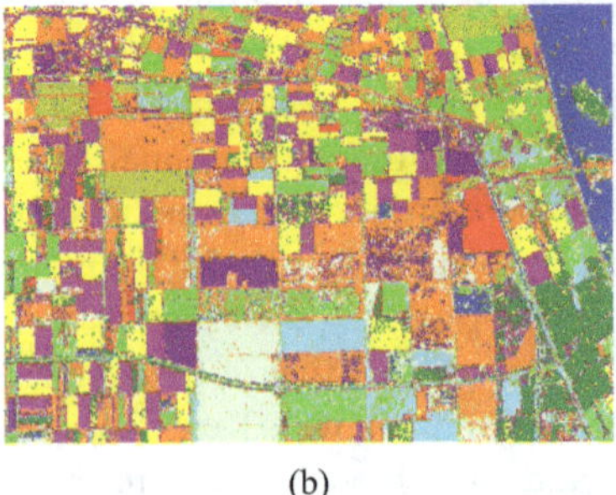
(b)

FIGURE 4.2.12 Classification results over the full-scene area of AIRSAR data using the DT classifier. (a) Conventional classification method, (b) polarimetric rotation domain classification method.

TABLE 4.2.3
Classification Accuracies for AIRSAR Data Using DT Classifier (%)

Classes	Conventional Classification Method	Polarimetric Rotation Domain Classification Method
Stembeans	96.56	**96.89**
Peas	96.95	**97.52**
Forest	92.16	**94.40**
Lucerne	94.01	**97.47**
Wheat	93.78	**97.49**
Beet	95.64	**96.64**
Potatoes	91.49	**93.68**
Bare soil	97.08	**97.09**
Grasses	84.56	**93.94**
Rapeseed	94.66	**96.04**
Water	**99.44**	99.39
OA	94.12	**96.38**

TABLE 4.2.4
Overall Classification Accuracies for AIRSAR Data with Different Speckle Filter Sliding Window Sizes (%)

Method		7 × 7	9 × 9	11 × 11	13 × 13	15 × 15	25 × 25
SVM classifier	Conventional classification method	89.07	91.17	92.43	93.21	93.87	95.17
	Polarimetric rotation domain classification method	**91.93**	**93.53**	**94.41**	**94.86**	**95.37**	**96.63**
DT classifier	Conventional classification method	88.66	91.04	92.39	93.50	94.12	95.57
	Polarimetric rotation domain classification method	**93.06**	**94.58**	**95.32**	**96.22**	**96.38**	**97.28**

the 94.12% of the conventional classification method. Moreover, for 10 of these 11 known land covers, the classification accuracies of the polarimetric rotation domain classification method are higher than those of the conventional classification method. In addition, it is worth noting that for both the conventional classification method and the polarimetric rotation domain classification method, the classification performance from the DT classifier is better than that from the SVM classifier.

Also, the overall classification accuracies in terms of different sliding window sizes (7 × 7, 9 × 9, 11 × 11, 13 × 13, 15 × 15, and 25 × 25) of the speckle filter are examined. The corresponding OAs are listed in Table 4.2.4. For the same classification method, it is clear that larger windows lead to higher OAs. According to studies on window size selection [28, 30], the 15 × 15 sliding window is adopted in this chapter. In addition, with different speckle filter window sizes, the performances from the polarimetric rotation domain classification method are still better than the conventional classification method, which further validates the advantage of the polarimetric rotation domain classification method.

4.2.3.4 Comparison with Multi-Temporal UAVSAR Data

With the SVM classifier, the classification results for the four temporal UAVSAR datasets over the 7 known land covers and the full-scene area are shown in Figures 4.2.13 and 4.2.14, respectively. The classification accuracies are listed in Table 4.2.5. It is clear that the performances of the polarimetric rotation domain classification method are much better than those of the conventional classification method. The mean OA from the polarimetric rotation domain classification method is 97.47%, which is much higher than 89.59% from the conventional classification method. Additionally, the OA increments for the four temporal datasets are 6.45% (DoY 169: from 90.19% to 96.64%), 6.30% (DoY 174: from 90.75% to 97.05%), 10.24% (DoY 187: from 88.03% to 98.27%), and 8.54% (DoY 199: from 89.39% to 97.93%). Moreover, the polarimetric rotation domain classification method exhibits

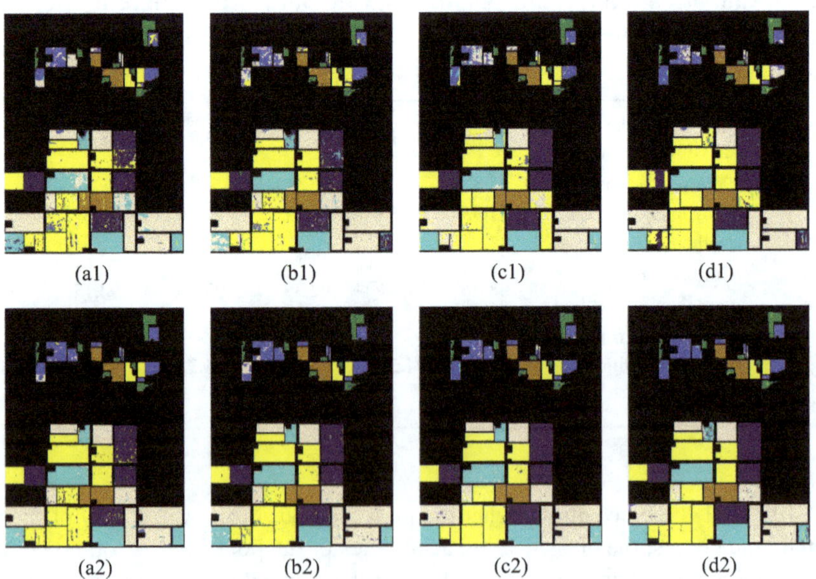

FIGURE 4.2.13 Classification results for multi-temporal UAVSAR data over 7 known land covers using the SVM classifier. (a1)–(d1) Conventional classification method for DoY 169, DoY 174, DoY 187, and DoY 199 datasets, respectively; (a2)–(d2) polarimetric rotation domain classification method for DoY 169, DoY 174, DoY 187, and DoY 199 datasets, respectively.

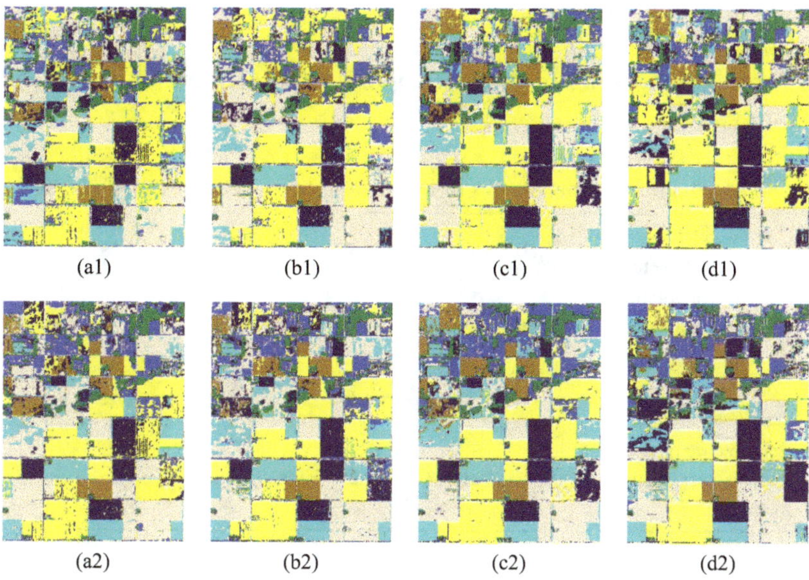

FIGURE 4.2.14 Classification results over the full-scene area of multi-temporal UAVSAR data using the SVM classifier. (a1)–(d1) Conventional classification method for DoY 169, DoY 174, DoY 187, and DoY 199 datasets, respectively; (a2)–(d2) polarimetric rotation domain classification method for DoY 169, DoY 174, DoY 187, and DoY 199 datasets, respectively.

TABLE 4.2.5
Classification Accuracies for Multi-Temporal UAVSAR Data Using the SVM Classifier (%)

DoY	Method	Broadleaf	Forage Crops	Soybeans	Corn	Wheat	Rapeseed	Oats	OA
169	Conventional classification method	98.47	62.24	92.64	96.12	93.63	91.70	86.37	90.19
	Polarimetric rotation domain classification method	**98.49**	**88.92**	**96.59**	**98.58**	**98.06**	**96.60**	**96.72**	**96.64**
174	Conventional classification method	**98.05**	61.38	94.14	97.30	97.89	93.82	77.29	90.75
	Polarimetric rotation domain classification method	97.75	**83.77**	**97.68**	**98.93**	**98.88**	**97.76**	**97.21**	**97.05**
187	Conventional classification method	**98.63**	56.36	92.31	**99.55**	76.85	99.24	94.61	88.03
	Polarimetric rotation domain classification method	98.60	**90.87**	**99.35**	99.45	**98.58**	**99.26**	**97.39**	**98.27**

(Continued)

TABLE 4.2.5 ***(Continued)***
Classification Accuracies for Multi-Temporal UAVSAR Data Using the SVM Classifier (%)

DoY	Method	Broadleaf	Forage Crops	Soybeans	Corn	Wheat	Rapeseed	Oats	OA
199	Conventional classification method	96.86	64.51	97.38	**99.78**	84.76	92.19	82.98	89.39
	Polarimetric rotation domain classification method	**97.20**	**94.16**	**99.47**	99.75	**97.79**	**99.74**	**94.09**	**97.93**
Mean	Conventional classification method	98.00	61.12	94.12	98.19	88.28	94.24	85.31	89.59
	Polarimetric rotation domain classification method	**98.01**	**89.43**	**98.27**	**99.18**	**98.33**	**98.34**	**96.35**	**97.47**

better robustness among different temporal PolSAR datasets. For the four temporal datasets, especially for the oats, wheat, and forage crops, the classification accuracy ranges from the conventional classification method are 77.29–94.61%, 76.85–97.89%, and 56.36–64.51%, while those from the polarimetric rotation domain classification method are 94.09–97.39%, 97.79–98.88%, and 83.77–94.16%, respectively, which further verifies the advantage of the polarimetric rotation domain classification method.

In addition, with the DT classifier, the classification results over 7 known land covers and the full-scene area are shown in Figures 4.2.15 and 4.2.16. The classification accuracies are summarized in Table 4.2.6. Similarly, the performance of the polarimetric rotation domain classification method is still better than the conventional classification method. For the four temporal datasets, the mean OA of the polarimetric rotation domain classification method is 99.39%, which is higher than 97.55% of the conventional classification method. In addition, the OA enhancements are 1.79% (DoY 169: from 97.48% to 99.27%), 1.65% (DoY 174: from 97.63% to 99.28%), 1.91% (DoY 187: from 97.65% to 99.56%), and 2.00% (DoY 199: from 97.45% to 99.45%). Moreover, the polarimetric rotation domain classification method still shows better robustness for the different temporal PolSAR data, especially in the areas of oats, rapeseed, and forage crops. Note that compared with the SVM classifier, the classification performances with the DT classifier are improved for both methods.

Also, based on the UAVSAR data of DoY 169, the OAs in terms of different speckle filter window sizes (7 × 7, 9 × 9, 11 × 11, 13 × 13, 15 × 15, and 25 × 25) are investigated and summarized in Table 4.2.7. The same conclusions can be obtained that larger windows usually lead to better classification accuracy. Meanwhile, with the different speckle filter window sizes, the performances from the polarimetric rotation domain classification method are still better than the conventional classification method.

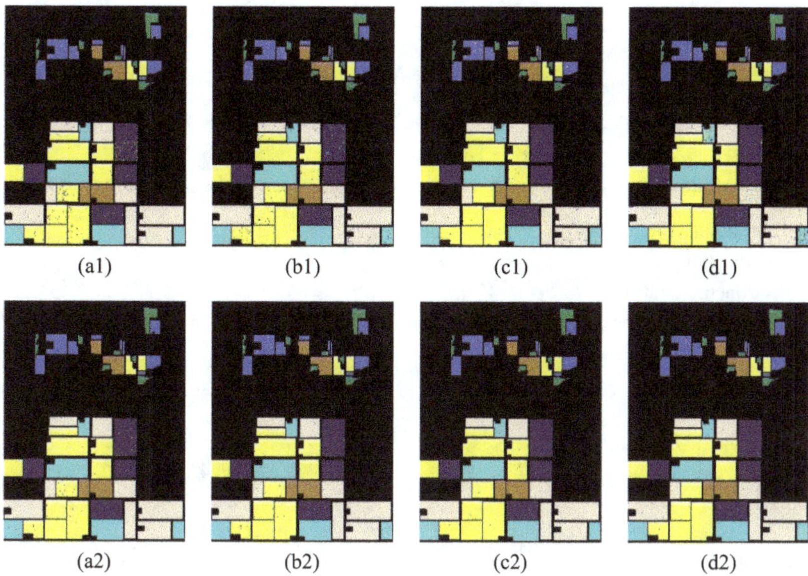

FIGURE 4.2.15 Classification results for multi-temporal UAVSAR data over 7 known land covers using the DT classifier. (a1)–(d1) Conventional classification method for DoY 169, DoY 174, DoY 187, and DoY 199 datasets, respectively; (a2)–(d2) polarimetric rotation domain classification method for DoY 169, DoY 174, DoY 187, and DoY 199 datasets, respectively.

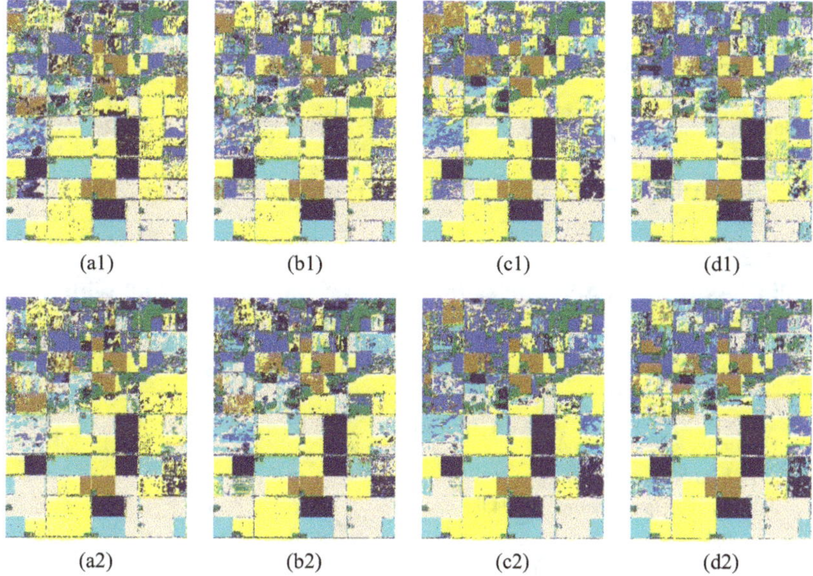

FIGURE 4.2.16 Classification results over the full-scene area of multi-temporal UAVSAR data using the DT classifier. (a1)–(d1) Conventional classification method for DoY 169, DoY 174, DoY 187, and DoY 199 datasets, respectively; (a2)–(d2) polarimetric rotation domain classification method for DoY 169, DoY 174, DoY 187, and DoY 199 datasets, respectively.

TABLE 4.2.6
Classification Accuracies for Multi-Temporal UAVSAR Data Using the DT Classifier (%)

DoY	Method	Broadleaf	Forage Crops	Soybeans	Corn	Wheat	Rapeseed	Oats	OA
169	Conventional classification method	98.71	94.42	97.09	98.28	98.59	95.79	98.98	97.48
	Polarimetric rotation domain classification method	**99.12**	**98.68**	**99.25**	**99.55**	**99.55**	**98.72**	**99.56**	**99.27**
174	Conventional classification method	98.46	92.72	98.24	99.00	97.87	97.45	97.66	97.63
	Polarimetric rotation domain classification method	**98.97**	**97.82**	**99.46**	**99.47**	**99.47**	**99.02**	**99.55**	**99.28**
187	Conventional classification method	98.64	91.62	97.64	99.80	97.16	99.46	98.39	97.65
	Polarimetric rotation domain classification method	**98.77**	**98.46**	**99.78**	**99.81**	**99.66**	**99.59**	**99.46**	**99.56**
199	Conventional classification method	97.94	95.74	98.61	99.78	97.06	97.77	94.88	97.45
	Polarimetric rotation domain classification method	**98.47**	**98.47**	**99.83**	**99.85**	**99.34**	**99.53**	**99.23**	**99.45**
Mean	Conventional classification method	98.44	93.63	97.90	99.22	97.67	97.62	97.48	97.55
	Polarimetric rotation domain classification method	**98.83**	**98.36**	**99.58**	**99.67**	**99.51**	**99.22**	**99.45**	**99.39**

TABLE 4.2.7
Overall Classification Accuracies for UAVSAR Data of DoY 169 with Different Speckle Filter Sliding Window Sizes (%)

	Method	7 × 7	9 × 9	11 × 11	13 × 13	15 × 15	25 × 25
SVM classifier	Conventional classification method	88.23	88.95	89.46	89.86	90.19	91.49
	Polarimetric rotation domain classification method	**94.88**	**95.57**	**96.06**	**96.38**	**96.64**	**97.31**
DT classifier	Conventional classification method	95.17	96.16	96.76	97.18	97.48	98.04
	Polarimetric rotation domain classification method	**98.44**	**98.84**	**99.06**	**99.16**	**99.27**	**99.44**

4.3 POLARIMETRIC ROTATION DOMAIN CLASSIFICATION WITH DEEP LEARNING

PolSAR image classification is an important application. Advanced deep learning techniques represented by deep CNN have been utilized to enhance the classification performance. One challenge is how to adapt a deep CNN classifier for PolSAR classification with limited training samples while keeping good generalization performance. This section introduces a polarimetric-feature–driven deep CNN classification scheme [15]. The core idea is to incorporate expert knowledge of target scattering mechanism interpretation and polarimetric feature mining to assist deep CNN classifier training and improve final classification performance. Both classical polarimetric roll-invariant features and polarimetric rotation domain features are used to drive the deep CNN model.

4.3.1 POLARIMETRIC-FEATURE-DRIVEN DEEP CNN CLASSIFIER

4.3.1.1 Selected Polarimetric Features with Expert Knowledge

Polarimetric roll-invariant features which are independent of target orientations are normally preferred in PolSAR data interpretation and utilization. Therefore, polarimetric entropy *H*, mean alpha angle $\bar{\alpha}$, polarimetric anisotropy *Ani*, and total backscattering power *SPAN*, which are commonly used in PolSAR land cover classification, can be adopted. In addition, according to Chapter 2, two polarimetric null angle features, $\theta_{\text{null}}\_\text{Re}[T_{12}]$ and $\theta_{\text{null}}\_\text{Im}[T_{12}]$, can indicate the scattering symmetry condition, the amplitude and phase differences between co-polarization channels, and their values have clear physical meanings. For convenience, their definitions are rewritten here

$$
\begin{aligned}
\theta_{\text{null}}\_\text{Re}[T_{12}] &= -\frac{1}{2}\text{Angle}\left\{\text{Re}[T_{13}] + \text{j}\text{Re}[T_{12}]\right\} \\
&= \frac{1}{2}\text{Angle}\left\{\text{Re}\left[\left\langle (S_{\text{HH}} + S_{\text{VV}}) S_{\text{HV}}^* \right\rangle\right] + \text{j}\frac{1}{2}\left(\left\langle |S_{\text{VV}}|^2 - |S_{\text{HH}}|^2 \right\rangle\right)\right\}
\end{aligned}
\tag{4.3.1}
$$

$$
\begin{aligned}
\theta_{\text{null}}\_\text{Im}[T_{12}] &= -\frac{1}{2}\text{Angle}\left\{\text{Im}[T_{13}] + \text{j}\text{Im}[T_{12}]\right\} \\
&= \frac{1}{2}\text{Angle}\left\{\text{Im}\left[\left\langle (S_{\text{HH}} + S_{\text{VV}}) S_{\text{HV}}^* \right\rangle\right] + \text{j}\text{Im}\left[\left\langle S_{\text{HH}} S_{\text{VV}}^* \right\rangle\right]\right\}
\end{aligned}
\tag{4.3.2}
$$

Therefore, the two null angle features $\theta_{\text{null}}\_\text{Re}[T_{12}]$ and $\theta_{\text{null}}\_\text{Im}[T_{12}]$ are sensitive to different land covers, which are expected to enhance the PolSAR land cover classification accuracy. In addition, from the polarimetric feature selection demonstrated in Section 4.2, it is clear that both null angle features, $\theta_{\text{null}}\_\text{Re}[T_{12}]$ and $\theta_{\text{null}}\_\text{Im}[T_{12}]$, are collected for both AIRSAR and UAVSAR sensors over different imaging scenes. In this vein, with expert knowledge incorporation, six polarimetric features of the polarimetric entropy *H*, mean alpha angle $\bar{\alpha}$, polarimetric anisotropy *Ani*, total backscattering power *SPAN*, and null angle features $\theta_{\text{null}}\_\text{Re}[T_{12}]$ and $\theta_{\text{null}}\_\text{Im}[T_{12}]$ are considered.

4.3.1.2 Architecture of the Deep CNN Classifier

The deep CNN classifier achieves great success in image classification. This work adopts this framework, and the architecture of the deep CNN model is shown in Figure 4.3.1. It contains three main convolutional layers interleaved with two max-pooling layers, two fully connected layers, and a Softmax classifier. The dimension of input data cubic is $15\times15\times m$, where m indicates the feature dimension. In detail, there are 30, 60, and 120 filters for the three convolutional layers while with the same convolutional size of 2×2 and the stride of 1. The two max-pooling layers have the same pooling size of 2×2 and the stride of 2. The rectified linear unit (ReLU) activation function is implemented for the three convolutional layers and the first fully connected layer. Furthermore, dropout processing with a 0.5 ratio is applied to the second fully connected layer to smooth the potential over-fitting effect.

From these investigations, the selected polarimetric features, including polarimetric roll-invariant features of polarimetric entropy *H*, mean alpha angle $\bar{\alpha}$, polarimetric anisotropy *Ani*, total backscattering power *SPAN*, and two null angle features $\theta_{\text{null}}\_\text{Re}[T_{12}]$ and $\theta_{\text{null}}\_\text{Im}[T_{12}]$ are used as the input. In this vein, a polarimetric-feature-driven deep CNN classifier is established. Meanwhile, another scheme is used for comparison. It uses the same deep CNN architecture shown in Figure 4.3.1 while using the nine elements (T_{11}, T_{22}, T_{33}, $\text{Re}[T_{12}]$, $\text{Im}[T_{12}]$, $\text{Re}[T_{13}]$, $\text{Im}[T_{13}]$, $\text{Re}[T_{23}]$, and $\text{Im}[T_{23}]$) of polarimetric coherency matrices as the input.

4.3.2 Classification Comparison Studies

In this section, comparison studies are carried out with the single-temporal AIRSAR PolSAR data and the multi-temporal UAVSAR PolSAR datasets. All these datasets are speckle filtered by the SimiTest approach [28] with a 15 × 15 sliding window. The polarimetric-feature-driven CNN classification method with selected polarimetric features as the input is named Selected Features+CNN (SF+CNN). The deep CNN

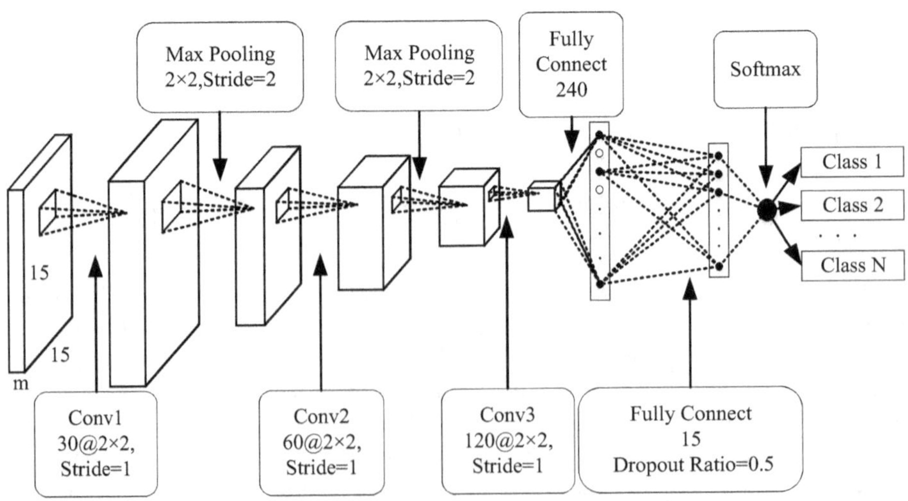

FIGURE 4.3.1 Polarimetric-feature-driven CNN classifier architecture.

classification method using only the nine elements of the polarimetric coherence matrix as input is named T+CNN. For both methods, the commonly used stochastic gradient descent and back-propagation algorithm with 64 minibatches is adopted for classifier training. The initial weights are randomly determined and updated with the learning rate 0.01, momentum parameter 0.9, and weight decay 0.0005. The classification performances of the two methods at different training rates (10%, 5%, and 1%) are compared, respectively. At the same time, each group of experiments is repeated ten times, the training sample is randomly selected each time, and the average classification accuracy is quantitatively compared.

4.3.2.1 Comparison with AIRSAR Data

The Pauli image of the single-temporal AIRSAR PolSAR data is shown in Figure 4.3.2 (a). Compared with the 11 known land covers shown in Figure 4.2.3, this section further divides the wheat into wheat 1, wheat 2, and wheat 3 and adds another two known land covers, barley and buildings. The ground-truth map for these 15 known land covers is shown in Figure 4.3.2 (b). The training epochs are 300 for the 10% and 5% training ratios and 1500 for the 1% training ratio situation. The training samples are randomly selected, and the remaining samples are used for validation for each experiment. There is no overlap between the training and validation samples. Classification results over the known land covers and full-scene area with 10% training ratio are given in Figures 4.3.3 and 4.3.4. Classification results over the known land covers and full-scene area with 5% training ratio are given in Figures 4.3.5 and 4.3.6. Classification results over the known land covers and full-scene area with 1% training ratio are given in Figures 4.3.7 and 4.3.8. Visually, the overall classification performances of the two methods are comparable.

Quantitative comparisons are summarized in Table 4.3.1. For more complete comparison, Table 4.3.1 also includes the classification accuracy of other three methods [7, 10, 13] over the same data. Compared with other traditional classification methods [7, 10] and the CNN method [13], both the SF+CNN and T+CNN methods achieve better classification accuracy. Meanwhile, the SF+CNN approach achieves the highest OAs of 99.3%, 98.83%, and 97.57%, respectively, for the training ratios of 10%, 5%, and 1%. The classification accuracies for each class from the SF+CNN method are

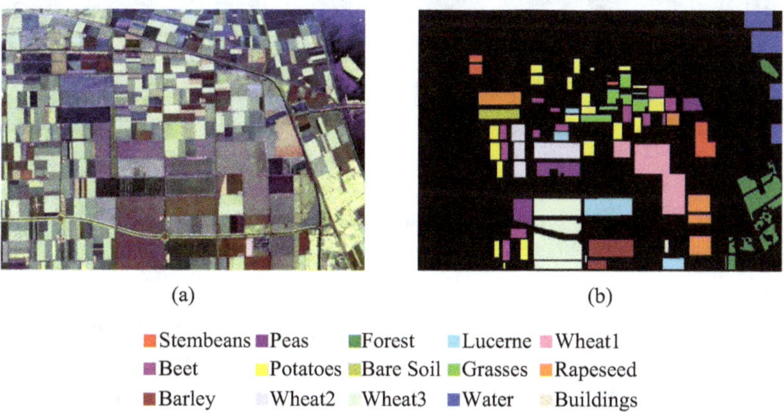

FIGURE 4.3.2 AIRSAR data. (a) Pauli image, (b) ground-truth map.

(a) (b)

FIGURE 4.3.3 Classification results over the known land covers with 10% training ratio. (a) T+CNN method, (b) SF+CNN method.

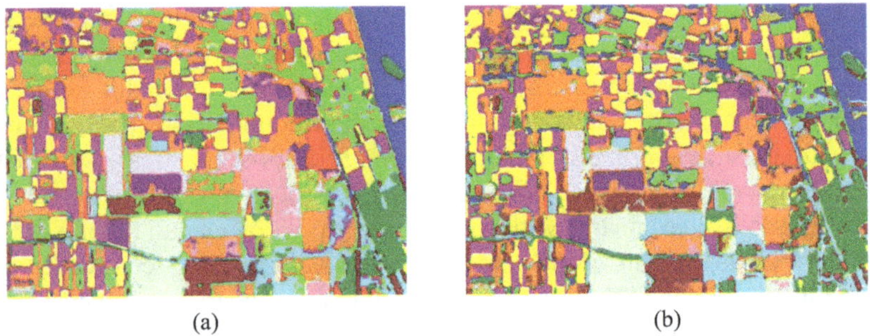

(a) (b)

FIGURE 4.3.4 Classification results over the full-scene area with 10% training ratio. (a) T+CNN method, (b) SF+CNN method.

(a) (b)

FIGURE 4.3.5 Classification results over the known land covers with 5% training ratio. (a) T+CNN method, (b) SF+CNN method.

given in Table 4.3.2. Except for the buildings, the classification accuracies are all above 90%. For training ratios of 5% and 1%, the classification accuracies of buildings are only 85.90% and 68.92%. The reason lies in that the total number of known building samples is only 342 (shown in Table 4.3.2). There are only 17 and 4 samples used during the model training for 5% and 1% training ratio cases. In comparison, for the

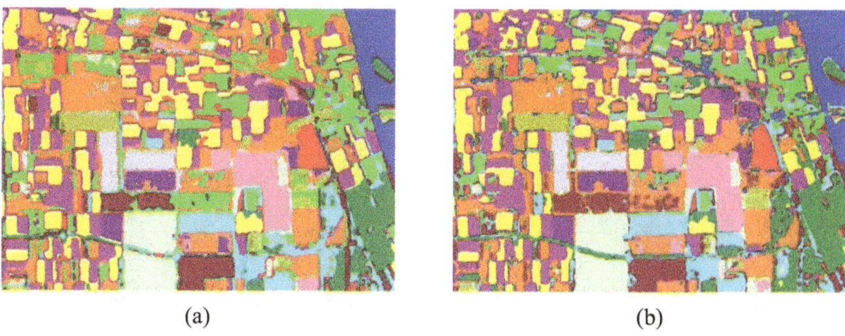

(a) (b)

FIGURE 4.3.6 Classification results over the full-scene area with 5% training ratio. (a) T+CNN method, (b) SF+CNN method.

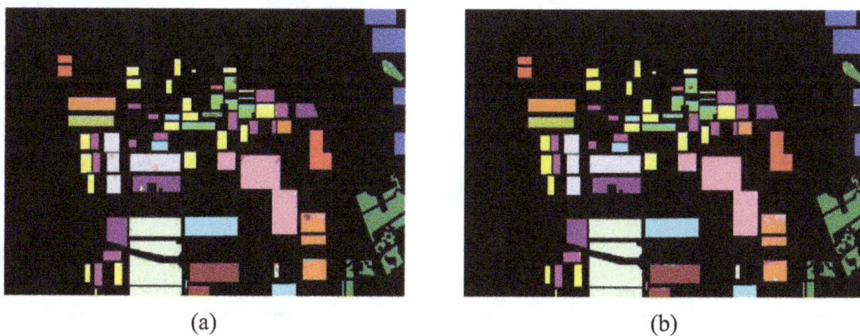

(a) (b)

FIGURE 4.3.7 Classification results over the known land covers with 1% training ratio. (a) T+CNN method, (b) SF+CNN method.

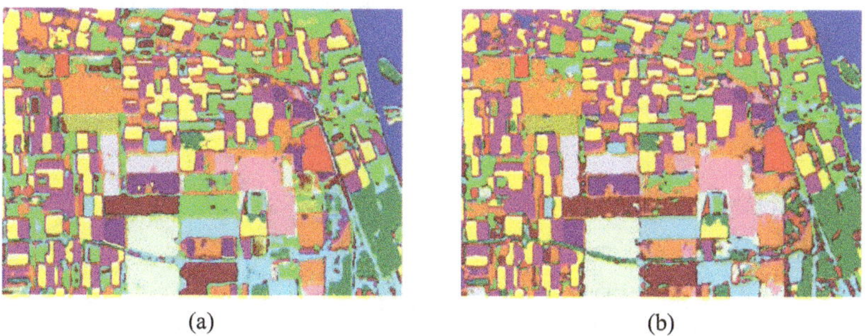

(a) (b)

FIGURE 4.3.8 Classification results over the full-scene with 1% training ratio. (a) T+CNN method, (b) SF+CNN method.

10% training ratio case, the classification accuracy is still as high as 93.08% for the buildings. These results validate the superiority of the SF+CNN classification scheme.

Furthermore, the model training speeds between the SF+CNN and T+CNN methods are plotted in Figure 4.3.9. For quantitative comparison, the convergence criteria that the training error is less than 0.5% and lasts for at least 5 epochs are adopted. For training ratios of 10%, 5%, and 1%, the averaged convergence epochs for the T+CNN method are respectively 190, 275, and 1201, while they are reduced to 82, 132, and

TABLE 4.3.1
Classification Results Comparison with AIRSAR Data

Method	Training Ratio (%)	Class Number	OA (%)
Lee et al. [7]	–	11	81.63
Wang et al. [10]	–	11	93.24
Zhou et al. [13]	–	11	93.38
	–	15	92.46
T+CNN	10	15	98.67
	5	15	98.64
	1	15	97.43
SF+CNN	10	15	**99.30**
	5	15	**98.83**
	1	15	**97.57**

TABLE 4.3.2
Classification Accuracy (%) of the SF+CNN Method

		Training Ratio		
Classes	Sample Number	10%	5%	1%
Stembeans	5484	99.42	99.11	98.48
Peas	8300	99.48	99.12	97.90
Forest	11758	99.58	99.49	99.25
Lucerne	8615	99.22	98.57	97.83
Beet	8596	98.98	98.67	97.43
Potatoes	13075	99.22	98.47	97.72
Bare soil	2800	99.75	99.48	98.79
Grasses	5342	96.87	96.10	91.61
Rapeseed	11632	99.17	98.23	95.31
Barley	6570	99.54	99.32	98.72
Wheat 1	16435	99.26	99.04	97.40
Wheat 2	9773	99.16	98.25	96.89
Wheat 3	19633	99.83	99.48	99.08
Water	10090	99.88	99.77	98.60
Buildings	342	93.08	85.90	68.92
OA		99.30	98.83	97.57

472 for the SF+CNN method. Generally, the SF+CNN method can be converged 2.3 times faster than the T+CNN method.

4.3.2.2 Comparison with Multi-Temporal UAVASAR Data

Multi-temporal PolSAR datasets are suitable for crop classification and growth monitoring. Due to the changes among multi-temporal acquisitions, developing a classifier with good robustness and high generalization is challenging [14]. In this

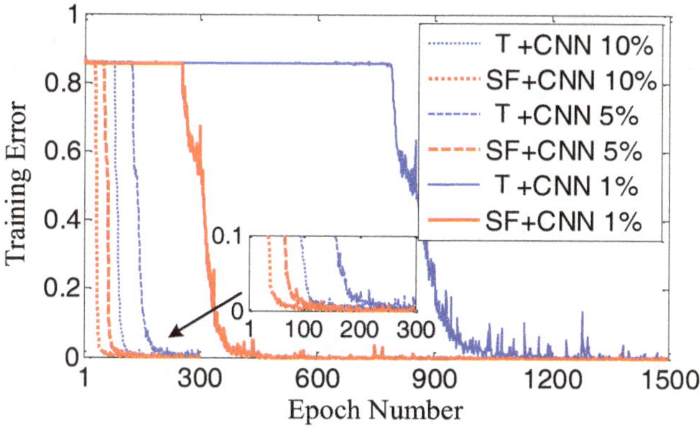

FIGURE 4.3.9 Model convergence speed comparison.

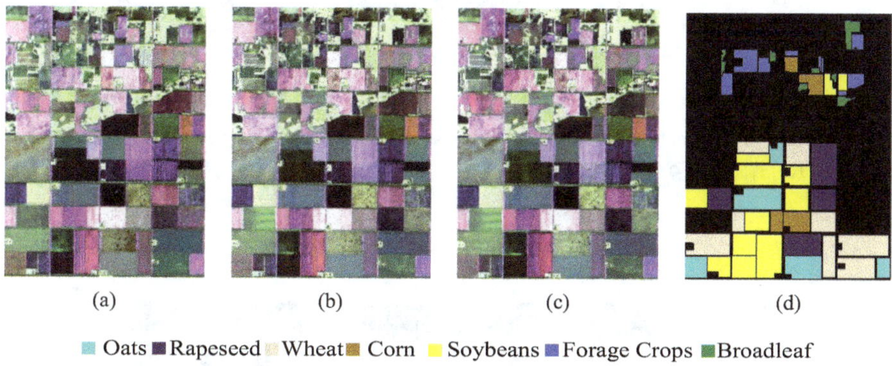

FIGURE 4.3.10 Multi-temporal UAVSAR data. Pauli images of (a) June 17, 2012 (DoY = 169), (b) June 22, 2012 (DoY = 174), and (c) June 23, 2012 (DoY = 175), (d) ground-truth map.

section, experimental studies with three temporal UAVSAR datasets obtained on July 17, July 22, and July 23, 2012, with DoYs of 169, 174, and 175, are carried out. The PolSAR Pauli images and ground-truth map are shown in Figure 4.3.10. The temporal changes due to crop growth can be observed, and the Pauli images of different temporal PolSAR datasets present variations. This section focuses on the comparative analysis of the SF+CNN method and the T+CNN method for multi-temporal PolSAR land cover classification. Meanwhile, in order to examine generalization performance, only samples from DoY 169 and 175 are used for training, while all three temporal PolSAR datasets are used for validation. This means that samples from DoY 174 will be relatively new for the trained classifiers. For DoY 174, the selected polarimetric features are illustrated in Figure 4.3.11.

The training epochs are 100 for 10% and 5% training ratios and 200 for the 1% training ratio case. Classification results over the known land covers and full-scene area with 10% training ratio are given in Figures 4.3.12 and 4.3.13. Classification results over the known land covers and full-scene area with 5% training ratio are given in Figures 4.3.14 and 4.3.15. Classification results over the known land covers

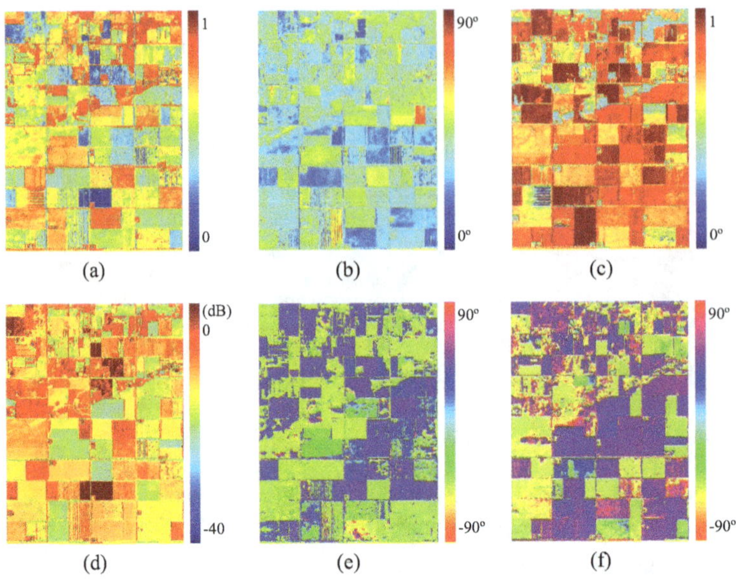

FIGURE 4.3.11 Selected polarimetric features (DoY = 174). (a) H, (b) $\overline{\alpha}$, (c) *Ani*, (d) *SPAN*, (e) $\theta_{\text{null}}\_\text{Re}[T_{12}]$, (f) $\theta_{\text{null}}\_\text{Im}[T_{12}]$.

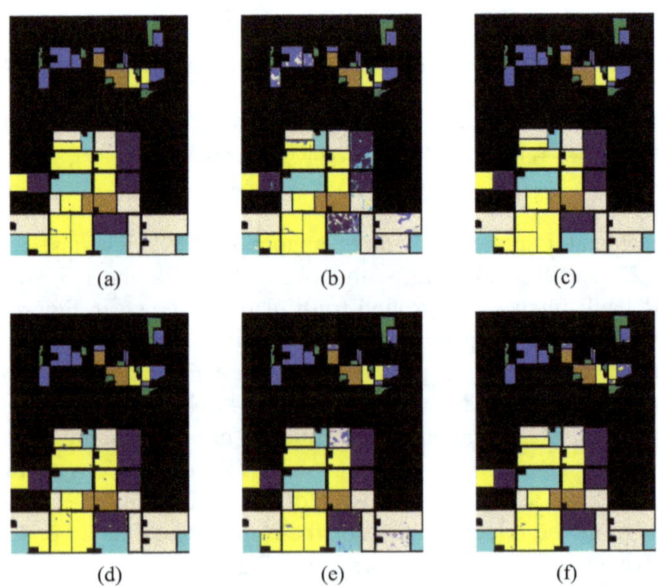

FIGURE 4.3.12 Classification results over the known land covers with 10% training ratio. (a)–(c) T+CNN method for DoYs 169, 174, and 175; (d)–(f) SF+CNN method for DoYs 169, 174, and 175, respectively.

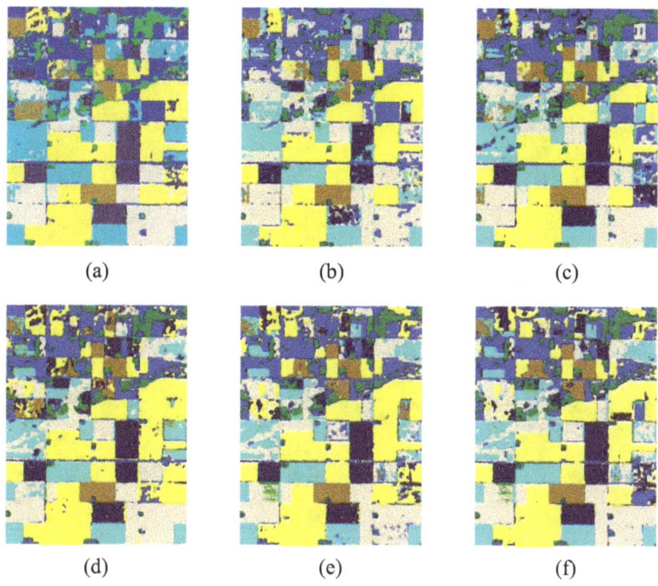

FIGURE 4.3.13 Classification results over the full-scene with 10% training ratio. (a)–(c) T+CNN method for DoYs 169, 174, and 175; (d)–(f) SF+CNN method for DoYs 169, 174, and 175, respectively.

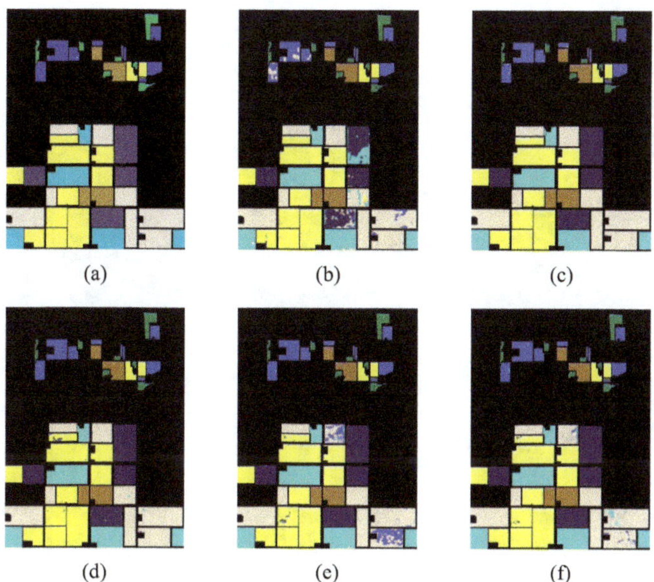

FIGURE 4.3.14 Classification results over the known land covers with 5% training ratio. (a)–(c) T+CNN method for DoYs 169, 174, and 175; (d)–(f) SF+CNN method for DoYs 169, 174, and 175, respectively.

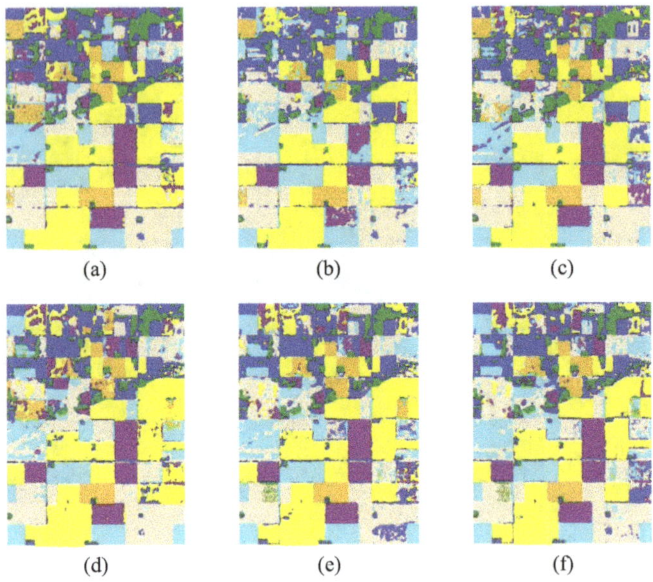

FIGURE 4.3.15 Classification results over the full-scene with 5% training ratio. (a)–(c) T+CNN method for DoYs 169, 174, and 175; (d)–(f) SF+CNN method for DoYs 169, 174, and 175, respectively.

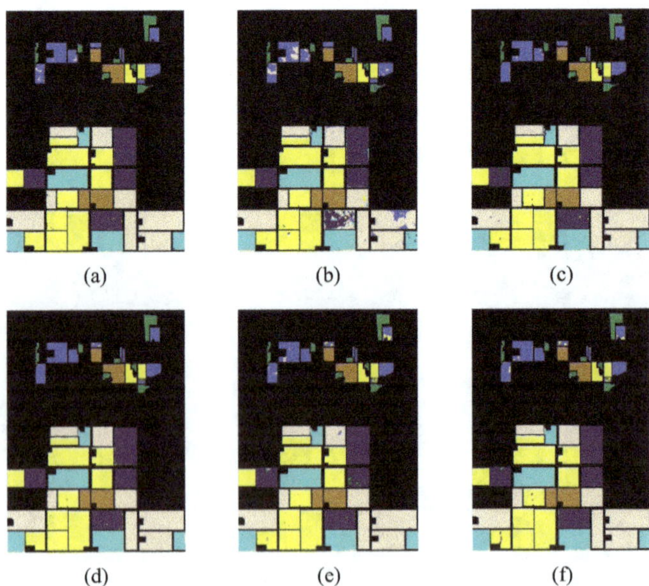

FIGURE 4.3.16 Classification results over the known land covers with 1% training ratio. (a)–(c) T+CNN method for DoYs 169, 174, and 175; (d)–(f) SF+CNN method for DoYs 169, 174, and 175, respectively.

and full-scene area with 1% training ratio are given in Figures 4.3.16 and 4.3.17. Quantitative classification performances are compared in Table 4.3.3. For training-used data of DoYs 169 and 175, both the T+CNN and SF+CNN methods achieve relatively equivalent classification performances. The obtained OAs are very high and stable, within 98.34% and 99.54% for three training ratio cases. The performance differences mainly lie in the data of DoY 174 not used for training. The SF+CNN method clearly outperforms the T+CNN method and obtains the higher OAs of 98.69%, 98.11%, and 97.04% for training ratios of 10%, 5%, and 1%, accordingly. In

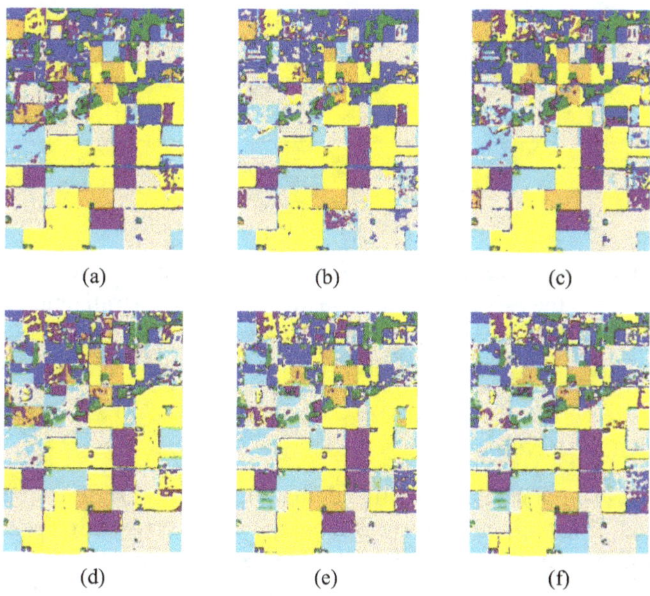

FIGURE 4.3.17 Classification results over the full-scene with 1% training ratio. (a)–(c) T+CNN method for DoYs 169, 174, and 175; (d)–(f) SF+CNN method for DoYs 169, 174, and 175, respectively.

TABLE 4.3.3
Classification Results Comparison with UAVSAR Data (%)

DoY	Method	Training Ratio 10%	5%	1%
169	T+CNN	99.35	99.38	98.92
	SF+CNN	99.54	99.43	98.86
174	T+CNN	93.18	93.70	92.37
	SF+CNN	**98.69**	**98.11**	**97.04**
175	T+CNN	99.53	99.43	98.87
	SF+CNN	99.42	99.23	98.34

TABLE 4.3.4
Classification Accuracy of the SF+CNN Method for Data Not Used in Training of DoY 174 (%)

Classes	Sample Number	Training Ratio		
		10%	5%	1%
Broadleaf	14149	99.21	98.79	98.46
Forage crops	32178	97.69	97.33	93.33
Soybeans	140697	99.56	98.94	98.90
Wheat	115305	97.28	95.95	93.65
Rapeseed	68613	99.22	98.78	97.91
Oats	64892	98.62	98.83	98.54
Corn	31463	99.82	99.83	99.41
OA	–	98.69	98.11	97.04

comparison, the SF+CNN method achieves 4.86% enhancement for the final OAs on average. Therefore, the SF+CNN scheme exhibits better generalization performance for multi-temporal PolSAR classification. The averaged classification accuracies for each class of DoY 174 from the SF+CNN method are given in Table 4.3.4.

4.4 CONCLUSIONS

In this chapter, two polarimetric rotation domain land cover classification approaches are introduced. The main idea is to combine the expert knowledge of target polarimetric scattering interpretation to drive traditional machine learning classifiers (e.g. SVM, DT) and deep learning classifiers (e.g. CNN) for PolSAR land cover classification by integrating traditional polarimetric roll-invariant features and selected polarimetric rotation domain features. Comparison studies are carried out with single-temporal and multi-temporal PolSAR datasets. The performance advantages of the polarimetric rotation domain land cover classification methods are clearly verified. Taking the polarimetric-feature-driven deep CNN method as an example, its performance advantages are mainly twofold. First, for single AIRSAR PolSAR data, the polarimetric-feature-driven deep CNN method achieves the best classification accuracy, while the model training convergence speed is increased by 2.3 times compared with the traditional deep CNN method. Second, for multi-temporal UAVSAR PolSAR datasets, both the polarimetric-feature-driven deep CNN method and the traditional deep CNN method achieve similar performance for train-used data. However, for data not used in training, the overall classification accuracy of the polarimetric-feature-driven deep CNN method is enhanced by 4.86% on average compared with the traditional deep CNN method. In this vein, it is validated that the deep CNN model with target scattering mechanism interpretation knowledge can equip better generalization capability. In addition, comparison experiments in terms of different training ratios further demonstrate that polarimetric rotation domain land cover classification methods can exhibit better robustness.

REFERENCES

1. J. S. Lee and E. Pottier, *Polarimetric Radar Imaging: From Basics to Applications*. Boca Raton: CRC Press, 2009.
2. S. R. Cloude, *Polarisation: Application in Remote Sensing*. New York: Oxford University Press, 2009.
3. J. J. Van Zyl and Y. Kim, *Synthetic Aperture Radar Polarimetry*. Hoboken: Wiley, 2011.
4. S. W. Chen, X. S. Wang, S. P. Xiao, and M. Sato, *Target Scattering Mechanism in Polarimetric Synthetic Aperture Radar-Interpretation and Application*. Singapore: Springer, 2018.
5. Y. Yamaguchi, *Polarimetric SAR Imaging: Theory and Applications*. Boca Raton: CRC Press, 2020.
6. J. Yang and J. J. Yin, *Polarimetric Radar Theory and Remote Sensing Applications*. Beijing: Science Press, 2020.
7. J. S. Lee, M. R. Grunes, and E. Pottier, "Quantitative comparison of classification capability: Fully polarimetric versus dual and single-polarization SAR," *IEEE Transactions on Geoscience and Remote Sensing,* vol. 39, no. 11, pp. 2343–2351, 2001.
8. S. R. Cloude and E. Pottier, "An entropy based classification scheme for land applications of polarimetric SAR," *IEEE Transactions on Geoscience and Remote Sensing,* vol. 35, no. 1, pp. 68–78, 1997.
9. J. S. Lee, M. R. Grunes, E. Pottier, and L. F. Famil, "Unsupervised terrain classification preserving polarimetric scattering characteristics," *IEEE Transactions on Geoscience and Remote Sensing,* vol. 42, no. 4, pp. 722–731, 2004.
10. H. Wang, Z. Zhou, J. Turnbull, Q. Song, and F. Qi, "Pol-SAR classification based on generalized polar decomposition of Mueller matrix," *IEEE Geoscience and Remote Sensing Letters,* vol. 13, no. 4, pp. 565–569, 2016.
11. H. Skriver, "Crop classification by multitemporal C- and L-band single- and dual-polarization and fully polarimetric SAR," *IEEE Transactions on Geoscience and Remote Sensing,* vol. 50, no. 6, pp. 2138–2149, 2012.
12. C. S. Tao, S. W. Chen, Y. Z. Li, and S. P. Xiao, "PolSAR land cover classification based on roll-invariant and selected hidden polarimetric features in the rotation domain," *Remote Sensing,* vol. 9, no. 7, p. 660, 2017.
13. Y. Zhou, H. Wang, F. Xu, and Y. Q. Jin, "Polarimetric SAR image classification using deep convolutional neural networks," *IEEE Geoscience and Remote Sensing Letters,* vol. 13, no. 12, pp. 1935–1939, 2016.
14. S. W. Chen and C. S. Tao, "Multi-temporal PolSAR crops classification using polarimetric-feature–driven deep convolutional neural network," in *International Workshop on Remote Sensing with Intelligent Processing*, Shanghai, China, 2017, pp. 1–4.
15. S. W. Chen and C. S. Tao, "PolSAR image classification using polarimetric-feature–driven deep convolutional neural network," *IEEE Geoscience and Remote Sensing Letters,* vol. 15, no. 4, pp. 627–631, 2018.
16. S. W. Chen, Y. Z. Li, X. S. Wang, S. P. Xiao, and M. Sato, "Modeling and interpretation of scattering mechanisms in polarimetric synthetic aperture radar: Advances and perspectives," *IEEE Signal Processing Magazine,* vol. 31, no. 4, pp. 79–89, 2014.
17. S. W. Chen, X. S. Wang, S. P. Xiao, and M. Sato, "General polarimetric model-based decomposition for coherency matrix," *IEEE Transactions on Geoscience and Remote Sensing,* vol. 52, no. 3, pp. 1843–1855, 2014.
18. S. W. Chen, X. S. Wang, Y. Z. Li, and M. Sato, "Adaptive model-based polarimetric decomposition using PolInSAR coherence," *IEEE Transactions on Geoscience and Remote Sensing,* vol. 52, no. 3, pp. 1705–1718, 2014.
19. S. R. Cloude and E. Pottier, "A review of target decomposition theorems in radar polarimetry," *IEEE Transactions on Geoscience and Remote Sensing,* vol. 34, no. 2, pp. 498–518, 1996.

20. L. F. Famil, E. Pottier, and J. S. Lee, "Unsupervised classification of multifrequency and fully polarimetric SAR images based on the H/A/Alpha-Wishart classifier," *IEEE Transactions on Geoscience and Remote Sensing,* vol. 39, no. 11, pp. 2332–2342, 2001.
21. A. Krizhevsky, I. Sutskever, and G. E. Hinton, "ImageNet classification with deep convolutional neural networks," *Communications of the ACM,* vol. 60, no. 6, pp. 84–90, 2017.
22. S. W. Chen, X. S. Wang, and M. Sato, "Uniform polarimetric matrix rotation theory and its applications," *IEEE Transactions on Geoscience and Remote Sensing,* vol. 52, no. 8, pp. 4756–4770, 2014.
23. X. S. Wang and S. W. Chen, "Polarimetric synthetic aperture radar interpretation and recognition: Advances and perspectives," *Journal of Radars,* vol. 9, no. 2, pp. 259–276, 2020.
24. C. S. Tao, S. W. Chen, Y. Z. Li, and S. P. Xiao, "Polarimetric SAR terrain classification using polarimetric features derived from rotation domain," *Journal of Radars,* vol. 6, no. 5, pp. 524–532, 2016.
25. S. W. Chen, "Polarimetric coherence pattern: A visualization and characterization tool for PolSAR data investigation," *IEEE Transactions on Geoscience and Remote Sensing,* vol. 56, no. 1, pp. 286–297, 2018.
26. C. C. Chang and C. J. Lin, "LIBSVM: A library for support vector machines," *ACM Transactions on Intelligent Systems and Technology,* vol. 2, no. 3, pp. 1–27, 2011.
27. A. Webb and K. Copsey, *Statistical Pattern Recognition.* Hoboken: Wiley, 2011.
28. S. W. Chen, X. S. Wang, and M. Sato, "PolInSAR complex coherence estimation based on covariance matrix similarity test," *IEEE Transactions on Geoscience and Remote Sensing,* vol. 50, no. 11, pp. 4699–4710, 2012.
29. H. McNairn, T. J. Jackson, G. Wiseman, S. Belair, A. Berg, P. Bullock, et al., "The soil moisture active passive validation experiment 2012 (SMAPVEX12): Prelaunch calibration and validation of the SMAP soil moisture algorithms," *IEEE Transactions on Geoscience and Remote Sensing,* vol. 53, no. 5, pp. 2784–2801, 2015.
30. S. W. Chen, "SAR image speckle filtering with context covariance matrix formulation and similarity test," *IEEE Transactions on Image Processing,* vol. 29, pp. 6641–6654, 2020.

5 Polarimetric Rotation Domain Target Detection

5.1 INTRODUCTION

Target detection is an important application for polarimetric radar. In particular, PolSAR manmade target detection, including ship detection, building detection, aircraft detection, and so on, plays an essential role in many application fields. Meanwhile, manmade targets are usually surrounded by complex environmental clutter, leading to relatively low TCRs. Moreover, manmade targets with different orientations usually exhibit quite varied scattering characteristics, leading to scattering mechanism interpretation ambiguity between manmade targets and background clutter. Therefore, PolSAR manmade target detection is still a challenging task.

As introduced in Chapter 2, compared with environmental clutter, manmade targets can exhibit quite different properties in the polarimetric rotation domain. In this vein, novel target detection techniques can be developed by mining and utilizing the scattering diversity between manmade targets and environmental clutter with polarimetric rotation domain interpretation tools. With PolSAR acquisition, this chapter mainly introduces polarimetric rotation domain ship detection, building detection, and aircraft detection approaches.

5.2 POLARIMETRIC ROTATION DOMAIN SHIP DETECTION

Ship detection is an important PolSAR application. This section will introduce two PolSAR ship detection methods based on the two-dimension and three-dimension polarimetric correlation pattern interpretation tools. The core ideas of both methods are to exploit the ship target scattering diversity in the polarimetric rotation domain and enhance the contrast between ship targets and sea clutter through polarimetric rotation domain feature extraction and selection. Thereby, PolSAR ship detection performance can be enhanced.

5.2.1 PolSAR Data Description

Three Radarsat-2 PolSAR datasets are adopted in this section for ship detection. These datasets were acquired on December 16, 2008; January 1, 2016; and September 8, 2011, and mainly cover the areas of Hong Kong, China; Strait of Gibraltar; and Bohai Sea, China, respectively. Three regions of interests are selected from these datasets for comparison study. The ROI information is listed in Table 5.2.1. The Pauli images are shown in Figure 5.2.1 (a1–c1), accordingly. Ground-truth images are given in

DOI: 10.1201/9781003461296-5

TABLE 5.2.1
Radarsat-2 PolSAR Data Information for Three ROIs

ROI	Acquisition Date	Image Size/Pixel	Pixel Spacing in Range and Azimuth Directions/m
Hong Kong, China	December 16, 2008	300 × 500	4.82 × 4.73
Strait of Gibraltar	January 1, 2016	700 × 1300	4.60 × 4.73
Bohai Sea, China	September 8, 2011	700 × 800	4.74 × 4.73

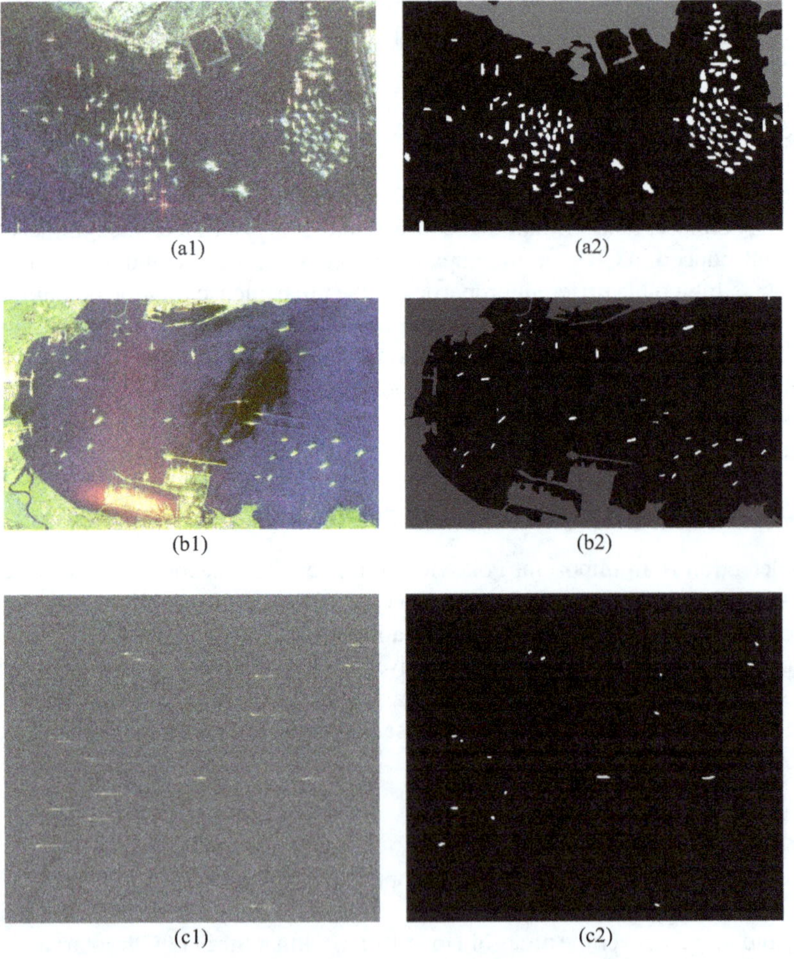

FIGURE 5.2.1 Radarsat-2 PolSAR datasets for three ROIs. (a1)–(c1) are Pauli images, and (a2)–(c2) are ground-truth images for the three ROIs of Hong Kong, China; Strait of Gibraltar; and Bohai Sea, China, respectively.

Figures 5.2.1 (a2–c2), which are acquired with the AIS information and the support of experts' confirmation. For ground-truth images, white, black, and gray pixels respectively indicate ships, sea clutter, and land areas. The ROIs from Hong Kong, China, and the Strait of Gibraltar are inshore areas and contain lots of ships, while the ROI from the Bohai Sea, China, is an offshore areas where ships are more sparsely appeared. The number of ships from the three ROIs are 137, 37, and 16, accordingly.

5.2.2 Ship Detection with Two-Dimension Polarimetric Correlation Pattern

5.2.2.1 Feature Investigation and Selection

From Chapter 2, ten polarimetric features can be derived from each two-dimension polarimetric correlation pattern. Among them, the maximum rotation angle, minimum rotation angle and correlation beamwidth are angle features, while the other 7 polarimetric features are amplitude features. This section mainly considers the amplitude features. In this vein, 28 amplitude polarimetric features from four typical two-dimension polarimetric correlation patterns, $\left|\hat{\gamma}_{\text{HH-HV}}(\theta)\right|$, $\left|\hat{\gamma}_{\text{HH-VV}}(\theta)\right|$, $\left|\hat{\gamma}_{\text{(HH+VV)-(HH-VV)}}(\theta)\right|$, and $\left|\hat{\gamma}_{\text{(HH-VV)-(HV)}}(\theta)\right|$, are considered.

Generally, the TCR index can reflect the detection potential of features. Usually, the higher the TCR, the better the detection performance [1]. Thus, the TCR index is adopted for quantitative analysis and feature selection. In detail, three ship regions with size 50×50 are randomly selected from the ROI of Hong Kong, China. For each polarimetric feature, the mean values of sea clutter pixels and ship pixels are denoted as $\text{sea}_{\text{mean}}^{\text{All}}$ and $\text{ship}_{\text{mean}}^{\text{All}}$. Therefore, the TCR index is calculated as

$$\text{TCR}=10\log_{10}\left\{\frac{\text{ship}_{\text{mean}}^{\text{All}}}{\text{sea}_{\text{mean}}^{\text{All}}}\right\} \tag{5.2.1}$$

TCR results are illustrated in Figure 5.2.2. It includes the TCR values of 28 two-dimension polarimetric correlation pattern features. There are four zones representing two-dimension polarimetric correlation patterns, $\left|\hat{\gamma}_{\text{HH-HV}}(\theta)\right|$, $\left|\hat{\gamma}_{\text{HH-VV}}(\theta)\right|$,

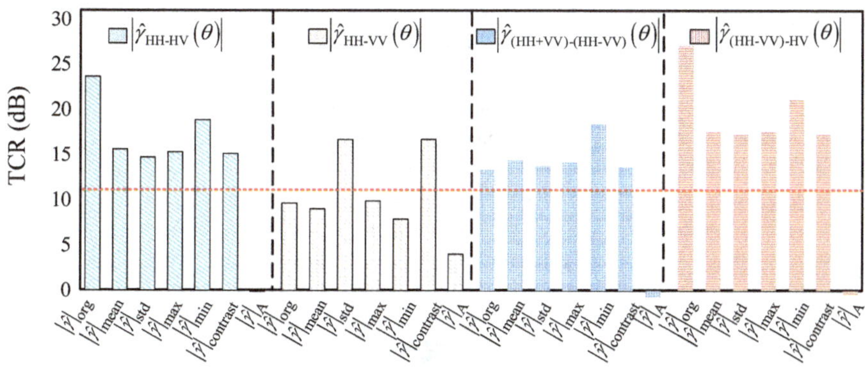

FIGURE 5.2.2 TCR comparison for polarimetric features between ships and sea clutter. The red dashed line indicates the TCR value of the *SPAN* feature.

$\left|\hat{\gamma}_{(\text{HH+VV})\text{-}(\text{HH-VV})}(\theta)\right|$, and $\left|\hat{\gamma}_{(\text{HH-VV})\text{-}(\text{HV})}(\theta)\right|$, respectively. For each zone, the TCR values of the original correlation $|\hat{\gamma}|_{\text{org}}$, the correlation degree $|\hat{\gamma}|_{\text{mean}}$, the correlation fluctuation $|\hat{\gamma}|_{\text{std}}$, the maximum correlation $|\hat{\gamma}|_{\text{max}}$, the minimum correlation $|\hat{\gamma}|_{\text{min}}$, the correlation contrast $|\hat{\gamma}|_{\text{contrast}}$, and the correlation anisotropy $|\hat{\gamma}|_{\text{A}}$ are given from left to right. Meanwhile, the red dashed line indicates the TCR value of the *SPAN* feature. Compared with the *SPAN* feature, most of the TCR values from two-dimension polarimetric correlation pattern features are obviously higher. Specially, polarimetric features $\left|\hat{\gamma}_{(\text{HH-VV})\text{-}(\text{HV})}(\theta)\right|_{\text{org}}$, $\left|\hat{\gamma}_{\text{HH-HV}}(\theta)\right|_{\text{org}}$, and $\left|\hat{\gamma}_{(\text{HH-VV})\text{-}(\text{HV})}(\theta)\right|_{\text{min}}$ exhibit the top three highest TCR values of 26.52 dB, 23.20 dB, and 20.75 dB, respectively. Therefore, these three polarimetric features are selected for ship detection thereafter.

5.2.2.2 Ship Detection with Two-Dimension Polarimetric Correlation Pattern Features

From the TCR investigations, three two-dimension polarimetric correlation pattern features, $\left|\hat{\gamma}_{(\text{HH-VV})\text{-}(\text{HV})}(\theta)\right|_{\text{org}}$, $\left|\hat{\gamma}_{\text{HH-HV}}(\theta)\right|_{\text{org}}$, and $\left|\hat{\gamma}_{(\text{HH-VV})\text{-}(\text{HV})}(\theta)\right|_{\text{min}}$, are selected. Considering the obvious differences between ships and sea clutter in terms of these three polarimetric features, ships can be detected with an adaptive thresholding procedure. In this vein, this section introduces a ship detection method with two-dimension polarimetric correlation pattern features [1, 2]. It contains three main steps. First, the two-dimension polarimetric correlation pattern features are extracted, and those with higher TCRs are selected. Second, an adaptive thresholding procedure is performed to obtain the ship candidates. Finally, morphological filtering is used to eliminate false alarms such as small isolated pixels, and the final ship detection results are obtained. The flowchart of the ship detection method is shown in Figure 5.2.3.

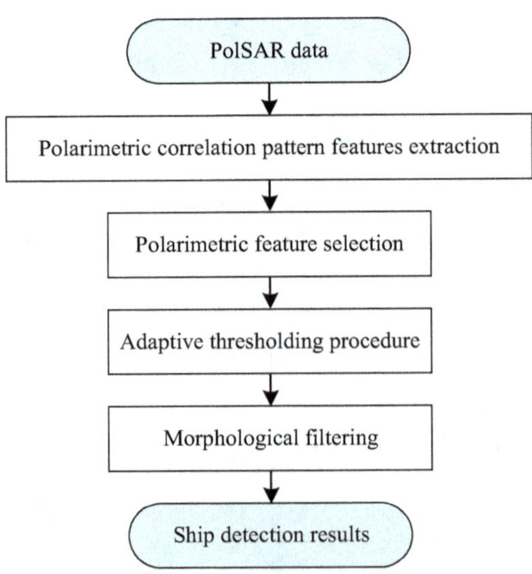

FIGURE 5.2.3 Ship detection flowchart with two-dimension polarimetric correlation pattern features.

5.2.2.3 Ship Detection Comparison

Experimental studies based on real PolSAR datasets are carried out with three comparison methods. The first comparison method is based on the *SPAN* feature and thresholding processing. The second comparative approach is the saliency method, which combines the superpixel technique and salient features for ship detection [3]. The third one is the R-filter method, which adopts the polarimetric correlation between the HH and HV polarization channels for ship detection [4]. Actually, the polarimetric feature used in [4] is the selected polarimetric feature $\left|\hat{\gamma}_{\text{HH-HV}}(\theta)\right|_{\text{org}}$. Note that the same thresholding procedure and morphological filtering procedure are fairly conducted for all these comparison methods.

The figure of merit (FoM) index is adopted to evaluate the performance of ship detection results:

$$\text{FoM} = \frac{N_{\text{C}}}{N_{\text{C}} + N_{\text{M}} + N_{\text{FA}}} \tag{5.2.2}$$

where N_{C}, N_{FA}, and N_{M} indicate the number of correct detections, false alarms, and missed detections, respectively. If more than $\alpha\%$ pixels of a ship are detected, the ship target is regarded as a correct detection. From the investigation of the parameter α, the ship detection performances of comparative methods are the best when α equals 10. Therefore, $\alpha = 10$ is adopted in this section. If there is no overlapping between a detected candidate and the ground-truth image, it is considered a false alarm. Otherwise, it is a missed detection.

For three ROIs of Radarsat-2 PolSAR data, the polarimetric feature maps of $\left|\hat{\gamma}_{\text{(HH-VV)-(HV)}}(\theta)\right|_{\text{org}}$, $\left|\hat{\gamma}_{\text{HH-HV}}(\theta)\right|_{\text{org}}$, $\left|\hat{\gamma}_{\text{(HH-VV)-(HV)}}(\theta)\right|_{\text{min}}$, and *SPAN* are illustrated in Figures 5.2.4–5.2.6, respectively. Note that each polarimetric feature is normalized independently, and the dynamic ranges are the same. For the *SPAN* feature, the feature values of ships and sea clutter are relatively close, especially from the ROI of the Bohai Sea, China, as shown in Figure 5.2.6. Compared with the *SPAN* feature, the selected two-dimension polarimetric correlation features exhibit better potential to distinguish ship targets from sea clutter environments, which is consistent with the TCR comparison shown in Figure 5.2.2.

For three ROIs of Radarsat-2 PolSAR data, the ship detection results are shown in Figures 5.2.7–5.2.9, respectively. The quantitative results in terms of the correct detection, missed detection, false alarm, and FoM are summarized in Table 5.2.2.

For the ROI of Hong Kong, China, ship targets are relative dense. From the detection results of the saliency method, many detected ships are connected with each other, which are hard to distinguish from the adjacent ships, as shown in Figure 5.2.7 (c). In comparison, the selected three two-dimension polarimetric correlation pattern features and the *SPAN* feature achieve better detection performances in terms of separating each individual ships. For quantitative comparison, the ship detection method with the $\left|\hat{\gamma}_{\text{(HH-VV)-(HV)}}(\theta)\right|_{\text{org}}$ feature achieves the highest FoM of 98.56% with no missed detections.

For the ROI of Strait of Gibraltar area, the detection results are shown in Figure 5.2.8. The ship detection method with the *SPAN* feature has 13 false alarms at

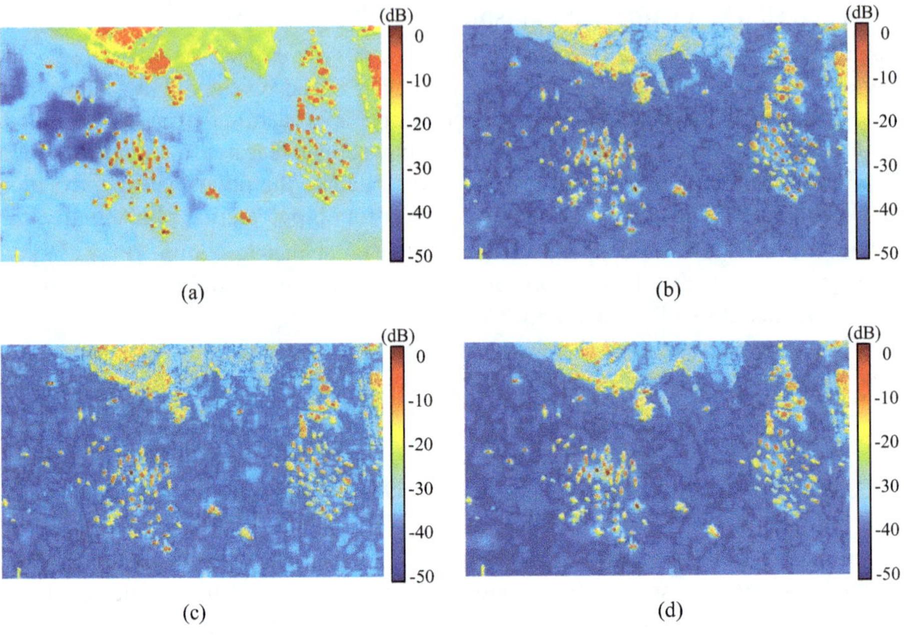

FIGURE 5.2.4 Polarimetric feature maps over the ROI of Hong Kong, China. (a) *SPAN*, (b) $\left|\hat{\gamma}_{\text{HH-HV}}(\theta)\right|_{\text{org}}$, (c) $\left|\hat{\gamma}_{\text{(HH-VV)-(HV)}}(\theta)\right|_{\text{min}}$, (d) $\left|\hat{\gamma}_{\text{(HH-VV)-(HV)}}(\theta)\right|_{\text{org}}$.

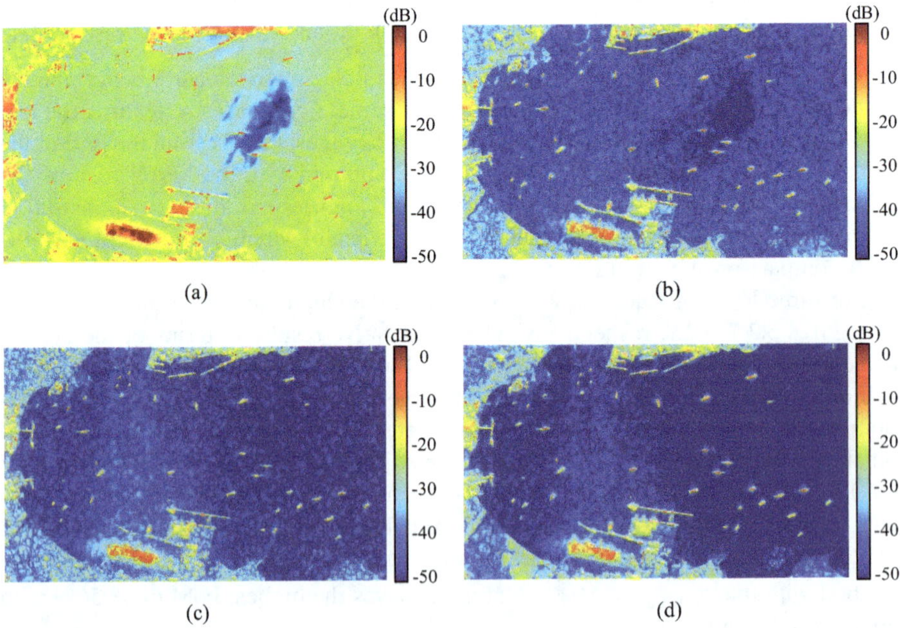

FIGURE 5.2.5 Polarimetric feature maps over the ROI of the Strait of Gibraltar. (a) *SPAN*, (b) $\left|\hat{\gamma}_{\text{HH-HV}}(\theta)\right|_{\text{org}}$, (c) $\left|\hat{\gamma}_{\text{(HH-VV)-(HV)}}(\theta)\right|_{\text{min}}$, (d) $\left|\hat{\gamma}_{\text{(HH-VV)-(HV)}}(\theta)\right|_{\text{org}}$.

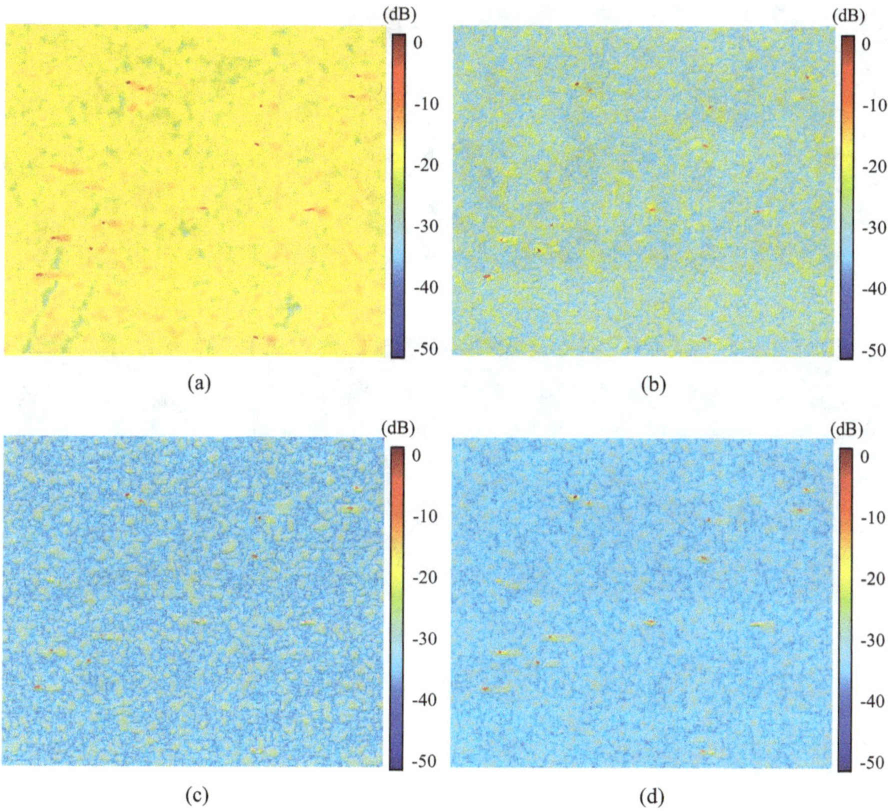

FIGURE 5.2.6 Polarimetric feature maps over the ROI of Bohai Sea, China. (a) *SPAN*, (b) $\left|\hat{\gamma}_{\text{HH-HV}}(\theta)\right|_{\text{org}}$, (c) $\left|\hat{\gamma}_{\text{(HH-VV)-(HV)}}(\theta)\right|_{\min}$, (d) $\left|\hat{\gamma}_{\text{(HH-VV)-(HV)}}(\theta)\right|_{\text{org}}$.

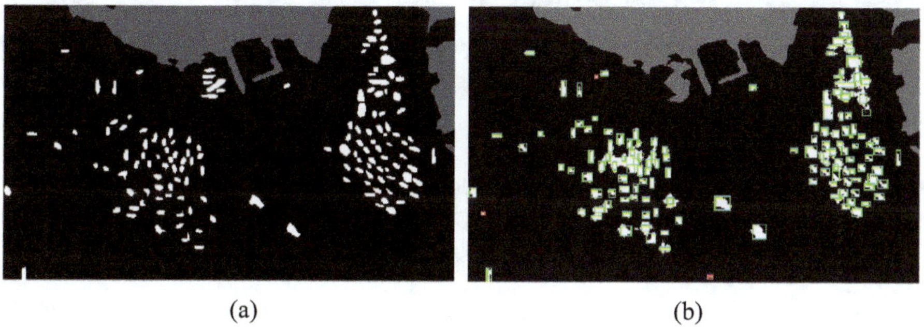

FIGURE 5.2.7 Ship detection results of Radarsat-2 PolSAR data over the ROI of Hong Kong, China. (a) Ground-truth image. Ship detection with (b) *SPAN*, (c) the saliency method, (d) $\left|\hat{\gamma}_{\text{HH-HV}}(\theta)\right|_{\text{org}}$ (R-filter), (e) $\left|\hat{\gamma}_{\text{(HH-VV)-(HV)}}(\theta)\right|_{\min}$, (f) $\left|\hat{\gamma}_{\text{(HH-VV)-(HV)}}(\theta)\right|_{\text{org}}$.

(Continued)

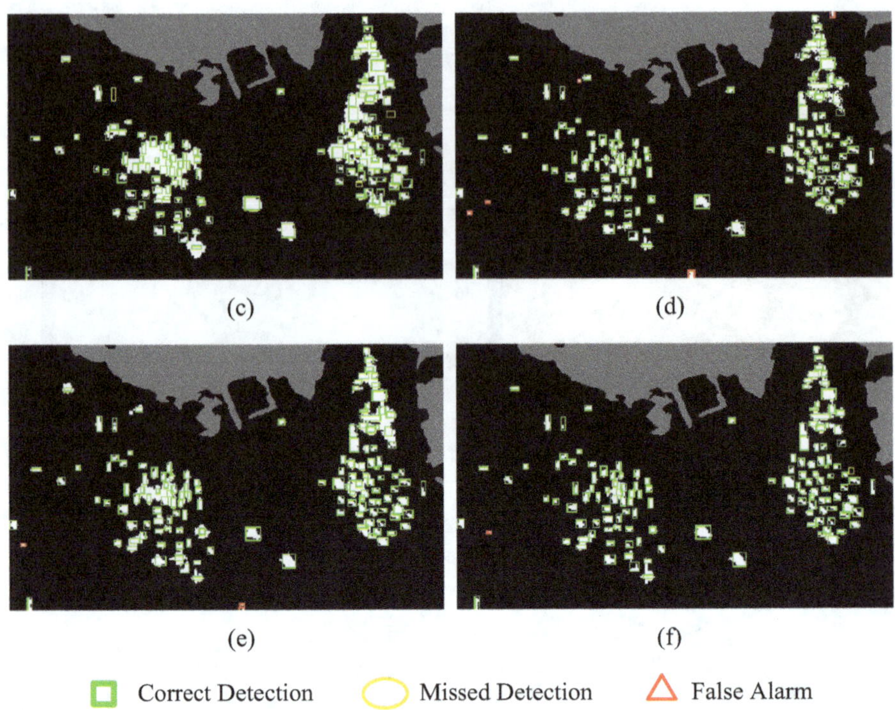

FIGURE 5.2.7 (Continued)

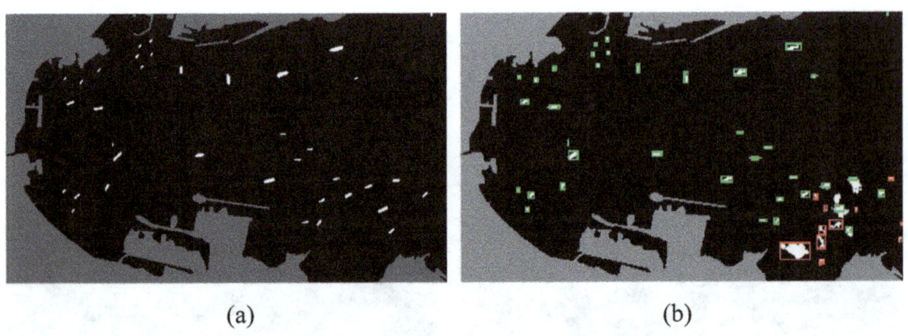

FIGURE 5.2.8 Ship detection results of Radarsat-2 PolSAR data over the ROI of the Strait of Gibraltar. (a) Ground-truth image. Ship detection with (b) *SPAN*, (c) the saliency method, (d) $\left|\hat{\gamma}_{\text{HH-HV}}(\theta)\right|_{\text{org}}$ (R-filter), (e) $\left|\hat{\gamma}_{\text{(HH-VV)-(HV)}}(\theta)\right|_{\text{min}}$, (f) $\left|\hat{\gamma}_{\text{(HH-VV)-(HV)}}(\theta)\right|_{\text{org}}$.

(Continued)

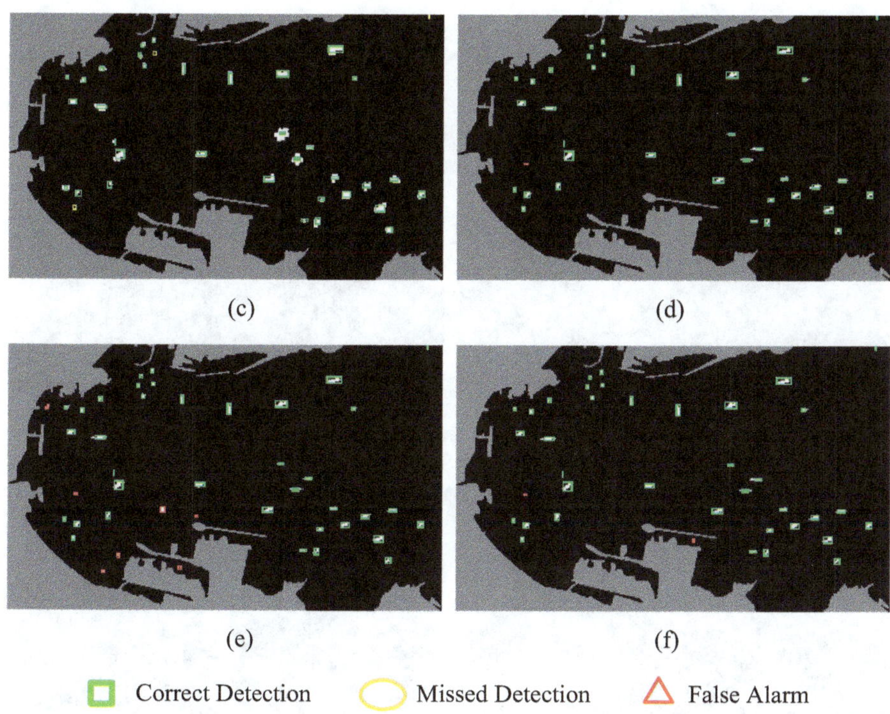

FIGURE 5.2.8 (Continued)

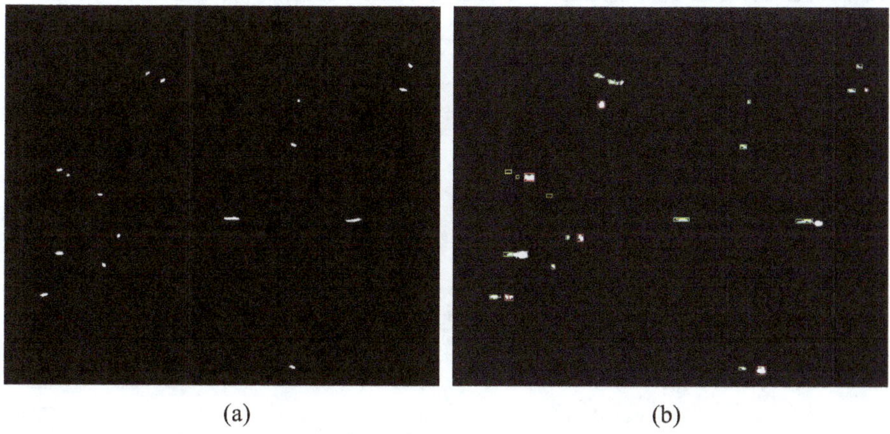

FIGURE 5.2.9 Ship detection results of Radarsat-2 PolSAR data over the ROI of Bohai Sea, China. (a) Ground-truth image. Ship detection with (b) *SPAN*, (c) the saliency method, (d) $\left|\hat{\gamma}_{\text{HH-HV}}(\theta)\right|_{\text{org}}$ (R-filter), (e) $\left|\hat{\gamma}_{\text{(HH-VV)-(HV)}}(\theta)\right|_{\min}$, (f) $\left|\hat{\gamma}_{\text{(HH-VV)-(HV)}}(\theta)\right|_{\text{org}}$.

(Continued)

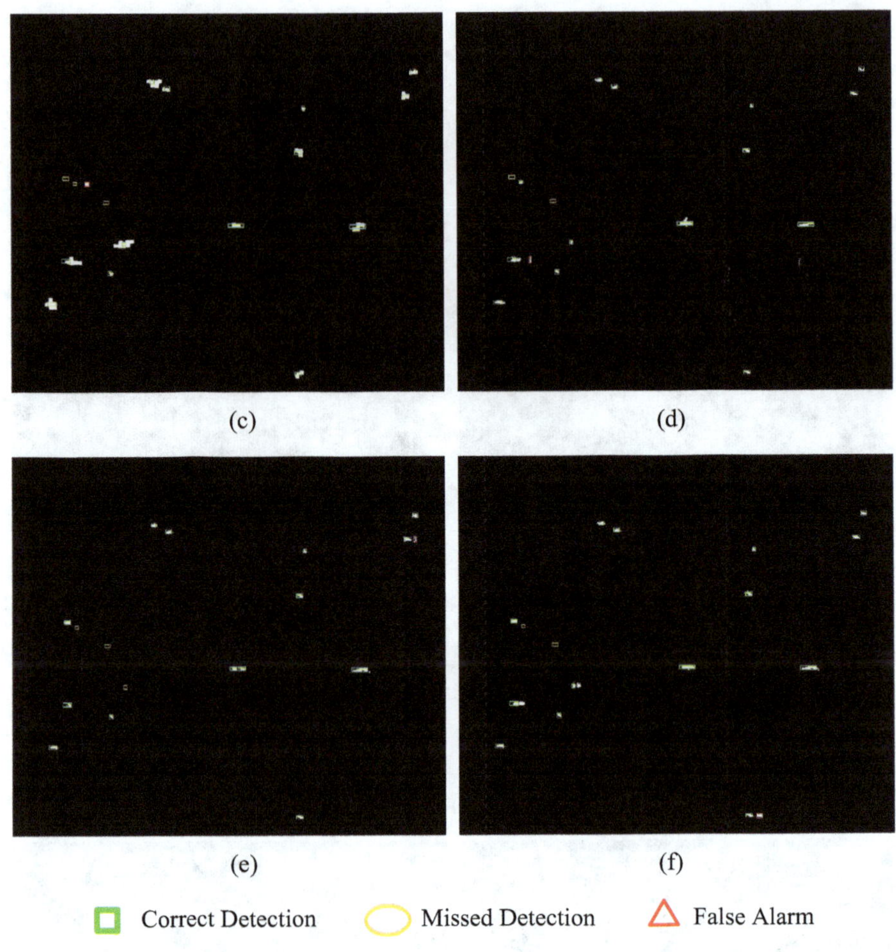

FIGURE 5.2.9 (Continued)

TABLE 5.2.2
Quantitative Comparison of Ship Detection Results

ROI	Method	N_C	N_M	N_{FA}	FoM
	SPAN	136	1	1	98.55%
	Saliency method	133	4	0	97.08%
Hong Kong, China	$\left\lvert\hat{\gamma}_{\text{HH-HV}}(\theta)\right\rvert_{\text{org}}$ (R-filter)	137	0	3	97.86%
	$\left\lvert\hat{\gamma}_{\text{(HH-VV)-(HV)}}(\theta)\right\rvert_{\text{min}}$	137	0	6	95.80%
	$\left\lvert\hat{\gamma}_{\text{(HH-VV)-(HV)}}(\theta)\right\rvert_{\text{org}}$	137	0	2	**98.56%**

(Continued)

TABLE 5.2.2 ***(Continued)***
Quantitative Comparison of Ship Detection Results

ROI	Method	N_C	N_M	N_{FA}	FoM
	SPAN	37	0	13	74.00%
	Saliency method	34	3	0	91.89%
Strait of Gibraltar	$\left\|\hat{\gamma}_{\text{HH-HV}}(\theta)\right\|_{\text{org}}$ (R-filter)	37	0	1	**97.37%**
	$\left\|\hat{\gamma}_{\text{(HH-VV)-(HV)}}(\theta)\right\|_{\text{min}}$	37	0	8	82.22%
	$\left\|\hat{\gamma}_{\text{(HH-VV)-(HV)}}(\theta)\right\|_{\text{org}}$	37	0	2	94.87%
	SPAN	13	3	7	56.52%
	Saliency method	13	3	1	76.47%
Bohai Sea, China	$\left\|\hat{\gamma}_{\text{HH-HV}}(\theta)\right\|_{\text{org}}$ (R-filter)	14	2	1	**82.35%**
	$\left\|\hat{\gamma}_{\text{(HH-VV)-(HV)}}(\theta)\right\|_{\text{min}}$	13	3	1	76.47%
	$\left\|\hat{\gamma}_{\text{(HH-VV)-(HV)}}(\theta)\right\|_{\text{org}}$	14	2	1	**82.35%**

the bottom right of the ROI, leading to a relatively low FoM of 74.00%. The saliency method produces 3 missed detections, as shown in Figure 5.2.8 (c). In comparison, ship detection methods with the selected three two-dimension polarimetric correlation features have much better detection performances. The ship detection method with the $\left|\hat{\gamma}_{\text{HH-HV}}(\theta)\right|_{\text{org}}$ feature can detect all ship targets, with only one false alarm, achieving the highest FoM value of 97.37%, as shown in Figure 5.2.8 (d). Meanwhile, the ship detection method with the $\left|\hat{\gamma}_{\text{(HH-VV)-(HV)}}(\theta)\right|_{\text{org}}$ feature has 2 false alarms and a FoM of 94.87%, which also outperforms comparison approaches.

For the ROI of Bohai Sea, China, since the ships show relatively low backscattering power, ship detection is a difficult task. The ship detection results are shown in Figure 5.2.9. It is clear that missed detections are observed for all these methods. In this vein, compared with the other two ROIs, the detection performances over the ROI of Bohai Sea area decrease. In detail, the ship detection method with the *SPAN* feature produces 3 missed detections and 7 false alarms. The corresponding FoM is the lowest among all these methods. Meanwhile, the saliency method and the method with the $\left|\hat{\gamma}_{\text{(HH-VV)-(HV)}}(\theta)\right|_{\text{min}}$ feature both have 3 missed detections and 1 false alarm. In comparison, ship detection methods respectively with the $\left|\hat{\gamma}_{\text{HH-HV}}(\theta)\right|_{\text{org}}$ and $\left|\hat{\gamma}_{\text{(HH-VV)-(HV)}}(\theta)\right|_{\text{org}}$ features only have 2 missed detections and 1 false alarm, which achieves the highest FoM of 83.25%.

In general, the ship detection approaches with the selected three two-dimension polarimetric correlation pattern features outperform the comparison methods for ship detection over the used PolSAR datasets, especially for dense ship detection and weak scattering ship detection.

5.2.3 Ship Detection with Three-Dimension Polarimetric Correlation Pattern

5.2.3.1 Visualization of Ship and Sea Areas

The three-dimension polarimetric correlation pattern interpretation tool [5–7] is introduced in Chapter 2. This section investigates the properties of ships and sea clutter in terms of three typical three-dimension polarimetric correlation patterns $\left|\hat{\gamma}_{\text{HH-HV}}(\theta,\tau)\right|$, $\left|\hat{\gamma}_{\text{HH-VV}}(\theta,\tau)\right|$, and $\left|\hat{\gamma}_{\text{(HH+VV)-(HH-VV)}}(\theta,\tau)\right|$. One pixel from a ship and one pixel from the sea area are randomly selected from Figure 5.2.1. The corresponding three-dimension polarimetric correlation patterns are generated and displayed in Figure 5.2.10. Compared with the typical scattering structures shown in Figure 2.6.2, the three-dimension polarimetric correlation patterns $\left|\hat{\gamma}_{\text{HH-HV}}(\theta,\tau)\right|$, $\left|\hat{\gamma}_{\text{HH-VV}}(\theta,\tau)\right|$, and $\left|\hat{\gamma}_{\text{(HH+VV)-(HH-VV)}}(\theta,\tau)\right|$ from the selected ship pixel are more similar to those of the

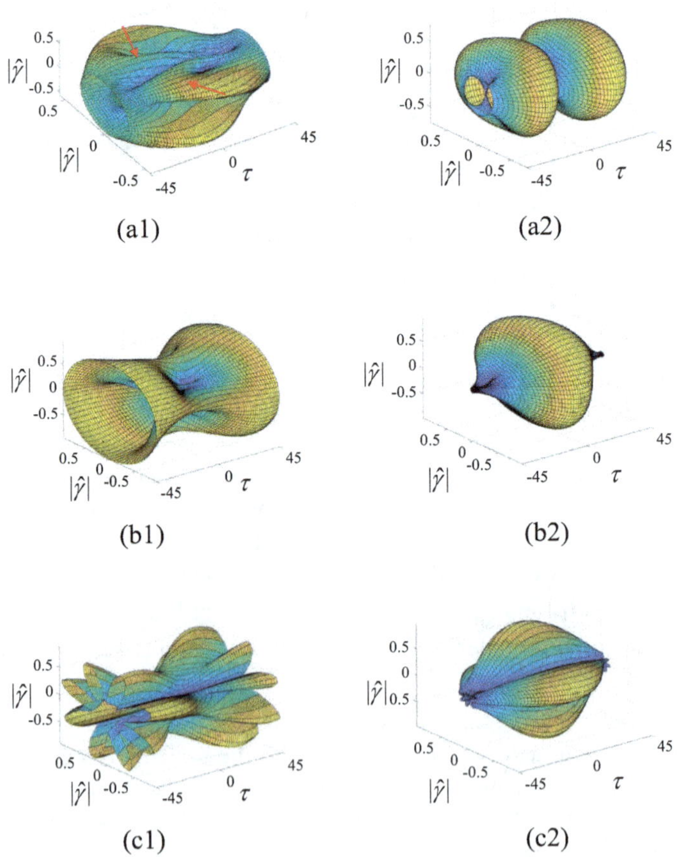

FIGURE 5.2.10 Three-dimension polarimetric correlation patterns for ship and sea pixels. (a1)–(a2) $\left|\hat{\gamma}_{\text{HH-HV}}(\theta,\tau)\right|$, (b1)–(b2) $\left|\hat{\gamma}_{\text{HH-VV}}(\theta,\tau)\right|$, and (c1)–(c2) $\left|\hat{\gamma}_{\text{(HH+VV)-(HH-VV)}}(\theta,\tau)\right|$. The numbers 1 and 2 indicate the ship pixel and the sea pixel, respectively.

dihedral structure. Meanwhile, the three-dimension polarimetric correlation patterns $\left|\hat{\gamma}_{\text{HH-HV}}(\theta,\tau)\right|$, $\left|\hat{\gamma}_{\text{HH-VV}}(\theta,\tau)\right|$, and $\left|\hat{\gamma}_{\text{(HH+VV)-(HH-VV)}}(\theta,\tau)\right|$ of the selected sea clutter pixel are more similar to those of the plane structure.

Furthermore, for the selected ship pixel and sea clutter pixel, the corresponding Gaussian curvature features can be extracted from three three-dimension polarimetric correlation patterns $\left|\hat{\gamma}_{\text{HH-HV}}(\theta,\tau)\right|$, $\left|\hat{\gamma}_{\text{HH-VV}}(\theta,\tau)\right|$, and $\left|\hat{\gamma}_{\text{(HH+VV)-(HH-VV)}}(\theta,\tau)\right|$, as shown in Figure 5.2.11. Several peaks can be observed from the Gaussian curvature features for the ship pixel, while those of the sea clutter pixel are close to zero. The

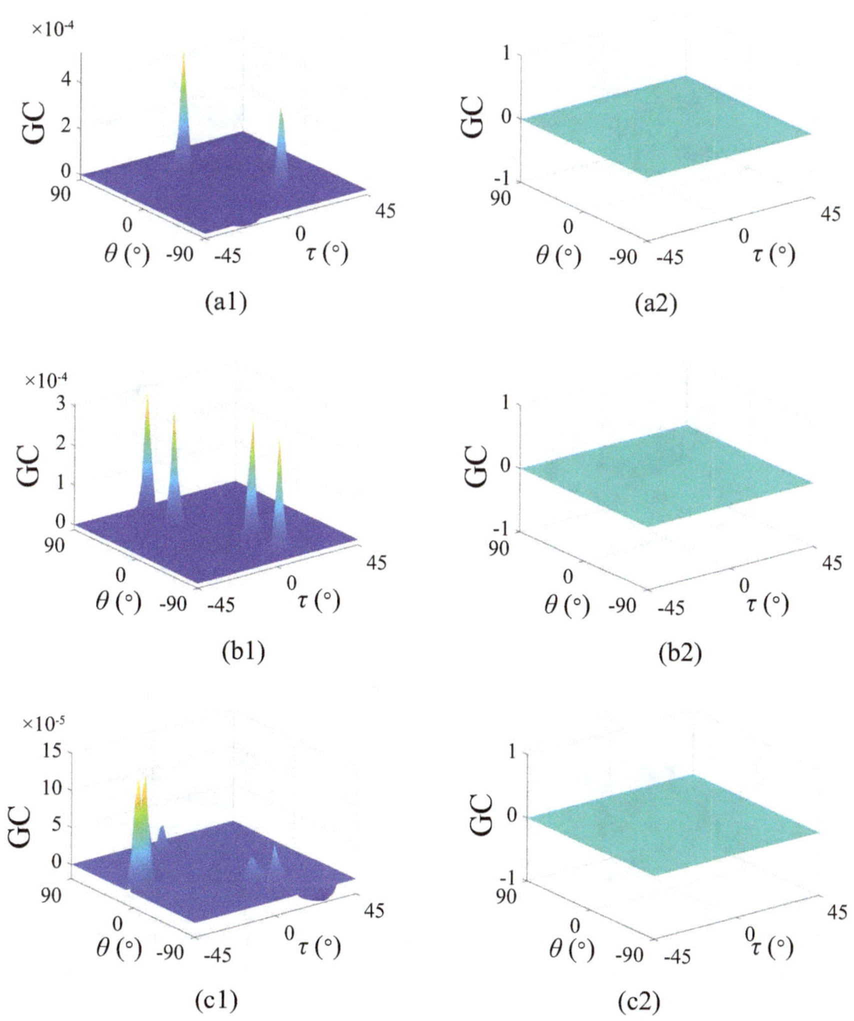

FIGURE 5.2.11 The Gaussian curvature features for ship and sea pixels from three-dimension polarimetric correlation patterns of (a1)–(a2) $\left|\hat{\gamma}_{\text{HH-HV}}(\theta,\tau)\right|$, (b1)–(b2) $\left|\hat{\gamma}_{\text{HH-VV}}(\theta,\tau)\right|$, and (c1)–(c2) $\left|\hat{\gamma}_{\text{(HH+VV)-(HH-VV)}}(\theta,\tau)\right|$, accordingly. The numbers 1 and 2 indicate the ship pixel and the sea pixel, respectively.

mean curvature and the maximum absolute value of the principal curvature features for the two selected pixels are displayed in Figures 5.2.12 and 5.2.13, respectively. In general, obvious fluctuations are observed from the mean curvature features and the principal curvature features for both ship and sea clutter pixels. In this vein, their differences need further study. In comparison, the corresponding Gaussian curvature features can distinguish ships and sea clutter more significantly.

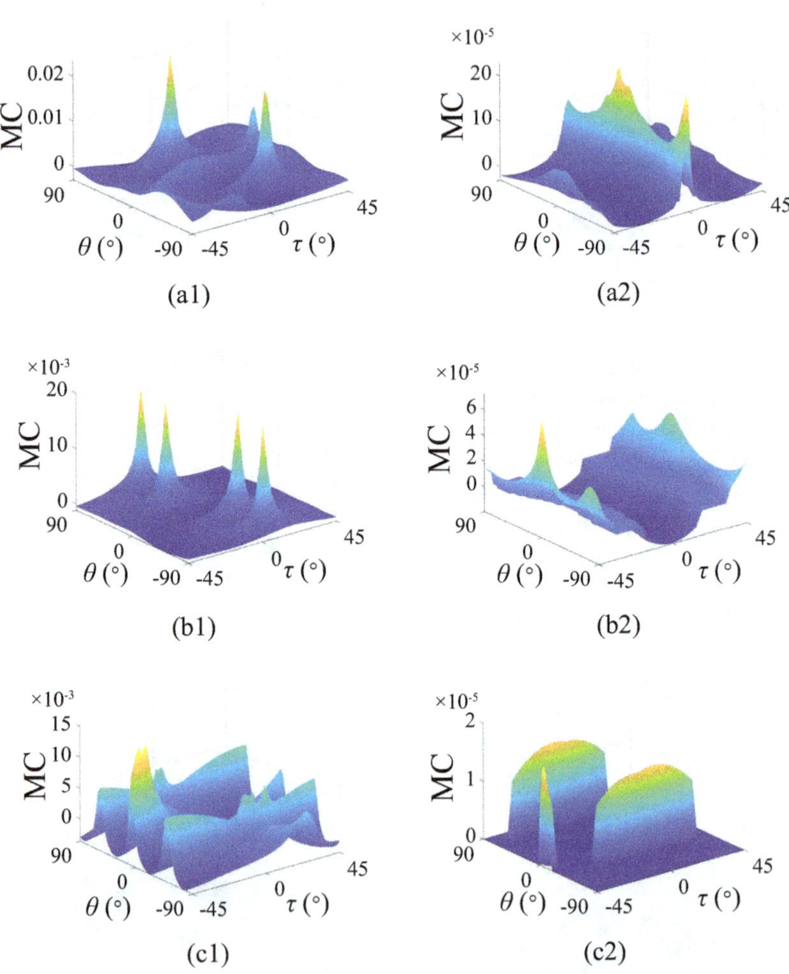

FIGURE 5.2.12 The mean curvature features for ship and sea pixels from three-dimension polarimetric correlation patterns of (a1)–(a2) $\left|\hat{\gamma}_{\text{HH-HV}}(\theta,\tau)\right|$, (b1)–(b2) $\left|\hat{\gamma}_{\text{HH-VV}}(\theta,\tau)\right|$, and (c1)–(c2) $\left|\hat{\gamma}_{\text{(HH+VV)-(HH-VV)}}(\theta,\tau)\right|$, accordingly. The numbers 1 and 2 indicate the ship pixel and the sea pixel, respectively.

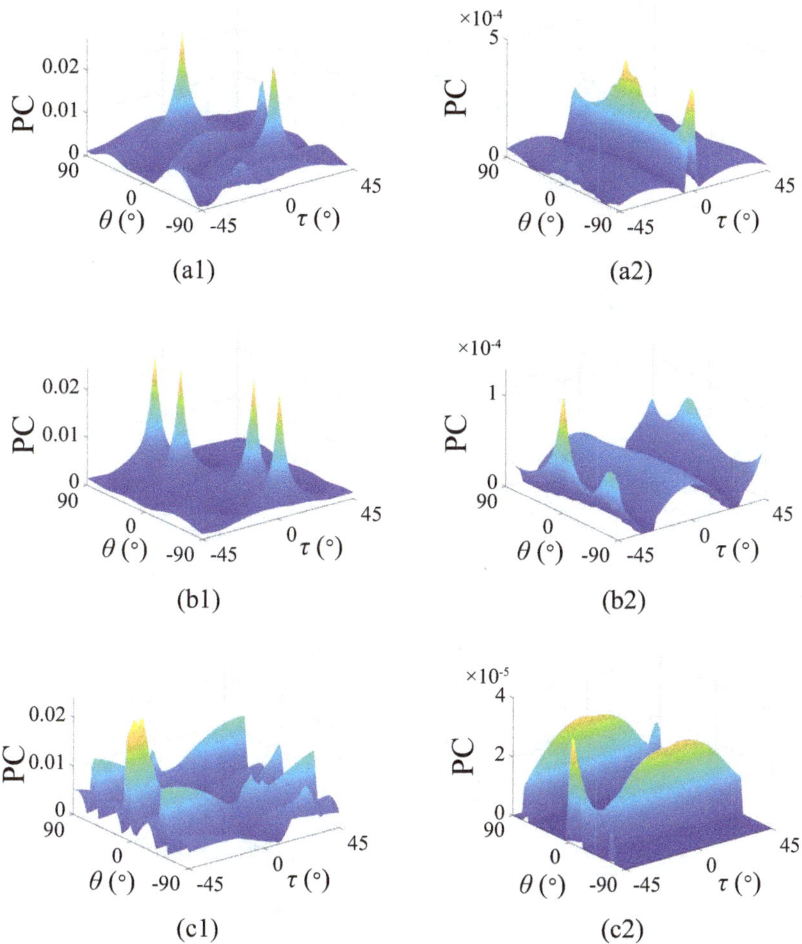

FIGURE 5.2.13 The principal curvature features for ship and sea pixels from three-dimension polarimetric correlation patterns of (a1)–(a2) $\left|\hat{\gamma}_{\text{HH-HV}}(\theta,\tau)\right|$, (b1)–(b2) $\left|\hat{\gamma}_{\text{HH-VV}}(\theta,\tau)\right|$, and (c1)–(c2) $\left|\hat{\gamma}_{\text{(HH+VV)-(HH-VV)}}(\theta,\tau)\right|$, accordingly. The numbers 1 and 2 indicate the ship pixel and the sea pixel, respectively.

5.2.3.2 Feature Investigation and Selection

As discussed in Chapter 2, 34 polarimetric features with clear physical meanings can be available for each independent three-dimension polarimetric correlation pattern. Therefore, for $\left|\hat{\gamma}_{\text{HH-HV}}(\theta,\tau)\right|$, $\left|\hat{\gamma}_{\text{HH-VV}}(\theta,\tau)\right|$ and $\left|\hat{\gamma}_{\text{(HH+VV)-(HH-VV)}}(\theta,\tau)\right|$, the total number of polarimetric features is 102. The TCR index is also adopted for quantitative comparison and feature selection.

For the ROI of Hong Kong, China, TCR results of these three-dimension polarimetric correlation pattern features are shown in Figure 5.2.14. There are three zones

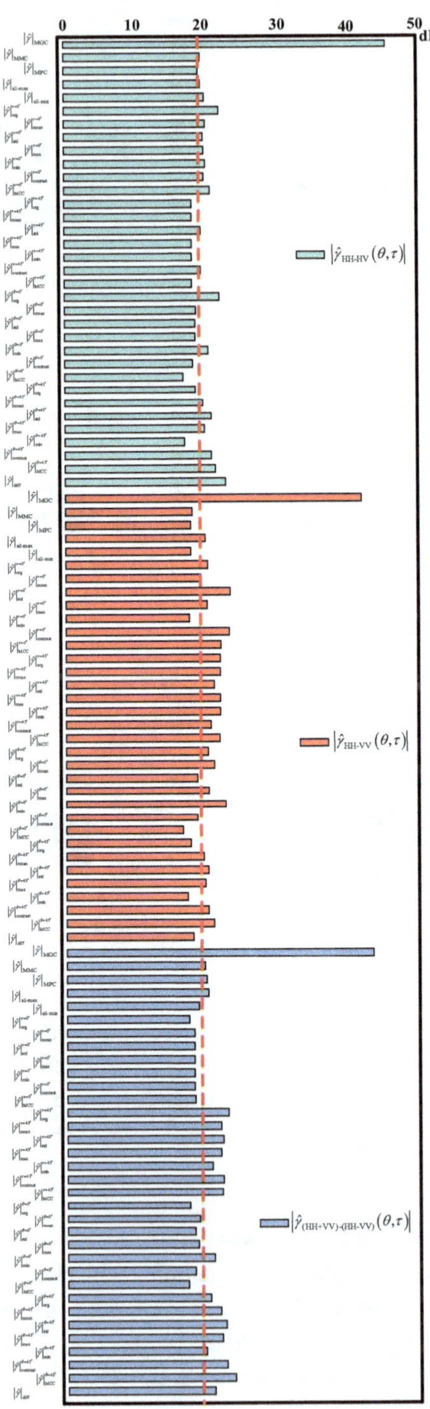

FIGURE 5.2.14 The TCR comparison results for Radarsat-2 data over the ROI of Hong Kong, China. The red dashed line indicates the TCR value of the *SPAN* feature. Different colored bars represent different three-dimension polarimetric correlation patterns, accordingly.

from the top to the bottom which correspond to the three-dimension polarimetric correlation patterns $\left|\hat{\gamma}_{\text{HH-HV}}(\theta,\tau)\right|$, $\left|\hat{\gamma}_{\text{HH-VV}}(\theta,\tau)\right|$, and $\left|\hat{\gamma}_{\text{(HH+VV)-(HH-VV)}}(\theta,\tau)\right|$, respectively. For each three-dimension polarimetric correlation pattern, the order of polarimetric features from the top to the bottom are global features: the maximum Gaussian curvature, the maximum mean curvature, the maximum principle curvature, the maximum correlation, and the minimum correlation and local features: the slice original correlation, the slice maximum correlation, the slice minimum correlation, the slice correlation degree, the slice correlation fluctuation, the slice correlation contrast, the slice maximum curve curvature, and the correlation difference for slices τ=0°, τ=45°, θ=0°, and θ=45°, respectively. For comparison, the dashed line from Figure 5.2.14 represents the TCR value of the *SPAN* feature. Then, the TCR comparison results of the *SPAN* and three-dimension polarimetric correlation pattern features with the top five highest TCRs are shown in Figure 5.2.15. Among them, the TCR values of the three maximum Gaussian curvature features are higher than those of other polarimetric features. The TCR value of the maximum Gaussian curvature feature $\left|\hat{\gamma}_{\text{HH-HV}}(\theta,\tau)\right|_{\text{MGC}}$ reaches a maximum of 45.49 dB, which is 26.14 dB higher than that of the *SPAN* feature. Meanwhile, the feature maps of these six polarimetric features are shown in Figure 5.2.16. Among them, for the three maximum Gaussian curvature features $\left|\hat{\gamma}_{\text{HH-HV}}(\theta,\tau)\right|_{\text{MGC}}$, $\left|\hat{\gamma}_{\text{(HH+VV)-(HH-VV)}}(\theta,\tau)\right|_{\text{MGC}}$, and $\hat{\gamma}_{\text{HH-VV}}(\theta,\tau)\big|_{\text{MGC}}$, the sea clutter is more homogeneous and the contrasts between ships and sea clutter are also more obvious.

For ROI of the Strait of Gibraltar area, TCR results of three-dimension polarimetric correlation pattern features are shown in Figure 5.2.17. As shown in Figure 5.2.14, there are also three zones from the top to the bottom which correspond to the three-dimension polarimetric correlation patterns $\left|\hat{\gamma}_{\text{HH-HV}}(\theta,\tau)\right|$, $\left|\hat{\gamma}_{\text{HH-VV}}(\theta,\tau)\right|$, and $\left|\hat{\gamma}_{\text{(HH+VV)-(HH-VV)}}(\theta,\tau)\right|$, respectively. For each three-dimension polarimetric correlation

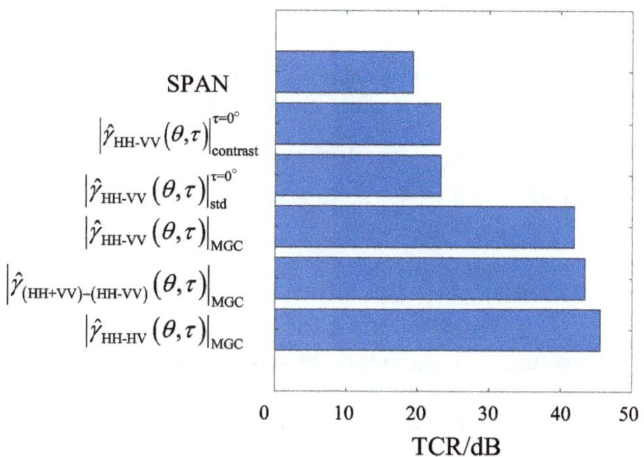

FIGURE 5.2.15 The TCR comparison results of the *SPAN* and three-dimension polarimetric correlation pattern features with the top five highest TCRs for Radarsat-2 data over the ROI of Hong Kong, China.

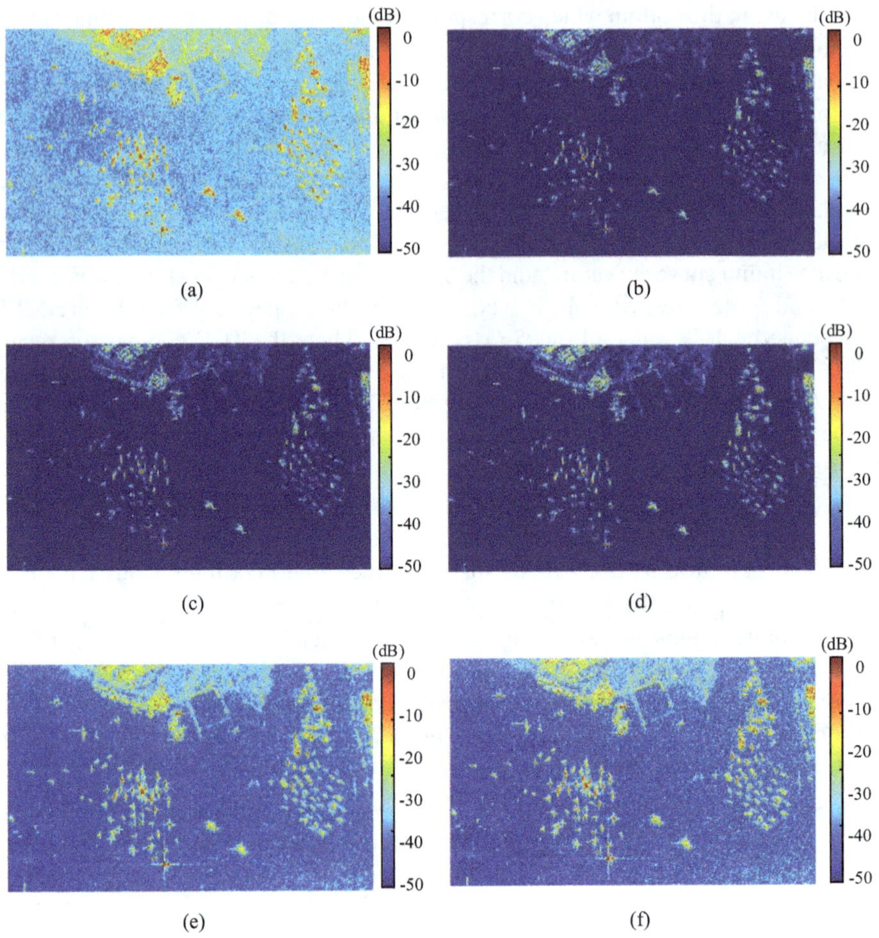

FIGURE 5.2.16 The *SPAN* and three-dimension polarimetric correlation pattern features with the top five highest TCRs for Radarsat-2 data over the ROI of Hong Kong, China. (a) *SPAN,* (b) $\left|\hat{\gamma}_{\text{HH-HV}}(\theta,\tau)\right|_{\text{MGC}}$, (c) $\left|\hat{\gamma}_{\text{(HH+VV)-(HH-VV)}}(\theta,\tau)\right|_{\text{MGC}}$, (d) $\left|\hat{\gamma}_{\text{HH-VV}}(\theta,\tau)\right|_{\text{MGC}}$, (e) $\left|\hat{\gamma}_{\text{HH-VV}}(\theta,\tau)\right|_{\text{std}}^{\tau=0^\circ}$, (f) $\left|\hat{\gamma}_{\text{HH-VV}}(\theta,\tau)\right|_{\text{contrast}}^{\tau=0^\circ}$.

pattern, the order of polarimetric features from the top to the bottom is global features: the maximum Gaussian curvature, the maximum mean curvature, the maximum principle curvature, the maximum correlation, and the minimum correlation and local features: the slice original correlation, the slice maximum correlation, the slice minimum correlation, the slice correlation degree, the slice correlation fluctuation, the slice correlation contrast, the slice maximum curve curvature, and the correlation difference for slices τ=0°, τ=45°, θ=0°, and θ=45°, respectively. The dashed line in Figure 5.2.17 indicates the TCR value of the *SPAN* feature. Compared with polarimetric features extracted from the PEA-plane at $\tau=0^\circ$, it can also be

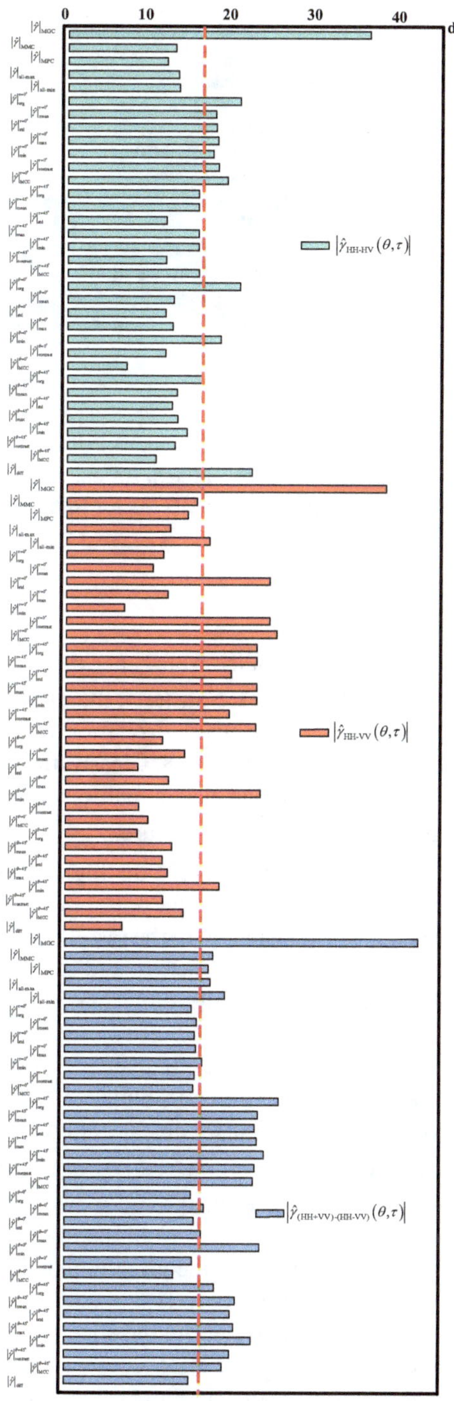

FIGURE 5.2.17 The TCR comparison results for Radarsat-2 data over the ROI of Strait of Gibraltar. The red dashed line indicates the TCR value of the *SPAN* feature. Different colored bars represent different three-dimension polarimetric correlation patterns, accordingly.

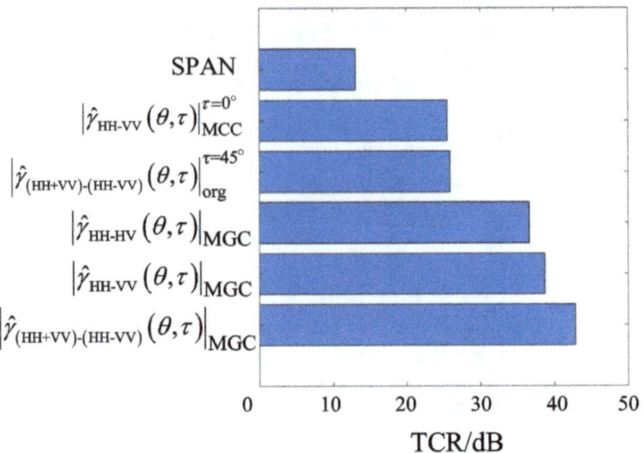

FIGURE 5.2.18 The TCR comparison results of the *SPAN* and three-dimension polarimetric correlation pattern features with the top five highest TCRs for Radarsat-2 data over the ROI of Strait of Gibraltar.

observed that polarimetric features extracted from other slices, such as the PEA-plane at $\tau = 45^\circ$, can achieve higher TCR values. Similarly, the TCR comparison results of the *SPAN* and three-dimension polarimetric correlation pattern features with the top five highest TCRs are shown in Figure 5.2.18. Clearly, three maximum Gaussian curvature features outperform other polarimetric features in term of TCR measure. Moreover, the maximum Gaussian curvature feature $\left|\hat{\gamma}_{\text{(HH+VV)-(HH-VV)}}(\theta,\tau)\right|_{\text{MGC}}$ exhibits the highest TCR of 42.83 dB, which is 26.77 dB higher than that of the *SPAN* feature. In addition, the feature maps of these six polarimetric features are shown in Figure 5.2.19. It can be seen that, consistent with the results from the ROI of the Hong Kong, China, the sea clutter is more homogeneous, and the contrasts between ships and ocean areas are clearly enhanced by the three maximum Gaussian curvature features.

For the ROI of the Bohai Sea, China, TCR results of three-dimension polarimetric correlation pattern features are shown in Figure 5.2.20. As in Figure 5.2.14, there are also three zones from the top to the bottom which correspond to the three-dimension polarimetric correlation patterns $\left|\hat{\gamma}_{\text{HH-HV}}(\theta,\tau)\right|$, $\left|\hat{\gamma}_{\text{HH-VV}}(\theta,\tau)\right|$, and $\left|\hat{\gamma}_{\text{(HH+VV)-(HH-VV)}}(\theta,\tau)\right|$, respectively. For each three-dimension polarimetric correlation pattern, the order of polarimetric features from the top to the bottom are global features: the maximum Gaussian curvature, the maximum mean curvature, the maximum principle curvature, the maximum correlation, and the minimum correlation and local features: the slice original correlation, the slice maximum correlation, the slice minimum correlation, the slice correlation degree, the slice correlation fluctuation, the slice correlation contrast, the slice maximum curve curvature, and the correlation difference for slices τ=0°, τ=45°, θ=0°, and θ=45°, respectively. Also,

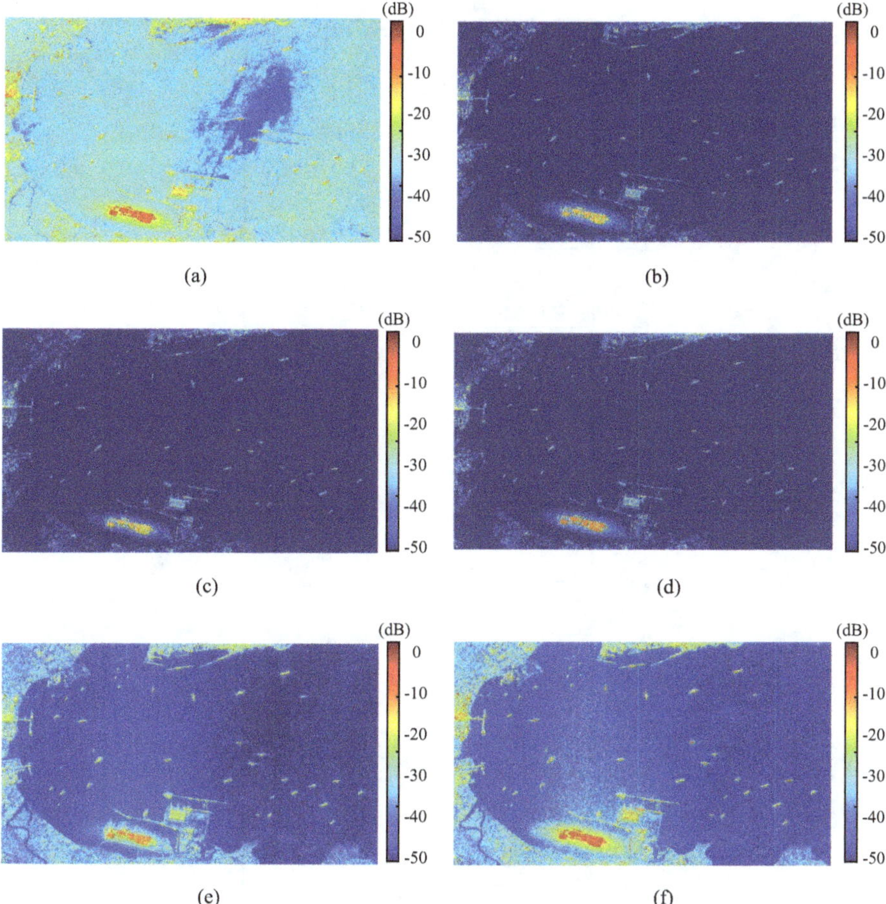

FIGURE 5.2.19 The *SPAN* and three-dimension polarimetric correlation pattern features with the top five highest TCRs for Radarsat-2 data over the ROI of Strait of Gibraltar. (a) *SPAN*, (b) $\left|\hat{\gamma}_{\text{(HH+VV)-(HH-VV)}}(\theta,\tau)\right|_{\text{MGC}}$, (c) $\left|\hat{\gamma}_{\text{HH-VV}}(\theta,\tau)\right|_{\text{MGC}}$, (d) $\left|\hat{\gamma}_{\text{HH-HV}}(\theta,\tau)\right|_{\text{MGC}}$, (e) $\left|\hat{\gamma}_{\text{(HH+VV)-(HH-VV)}}(\theta,\tau)\right|_{\text{org}}^{\tau=45^\circ}$, (f) $\left|\hat{\gamma}_{\text{HH-VV}}(\theta,\tau)\right|_{\text{MCC}}^{\tau=0^\circ}$.

the TCR value of the *SPAN* feature is given and indicated by the red dashed line. Meanwhile, the TCR comparison results of the *SPAN* and three-dimension polarimetric correlation pattern features with the top five highest TCRs are shown in Figure 5.2.21. Three maximum Gaussian curvature features still outperform the other polarimetric features, which is consistent with the other two ROIs. Furthermore, the maximum Gaussian curvature feature $\left|\hat{\gamma}_{\text{(HH+VV)-(HH-VV)}}(\theta,\tau)\right|_{\text{MGC}}$ achieves the highest TCR of 28.10 dB, which is 20.36 dB higher than that of the *SPAN* feature. In addition, the feature maps of these six polarimetric features are shown in Figure 5.2.22. It

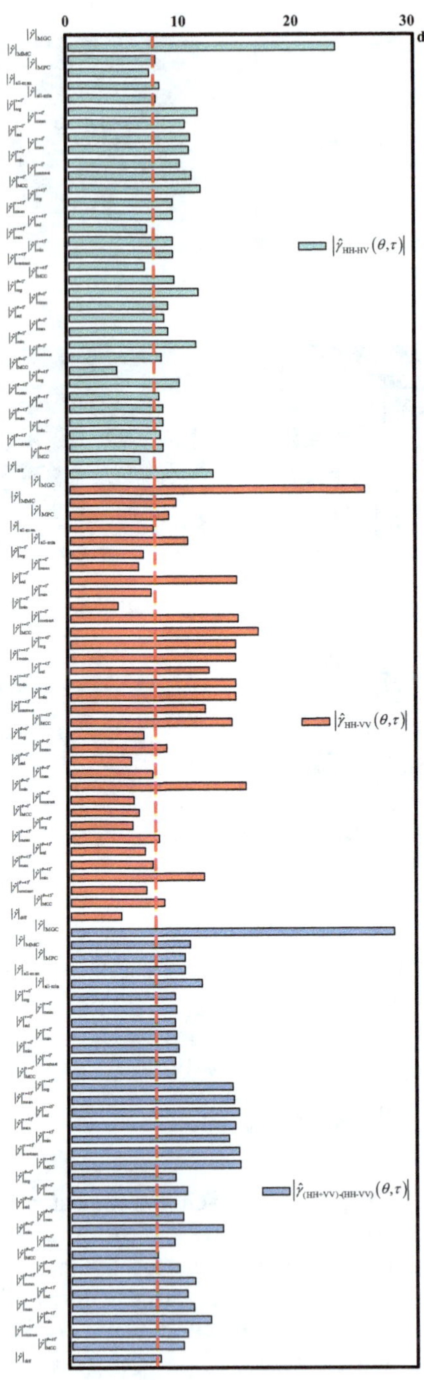

FIGURE 5.2.20 The TCR comparison results for Radarsat-2 data over the ROI of Bohai, China. The red dashed line indicates the TCR value of the *SPAN* feature. Different colored bars represent different three-dimension polarimetric correlation patterns, accordingly.

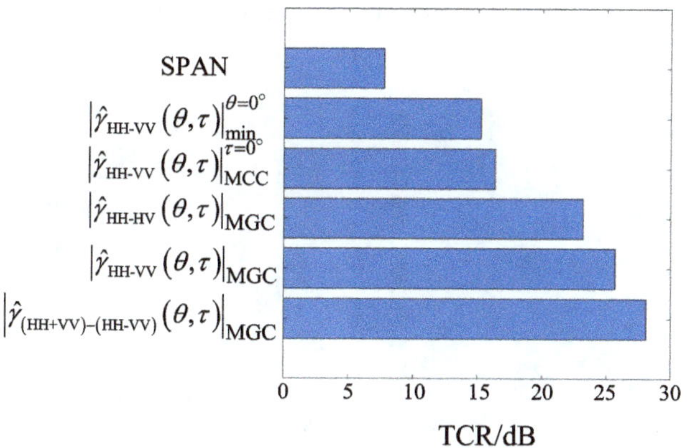

FIGURE 5.2.21 The TCR comparison results of the *SPAN* and three-dimension polarimetric correlation pattern features with the top five highest TCRs for Radarsat-2 data over the ROI of Bohai Sea, China.

(a) (b)

(c) (d)

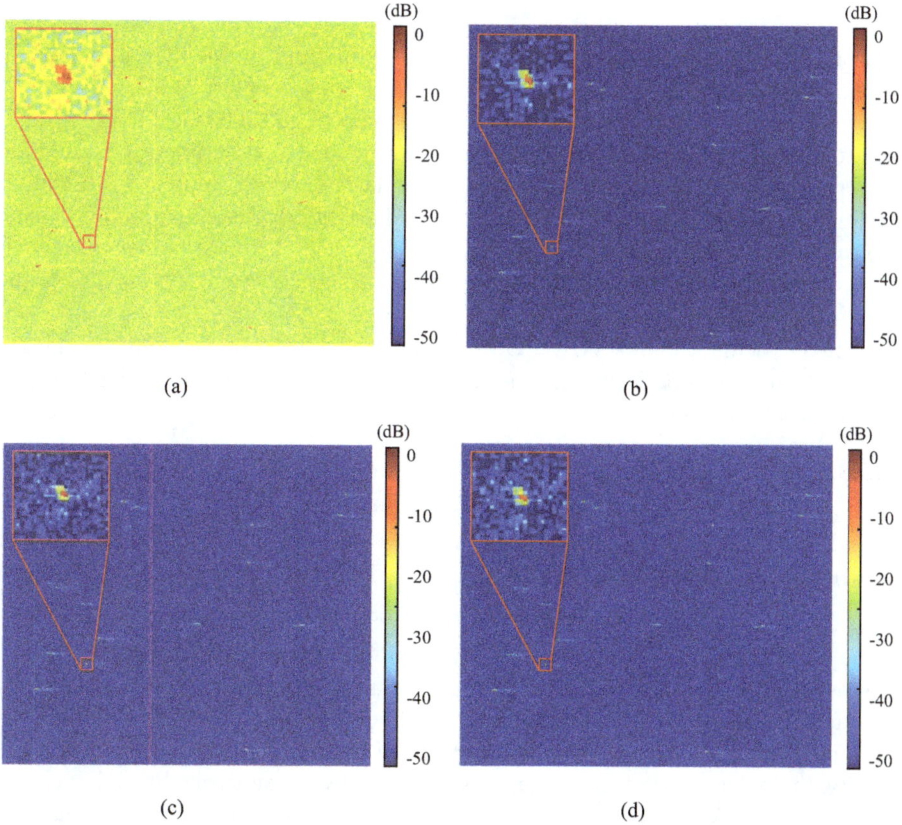

FIGURE 5.2.22 The *SPAN* and three-dimension polarimetric correlation pattern features with the top five highest TCRs for Radarsat-2 data over the ROI of Bohai Sea, China. (a) *SPAN*, (b) $\left|\hat{\gamma}_{\text{(HH+VV)-(HH-VV)}}(\theta,\tau)\right|_{\text{MGC}}$, (c) $\left|\hat{\gamma}_{\text{HH-VV}}(\theta,\tau)\right|_{\text{MGC}}$, (d) $\left|\hat{\gamma}_{\text{HH-HV}}(\theta,\tau)\right|_{\text{MGC}}$, (e) $\left|\hat{\gamma}_{\text{(HH+VV)-(HH-VV)}}(\theta,\tau)\right|_{\text{org}}^{\tau=45^\circ}$, (f) $\left|\hat{\gamma}_{\text{HH-VV}}(\theta,\tau)\right|_{\text{MCC}}^{\tau=0^\circ}$.

(Continued)

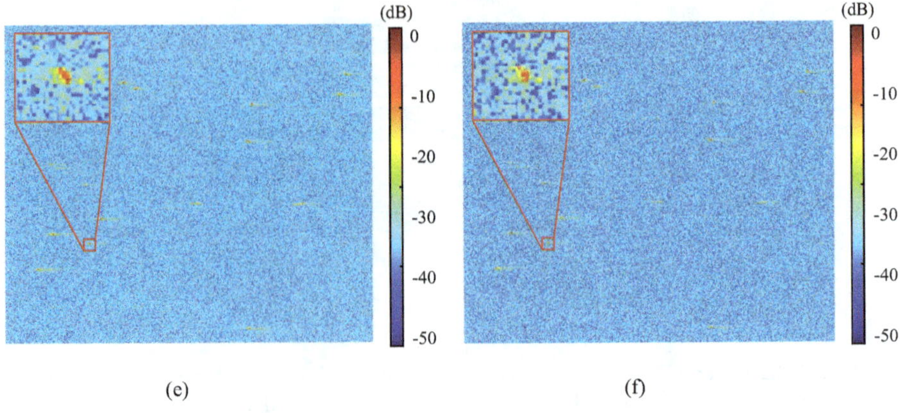

FIGURE 5.2.22 (Continued)

is worth noting that the backscattering powers of the ships in this ROI are relatively weak. For example, the TCR value of the *SPAN* feature is only 7.74 dB.

In summary, the experimental comparisons with three Radarsat-2 PolSAR datasets demonstrate and validate that the three-dimension polarimetric correlation pattern features can significantly enhance the TCR between ships and sea clutter, compared with the traditional *SPAN* features. Particularly, the maximum Gaussian curvature features can generally produce much higher TCRs than other polarimetric features. In addition, compared with the polarimetric features obtained from the PEA-plane $\tau = 0^{\circ}$, polarimetric features from other PEA-planes such as $\tau = 45^{\circ}$ are expected to achieve higher TCRs. This further confirms the information advantages and performance superiorities of the three-dimension polarimetric correlation pattern interpretation tool.

5.2.3.3 Ship Detection with Three-Dimension Polarimetric Correlation Pattern Features

Based on the TCR investigations with the three Radarsat-2 PolSAR datasets, three maximum Gaussian curvature features $\left|\hat{\gamma}_{\text{HH-HV}}(\theta,\tau)\right|_{\text{MGC}}$, $\left|\hat{\gamma}_{\text{HH-VV}}(\theta,\tau)\right|_{\text{MGC}}$, and $\left|\hat{\gamma}_{\text{(HH+VV)-(HH-VV)}}(\theta,\tau)\right|_{\text{MGC}}$ generally achieve the best performance. In this vein, these three polarimetric features are selected for ship detection in this section. To make full use of these selected polarimetric features, a superpixel-level non-local contrast method, named the non-local superpixel-level contrast measure method (NSLCM), is introduced and adopted [8]. Since superpixel usually has good edge reservation ability, the polarimetric feature map is segmented by the superpixel technique first. Then, a new background pixel searching strategy is adopted to determine the non-local pure samples. Then, a salient map is constructed using the non-local pure sea clutter samples. Meanwhile, the selected three maximum Gaussian curvature features are integrated. Finally, ship detection results can be obtained by an adaptive thresholding procedure. The ship detection flowchart with three-dimension polarimetric correlation pattern features is displayed in Figure 5.2.23.

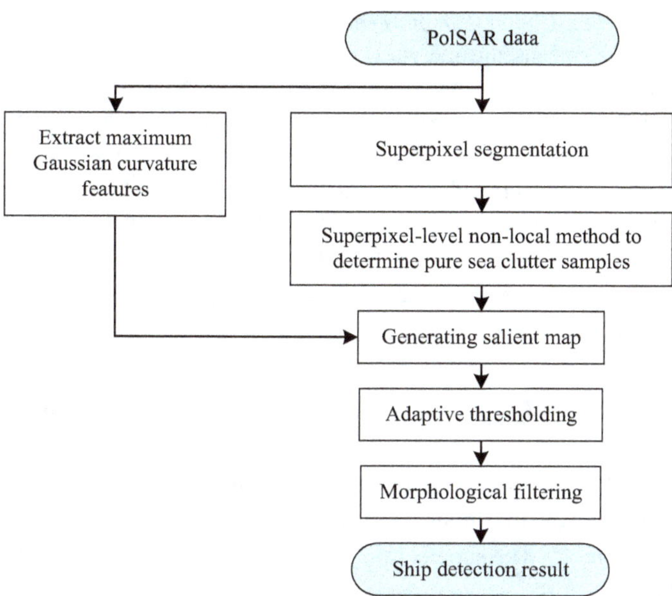

FIGURE 5.2.23 Ship detection flowchart with three-dimension polarimetric correlation pattern features.

The introduced NSLCM mainly contains superpixel segmentation, non-local sea clutter samples determination, maximum Gaussian curvature feature extraction, salient map generation, adaptive thresholding procedure, and morphological filtering.

1) *Superpixel Segmentation.* Both the *SPAN* feature and the SLIC method [9] are adopted for PolSAR image superpixel segmentation. The distance measure D in the SLIC method is defined as:

$$\begin{aligned} D &= \sqrt{\left(\frac{d_c}{N_c}\right)^2 + \left(\frac{d_o}{N_s}\right)^2} \\ d_c &= \sqrt{\left(l_p - l_q\right)^2} \\ d_o &= \sqrt{\left(x_p - x_q\right)^2 + \left(y_p - y_q\right)^2} \end{aligned} \tag{5.2.3}$$

where l_p and l_q are the feature values; $\left(x_p, y_p\right)$ and $\left(x_q, y_q\right)$ are the coordinates of pixels; p and q are the indexes of different pixels; N_c is the compactness coefficient within the range of 2~10. In this section, $N_c = 10$ is adopted. N_s is the desired number of superpixels, which can be predefined according to the smallest ship. Then superpixel set $\{\bullet\}_{\text{All}}$ can be collected.

2) *Non-Local Sea Clutter Sample Determination.* After the superpixel segmentation, the K-means unsupervised clustering method is implemented. Then, superpixels are divided into two categories. The one with higher variance is denoted as the target candidate superpixel set $\{\bullet\}_{\text{candidate}}$. For every superpixel sp^i_{target} in the superpixel set $\{\bullet\}_{\text{All}}$, its clutter superpixels are searched outward from the central superpixel, where i indicates the index of the target superpixel. A schematic illustration for non-local sea clutter samples determination is shown in Figure 5.2.24. The yellow arrows stand for the search directions. The collected clutter superpixels form the non-local sea clutter sample set $\{\bullet\}^i_{\text{clutter}}$.
3) *Maximum Gaussian Curvature Feature Extraction.* The maximum Gaussian curvature features $\left|\hat{\gamma}_{\text{HH-HV}}(\theta,\tau)\right|_{\text{MGC}}$, $\left|\hat{\gamma}_{\text{HH-VV}}(\theta,\tau)\right|_{\text{MGC}}$, and $\left|\hat{\gamma}_{\text{(HH+VV)-(HH-VV)}}(\theta,\tau)\right|_{\text{MGC}}$ are extracted, which are denoted as M_1, M_2, and M_3, respectively.
4) *Salient Map Generation.* Taking the M_1 feature, for example, the mean feature values of every clutter superpixel sp^j_{clutter} in $\{\bullet\}^i_{\text{clutter}}$ are calculated and denoted as $m^j_{i\text{-M1}}$. The superscript j represents the index of the clutter superpixel in $\{\bullet\}^i_{\text{clutter}}$. For the kth pixel in sp^i_{target}, the M_1 feature value is $f^k_{i\text{-M1}}$. The number of clutter superpixels for the target superpixel sp^i_{target} is N_i. The salient value $C^k_{i\text{-M}_1}$ of the kth pixel in sp^i_{target} is

$$C^k_{i\text{-M}_1} = \frac{1}{N_i}\sum_{j=1}^{N_i} f^k_{i\text{-M}_1} \Big/ m^j_{i\text{-M}_1} \tag{5.2.4}$$

Similarly, salient maps of M_2 and M_3 features can be obtained as $C^k_{i\text{-M}_2}$ and $C^k_{i\text{-M}_3}$.

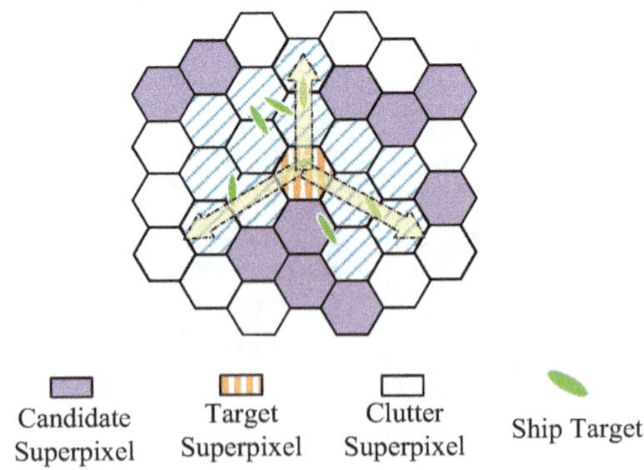

FIGURE 5.2.24 The schematic illustration for non-local sea clutter sample determination. The yellow arrows indicate the search direction for clutter superpixels.

The final fused salient value $C_{i-fusion}^{k}$ is calculated as

$$C_{i-\text{fusion}}^{k} = \max\left\{C_{i-\text{M}_1}^{k}, C_{i-\text{M}_2}^{k}, C_{i-\text{M}_3}^{k}\right\} \tag{5.2.5}$$

After the calculation of fused salient values of all target superpixels in $\{\cdot\}_{\text{All}}$, the salient contrast map M_{fusion} for one PolSAR dataset is constructed.

5) *Adaptive Thresholding.* For the salient map M_{fusion}, the mean and standard values of all pixels are denoted as I and σ. The adaptive threshold is

$$Th = I + w\sigma \tag{5.2.6}$$

where w is the adjustment parameter and can be set according to different scenes. In general, $w \in [10,20]$ is recommended for weak ship areas and $w \in [80,105]$ is recommended for other scenes. In this work, adjustment parameter w for the ROIs of Hong Kong, China; the Strait of Gibraltar; and the Bohai Sea, China, are 90, 90, and 15, respectively.

6) *Morphological Filtering.* The detection results are processed by the morphological filter to exclude isolated small false alarms.

With this procedure, the final ship detection results can be obtained.

5.2.3.4 Ship Detection Comparison

In this experimental study, the smallest of-constant false alarm rate (SO-CFAR) detector [10], a new polarimetric notch filter (NPNF) method [11], and the superpixel-based local contrast measure (SLCM) detector [12] are adopted for comparison. For the SO-CFAR method, the *SPAN* feature is utilized. The target, guard, and clutter windows are 3×3, 9×9, and 19×19, respectively. For the NPNF method, a 50×50 region is selected from each ROI as the prior information of sea clutter. For SLCM method, the *SPAN* feature is also utilized. The superpixel segmentation parameters and adaptive thresholding procedure are the same as the introduced NSLCM method. The FoM index is used to evaluate the detection performances quantitatively.

For the ROIs of the three Radarsat-2 PolSAR datasets introduced in Section 5.2.1, the corresponding ship detection results are shown in Figures 5.2.25, 5.2.26, and 5.2.27, respectively. The quantitative ship detection results are summarized in Table 5.2.3. For these three ROIs, the NSLCM method achieves the best ship detection performance. In detail, for the ROI of Hong Kong, China, the FoM value of the NSLCM method is 96.38%, which is 0.58% higher than the best comparison method, NPNF. Although ship distributions in this ROI are very dense, the NSLCM method has only one false alarm. For the ROI of Strait of Gibraltar area, both the NSLCM method and the SLCM method achieve the best detection performance, without missed detections and false alarms. Their FoM values are both 100%. However, compared with the SLCM method, the NSLCM method can detect more

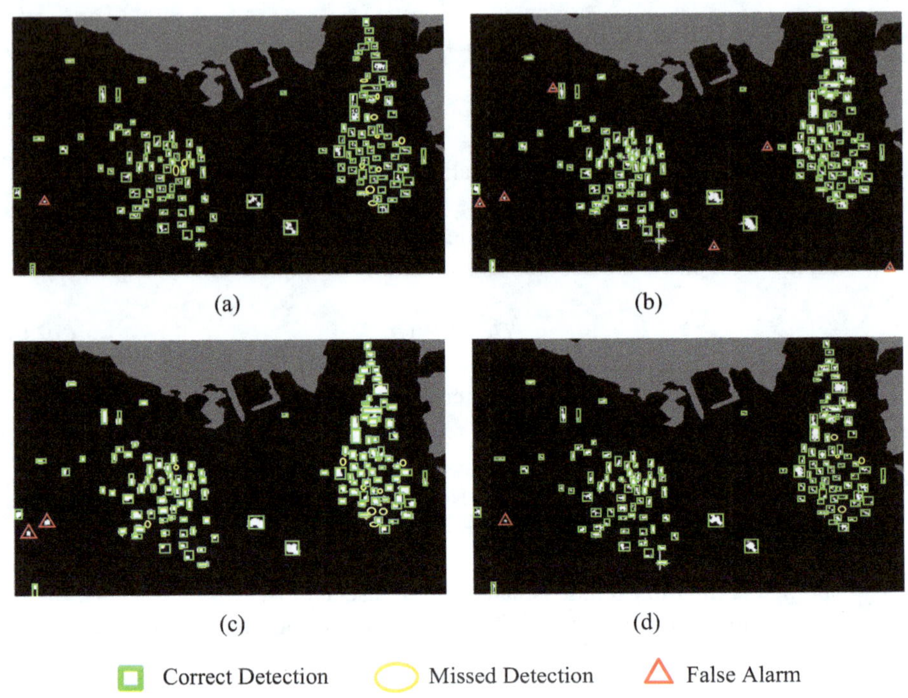

FIGURE 5.2.25 Ship detection results of Radarsat-2 PolSAR data over the ROI of Hong Kong, China. (a) SO-CFAR method, (b) NPNF method, (c) SLCM method, (d) NSLCM method.

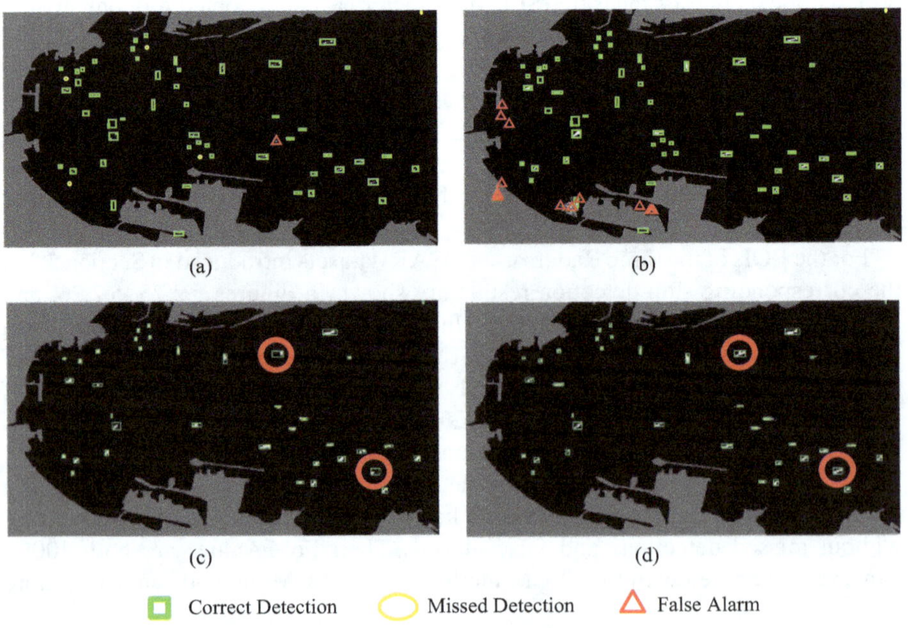

FIGURE 5.2.26 Ship detection results of Radarsat-2 PolSAR data over the ROI of the Strait of Gibraltar. (a) SO-CFAR method, (b) NPNF method, (c) SLCM method, (d) NSLCM method.

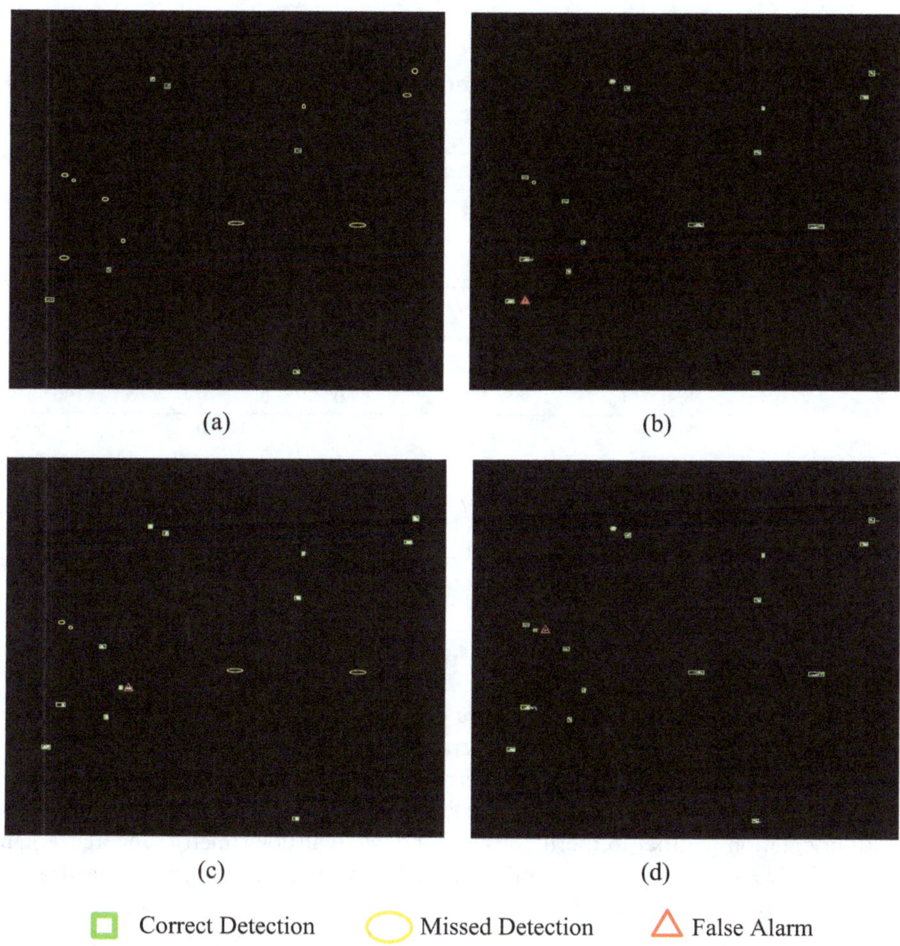

FIGURE 5.2.27 Ship detection results of Radarsat-2 PolSAR data over the ROI of Bohai Sea, China. (a) SO-CFAR method, (b) NPNF method, (c) SLCM method, (d) NSLCM method.

ship pixels, and the ship contours are more complete, as indicated by the red circles in Figure 5.2.26. For the ROI of Bohai Sea, China, the ship detection performance of the NSLCM method is better than all other comparison methods. Its FoM value reaches 94.12%, with only one false alarm and without any missed detections. Among the comparison methods, the NPNF method shows the best detection performance, with a FoM value of 88.24%, which is still 5.88% lower than that of the NSLCM method. It is worth noting that the backscattering powers of the ship targets in this ROI are relatively weak. There are many missed detections from the SO-CFAR method.

TABLE 5.2.3
Quantitative Comparison of Ship Detection Results

ROI	Method	N_C	N_M	N_{FA}	FoM
Hong Kong, China	SO-CFAR	126	11	1	91.30%
	NPNF	137	0	6	95.80%
	SLCM	128	9	2	92.09%
	NSLCM	133	4	1	**96.38%**
Strait of Gibraltar	SO-CFAR	34	3	1	89.47%
	NPNF	36	1	7	81.82%
	SLCM	37	0	0	**100.00%**
	NSLCM	37	0	0	**100.00%**
Bohai Sea, China	SO-CFAR	6	10	0	37.50%
	NPNF	15	1	1	88.24%
	SLCM	12	4	1	70.59%
	NSLCM	16	0	1	**94.12%**

5.3 POLARIMETRIC ROTATION DOMAIN BUILDING DETECTION

As introduced in Chapter 2, polarimetric coherence features are strongly related to the types and orientations of the scatterers. For scatterers with reflection symmetry and rotation symmetry properties, the corresponding polarimetric coherence may be relatively unchanged in the polarimetric rotation domain. Meanwhile, for buildings with orientation parallel to the flight pass, the original polarimetric coherence can be relatively high. For oriented buildings, the corresponding original polarimetric coherence is obviously reduced and can be as low as those of vegetated areas such as forests. However, the reflection symmetry and rotation symmetry properties are violated in oriented building areas. Thereby, their polarimetric coherence can be enhanced in the polarimetric rotation domain [13–16], which provides the physical basis for building detection.

5.3.1 Building Enhancement with Polarimetric Coherence Pattern

PiSAR X-band PolSAR data, acquired over Sendai, Japan, is used for demonstration. The details of this data have been described in Section 3.4.1. The corresponding optical image, PolSAR Pauli image, and *SPAN* feature image are shown in Figure 5.3.1.

For four polarimetric coherence patterns $\left|\gamma_{\text{HH-VV}}(\theta)\right|$, $\left|\gamma_{\text{HH-HV}}(\theta)\right|$, $\left|\gamma_{\text{(HH+VV)-(HV)}}(\theta)\right|$, and $\left|\gamma_{\text{(HH-VV)-(HV)}}(\theta)\right|$, their original coherence and maximum coherence features are shown in Figure 5.3.2, while the corresponding histograms are given in Figure 5.3.3. Compared with the original coherence features, it is observed that the maximum coherence features are obviously enhanced for most of these land covers. Quantitative

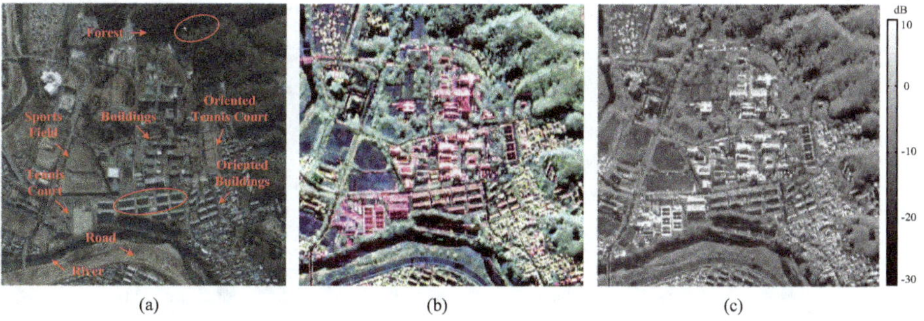

(a) (b) (c)

FIGURE 5.3.1 Study area of PiSAR data. (a) Optical image, (b) PolSAR Pauli image, (c) *SPAN*.

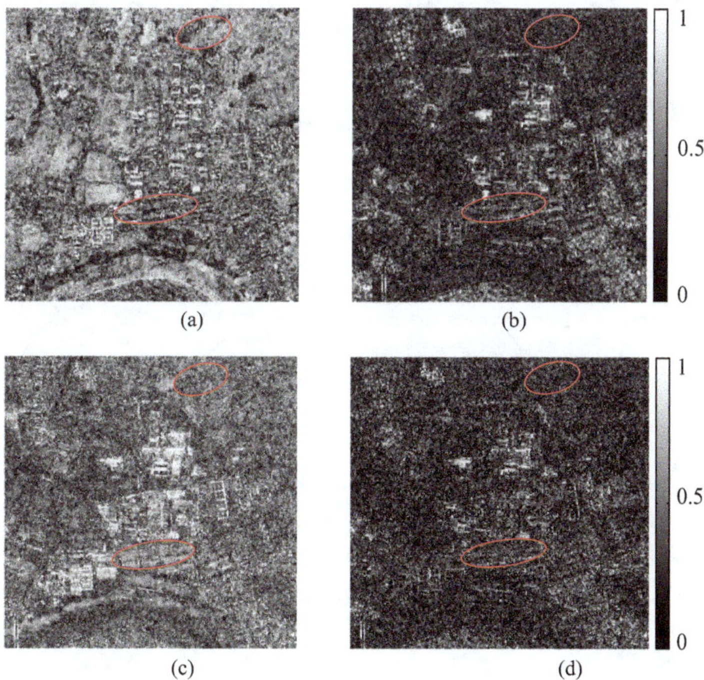

(a) (b)

(c) (d)

FIGURE 5.3.2 Polarimetric coherence pattern features of PiSAR PolSAR data. (a) Original coherence $\left|\gamma_{\text{HH-VV}}(\theta)\right|_{\text{org}}$, (b) original coherence $\left|\gamma_{\text{HH-HV}}(\theta)\right|_{\text{org}}$, (c) original coherence $\left|\gamma_{\text{(HH+VV)-(HV)}}(\theta)\right|_{\text{org}}$, (d) original coherence $\left|\gamma_{\text{(HH-VV)-(HV)}}(\theta)\right|_{\text{org}}$, (e) maximum coherence $\left|\gamma_{\text{HH-VV}}(\theta)\right|_{\max}$, (f) maximum coherence $\left|\gamma_{\text{HH-HV}}(\theta)\right|_{\max}$, (g) maximum coherence $\left|\gamma_{\text{(HH+VV)-(HV)}}(\theta)\right|_{\max}$, (h) maximum coherence $\left|\gamma_{\text{(HH-VV)-(HV)}}(\theta)\right|_{\max}$.

(Continued)

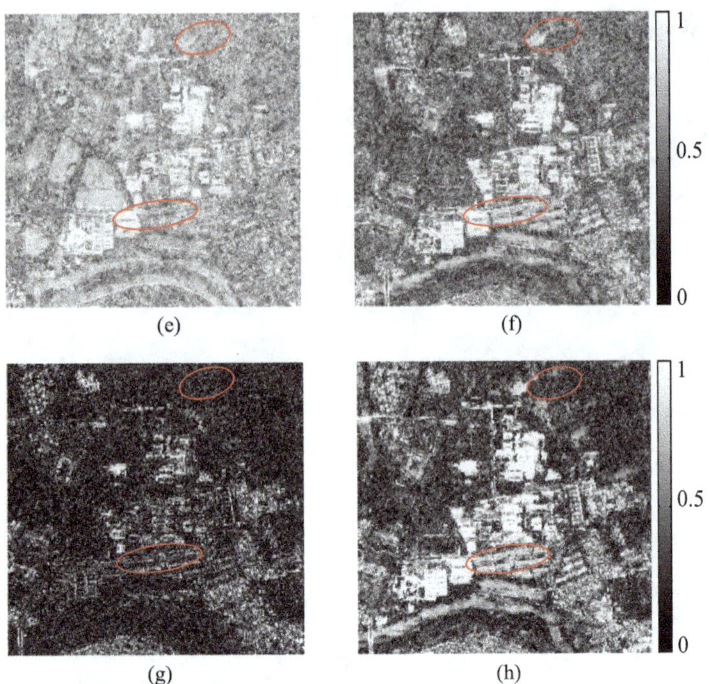

FIGURE 5.3.2 (Continued)

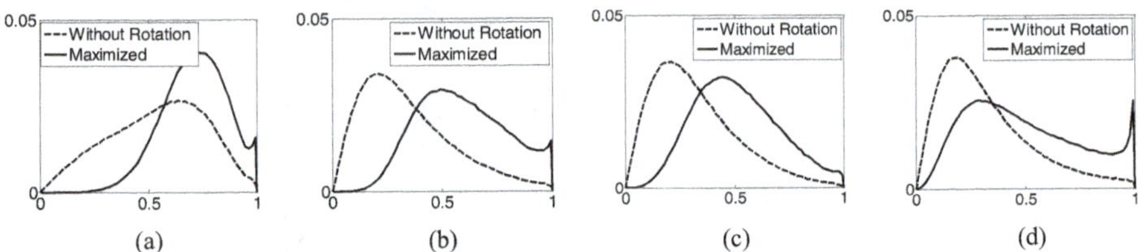

FIGURE 5.3.3 Histograms of polarimetric coherence magnitudes in term of the original coherence and the maximum coherence for polarimetric coherence patterns (a) $\left|\gamma_{\text{HH-VV}}(\theta)\right|$, (b) $\left|\gamma_{\text{HH-HV}}(\theta)\right|$, (c) $\left|\gamma_{\text{(HH+VV)-(HV)}}(\theta)\right|$, (d) $\left|\gamma_{\text{(HH-VV)-(HV)}}(\theta)\right|$.

comparison results are summarized in Table 5.3.1 in terms of the mean values of polarimetric coherence without and with maximization for the full-scene. The mean values of the original coherence $\left|\gamma_{\text{HH-VV}}(\theta)\right|_{\text{org}}$, $\left|\gamma_{\text{HH-HV}}(\theta)\right|_{\text{org}}$, $\left|\gamma_{\text{(HH+VV)-(HV)}}(\theta)\right|_{\text{org}}$, and $\left|\gamma_{\text{(HH-VV)-(HV)}}(\theta)\right|_{\text{org}}$ are respectively 0.54, 0.34, 0.32, and 0.32, while they are enhanced up to 0.71, 0.59, 0.51, and 0.49 accordingly by the maximization coherence $\left|\gamma_{\text{HH-VV}}(\theta)\right|_{\text{max}}$, $\left|\gamma_{\text{HH-HV}}(\theta)\right|_{\text{max}}$, $\left|\gamma_{\text{(HH+VV)-(HV)}}(\theta)\right|_{\text{max}}$, and $\left|\gamma_{\text{(HH-VV)-(HV)}}(\theta)\right|_{\text{max}}$. The corresponding enhanced percentages are 31.48%, 73.53%, 59.38%, and 53.13%. In particular, the enhancement effect from the maximum polarimetric coherence

TABLE 5.3.1
Quantitative Comparison of Polarimetric Original Coherence and Maximum Coherence for PiSAR PolSAR Data

Polarimetric Coherence Patterns	Means of Original Coherence	Means of Maximum Coherence	Enhanced Percentage
$\lvert\gamma_{\text{HH-VV}}(\theta)\rvert$	0.54	0.71	31.48%
$\lvert\gamma_{\text{HH-HV}}(\theta)\rvert$	0.34	0.59	73.53%
$\lvert\gamma_{\text{(HH+VV)-(HV)}}(\theta)\rvert$	0.32	0.51	59.38%
$\lvert\gamma_{\text{(HH-VV)-(HV)}}(\theta)\rvert$	0.32	0.49	53.13%

$\left|\gamma_{\text{(HH-VV)-(HV)}}(\theta)\right|_{\max}$ is quite obvious for buildings (especially the oriented buildings). According to the uniform polarimetric matrix rotation theory introduced in Chapter 2, such an enhancement effect can be explainable. The polarimetric coherence pattern of $\left|\gamma_{\text{(HH-VV)-(HV)}}(\theta)\right|$ mainly relates to the polarimetric coherency matrix element $\left|T_{23}(\theta)\right|^2$, which shows good sensitivity to the reflection symmetry condition. The definition of $\left|T_{23}(\theta)\right|^2$ is rewritten as

$$\begin{aligned}\left|T_{23}(\theta)\right|^2 &= \frac{1}{4}(T_{33}-T_{22})^2\sin^2 4\theta + \text{Re}^2\left[T_{23}\right]\cos^2 4\theta + \\ &\quad \frac{1}{2}(T_{33}-T_{22})\text{Re}\left[T_{23}\right]\sin 8\theta + \text{Im}^2\left[T_{23}\right] \\ &= \frac{1}{4}A^2\sin\left[8(\theta+\theta_0)\right] + \frac{1}{2}A + \text{Im}^2\left[T_{23}\right]\end{aligned} \tag{5.3.1}$$

where $A = \frac{1}{4}(T_{33}-T_{22})^2 + \text{Re}^2\left[T_{23}\right]$ is the oscillation amplitude feature and $\theta_0 = \frac{1}{8}\text{Angle}\left\{\frac{1}{2}(T_{33}-T_{22})\text{Re}\left[T_{23}\right] + j\frac{1}{2}\left[\text{Re}^2\left[T_{23}\right] - \frac{1}{4}(T_{33}-T_{22})^2\right]\right\}$ is the initial angle feature.

If the polarimetric oscillation amplitude feature A is rewritten with the elements of the polarimetric scattering matrix, it becomes

$$\begin{aligned}A &= \frac{1}{4}(T_{33}-T_{22})^2 + \text{Re}^2\left[T_{23}\right] \\ &= \frac{1}{4}\left(\left\langle\left|S_{\text{HH}}-S_{\text{VV}}\right|^2 - 4\left|S_{\text{HV}}\right|^2\right\rangle\right)^2 + 4\left\{\text{Re}\left[\left\langle(S_{\text{HH}}-S_{\text{VV}})S_{\text{HV}}^*\right\rangle\right]\right\}^2\end{aligned} \tag{5.3.2}$$

Therefore, the polarimetric oscillation amplitude feature A is sensitive to the reflection symmetry condition. In this vein, the polarimetric coherence pattern

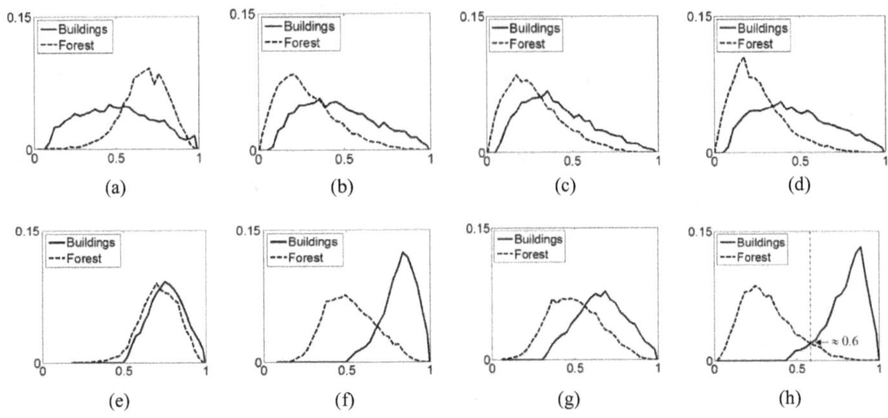

FIGURE 5.3.4 Histogram comparisons of polarimetric coherence magnitudes for selected building and forest areas. (a) Original coherence $\left|\gamma_{\text{HH-VV}}(\theta)\right|_{\text{org}}$, (b) original coherence $\left|\gamma_{\text{HH-HV}}(\theta)\right|_{\text{org}}$, (c) original coherence $\left|\gamma_{\text{(HH+VV)-(HV)}}(\theta)\right|_{\text{org}}$, (d) original coherence $\left|\gamma_{\text{(HH-VV)-(HV)}}(\theta)\right|_{\text{org}}$, (e) maximum coherence $\left|\gamma_{\text{HH-VV}}(\theta)\right|_{\text{max}}$, (f) maximum coherence $\left|\gamma_{\text{HH-HV}}(\theta)\right|_{\text{max}}$, (g) maximum coherence $\left|\gamma_{\text{(HH+VV)-(HV)}}(\theta)\right|_{\text{max}}$, (h) maximum coherence $\left|\gamma_{\text{(HH-VV)-(HV)}}(\theta)\right|_{\text{max}}$.

$\left|\gamma_{\text{(HH-VV)-(HV)}}(\theta)\right|$ and the maximum coherence $\left|\gamma_{\text{(HH-VV)-(HV)}}(\theta)\right|_{\text{max}}$ are also sensitive to the reflection symmetry condition.

Furthermore, an oriented building area and a forest area are selected from Figure 5.3.1, indicated by the ellipses. Their histogram comparisons in terms of original and maximum coherence features are illustrated in Figure 5.3.4. These comparisons further confirm that the maximum coherence $\left|\gamma_{\text{(HH-VV)-(HV)}}(\theta)\right|_{\text{max}}$ has the better performance to discriminate oriented buildings and forests, as shown in Figure 5.3.4 (h). Besides, the best threshold to discriminate them is around 0.6, according to the histogram shown in Figure 5.3.4 (h).

5.3.2 Building Detection Approach and Demonstration

A building detection approach based on the maximum coherence $\left|\gamma_{\text{(HH-VV)-(HV)}}(\theta)\right|_{\text{max}}$ is introduced in this section and demonstrated with PiSAR PolSAR data. The flowchart of building detection approach is shown in Figure 5.3.5. First, the PolSAR data are speckle filtered, and the maximum coherence feature $\left|\gamma_{\text{(HH-VV)-(HV)}}(\theta)\right|_{\text{max}}$ and the *SPAN* feature are extracted accordingly. Then, the building extraction scheme includes two main thresholding procedures. The first thresholding procedure is conducted with the maximum coherence $\left|\gamma_{\text{(HH-VV)-(HV)}}(\theta)\right|_{\text{max}}$, and building candidates are determined. As shown in Figure 5.3.4 (h), building candidates can be optimally

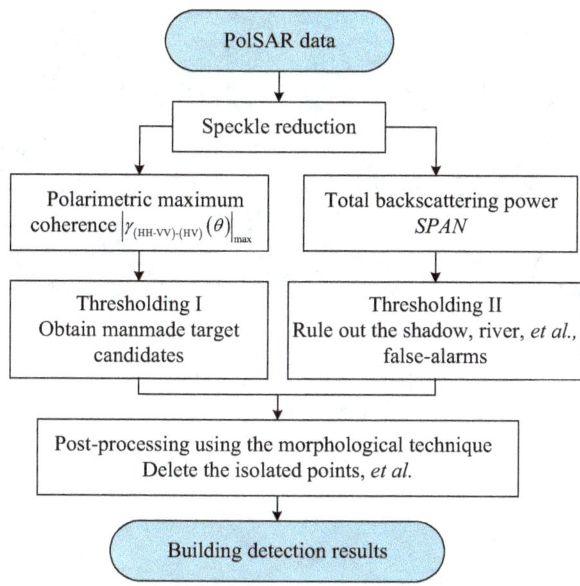

FIGURE 5.3.5 Flowchart of the building detection method.

identified when the maximum coherence $|\gamma_{(HH-VV)-(HV)}(\theta)|_{max}$ is around 0.6. Thereby, the first threshold is determined as 0.6 for this PiSAR data. The building candidate detection result is shown in Figure 5.3.6 (a). Compared with the optical image shown in Figure 5.3.1 (a), buildings and oriented buildings are well detected. Meanwhile, some very low backscattering areas of the shadow areas and river areas are incorrectly detected and become false alarms. Principally, the shadow areas without signal illumination and the river areas with mainly a specular reflection mechanism exhibit null or extremely low backscattering powers. The maximum coherence values of these areas are enhanced as well. These false alarm areas should be ruled out and can be easily accomplished using the *SPAN* feature, which leads to the second thresholding processing. Note that the noise floor of the SAR system is a good measure for shadow areas. For river areas, the backscattered power depends on the surface roughness. Basically, the backscattered power is higher than the noise floor of a SAR system. With the investigation of Figure 5.3.1 (c), the second threshold is chosen as –10 dB. The mask of low backscattering areas is shown in Figure 5.3.6 (b), which can well rule out the shadow and river areas. It should be noted that these two thresholds are tunable in practical applications. Combing the detected building candidates and the false alarm mask, the final building detection results can be generated with a morphological filtering technique to exclude small isolated areas. The final building detection result is shown in Figure 5.3.6 (c). For convenient visual comparison, the building detection results (red) have been superimposed on the original PolSAR Pauli RGB image, as shown in Figure 5.3.7. It is clear that the introduced method can detected buildings with various orientations and the detection results are accurate.

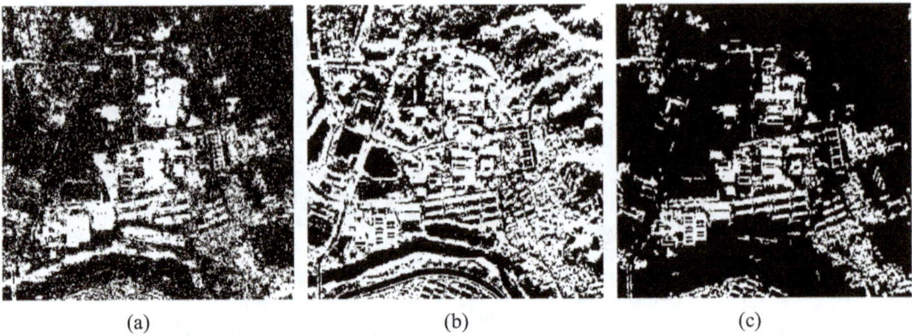

(a) (b) (c)

FIGURE 5.3.6 Building detection results. (a) Building candidates (white pixels) after thresholding I, (b) false alarm (black pixels) ruled out result after thresholding II, (c) final building detection result.

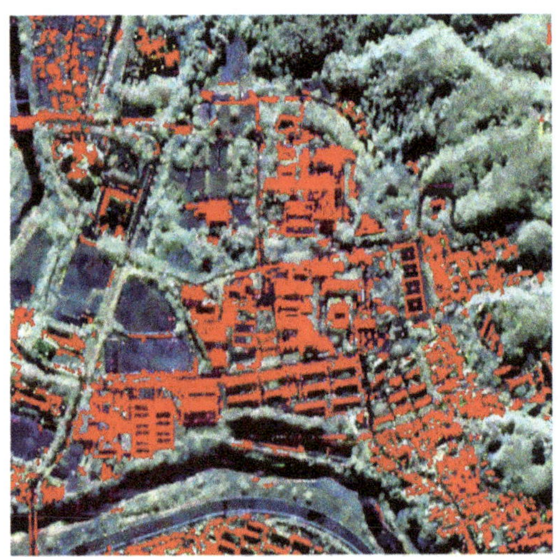

FIGURE 5.3.7 Superimposed building detection results (red) on the original PolSAR Pauli RGB image.

5.4 POLARIMETRIC ROTATION DOMAIN AIRCRAFT DETECTION

Aircraft detection is an important application for PolSAR data. PolSAR aircraft detection is challenging due to the complex structures of aircraft. An aircraft detection method based on the two-dimension polarimetric correlation pattern and superpixel technology is introduced in this section. The polarimetric scattering characteristics of aircrafts are investigated with the two-dimension polarimetric correlation pattern interpretation tool. Aircraft detection is realized by combining the selected polarimetric rotation domain features and the superpixel-level CFAR detector.

5.4.1 Feature Investigation and Selection

The Radarsat-2 data acquired over Tucson, Arizona, USA, on May 3, 2009, are used for investigation. The Pauli image and the ground-truth image are shown in Figures 5.4.1, respectively. One aircraft pixel and one ground clutter pixel are randomly selected. The corresponding two-dimension polarimetric correlation patterns $\left|\hat{\gamma}_{\text{(HH-VV)-(HV)}}(\theta)\right|$ are shown in Figure 5.4.2. It is clear that the two-dimension polarimetric correlation patterns are quite different for the selected aircraft pixel and ground clutter pixel, which can be used for aircraft detection. Similarly to Section 5.2.2, amplitude features deduced from the two-dimension polarimetric correlation pattern are adopted for investigation. For four typical two-dimension polarimetric correlation patterns

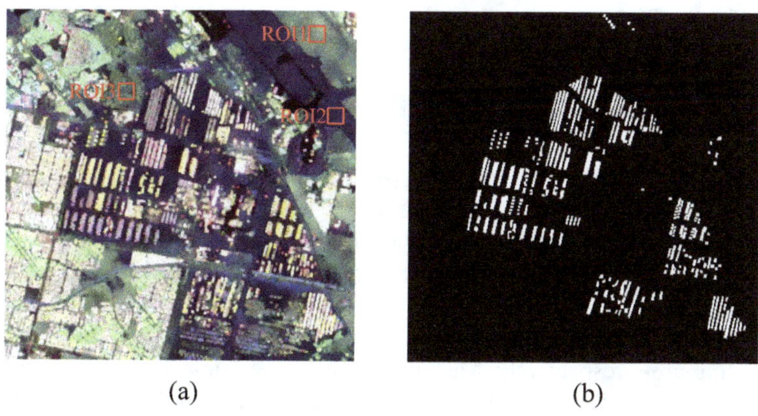

FIGURE 5.4.1 Study area of Radarsat-2 data, (a) PolSAR Pauli image, (b) ground-truth image.

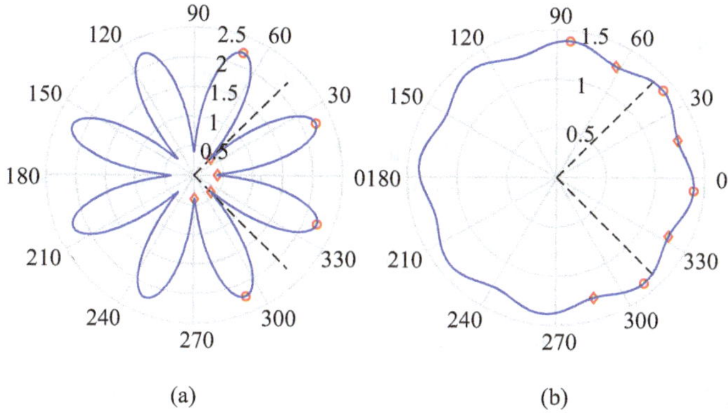

FIGURE 5.4.2 Two-dimension polarimetric correlation pattern $\left|\hat{\gamma}_{\text{(HH-VV)-(HV)}}(\theta)\right|$ for the (a) aircraft pixel, (b) clutter pixel.

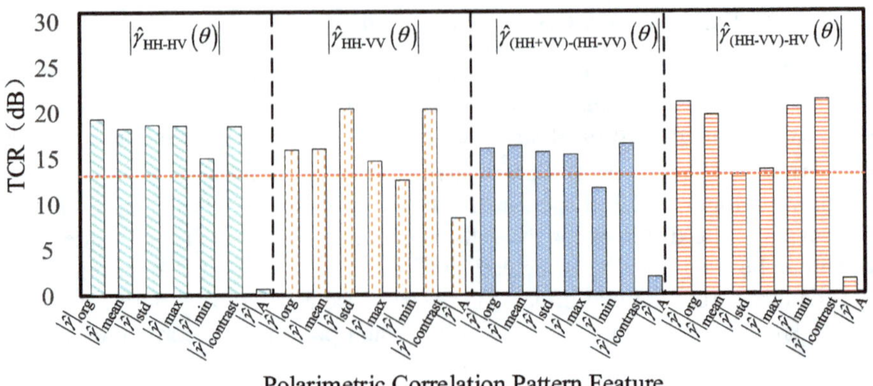

FIGURE 5.4.3 TCR comparison for polarimetric features between aircraft and clutter. The red dashed line indicates the TCR value of the *SPAN* feature.

$\left|\hat{\gamma}_{\text{HH-HV}}(\theta)\right|$, $\left|\hat{\gamma}_{\text{HH-VV}}(\theta)\right|$, $\left|\hat{\gamma}_{\text{(HH+VV)-(HH-VV)}}(\theta)\right|$, and $\left|\hat{\gamma}_{\text{(HH-VV)-(HV)}}(\theta)\right|$, seven kinds of amplitude features can be obtained. The *SPAN* feature is adopted for comparison.

TCR comparison and feature selection are conducted with the 28 two-dimension polarimetric correlation pattern features and *SPAN* feature. In Figure 5.4.1, three typical clutter regions with a size of $20\times20$are randomly selected and marked with red boxes. Among them, ROIs 1, 2, and 3 represent forest area, grass area, and building area, respectively. Meanwhile, the same number of aircraft pixels are randomly selected from the aircraft area for TCR calculation. For 28 two-dimension polarimetric correlation pattern features and the *SPAN* feature, the TCR results are shown in Figure 5.4.3. There are four zones which represent the four independent two-dimension polarimetric correlation patterns $\left|\hat{\gamma}_{\text{HH-HV}}(\theta)\right|$, $\left|\hat{\gamma}_{\text{HH-VV}}(\theta)\right|$, $\left|\hat{\gamma}_{\text{(HH+VV)-(HH-VV)}}(\theta)\right|$, and $\left|\hat{\gamma}_{\text{(HH-VV)-(HV)}}(\theta)\right|$. For each zone, TCR values of the original correlation $|\hat{\gamma}|_{\text{org}}$, the correlation degree $|\hat{\gamma}|_{\text{mean}}$, the correlation fluctuation $|\hat{\gamma}|_{\text{std}}$, the maximum correlation $|\hat{\gamma}|_{\text{max}}$, the minimum correlation $|\hat{\gamma}|_{\text{min}}$, the correlation contrast $|\hat{\gamma}|_{\text{contrast}}$, and the correlation anisotropy $|\hat{\gamma}|_{\text{A}}$ are given from left to right. Meanwhile, the dashed line indicates the TCR value of the *SPAN* feature. It can be seen that the polarimetric correlation contrast feature $\left|\hat{\gamma}_{\text{(HH-VV)-(HV)}}(\theta)\right|_{\text{contrast}}$ achieves the highest TCR and is adopted for aircraft detection. It is worth noting that other two-dimension polarimetric correlation pattern features and their combinations also have the potential for aircraft detection, which is not demonstrated in this section.

5.4.2 Aircraft Detection Approach and Comparison

Based on TCR investigations of polarimetric features, an aircraft detection method combining selected two-dimension polarimetric correlation pattern features and superpixel-level CFAR detector is introduced in this section, named superpixel-level

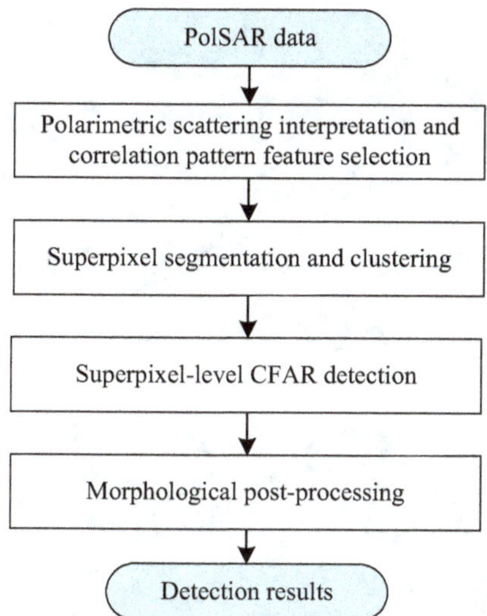

FIGURE 5.4.4 Flowchart of the aircraft detection method.

CFAR (SP-CFAR). The flowchart of the aircraft detection method is shown in Figure 5.4.4.

First, the two-dimension polarimetric correlation pattern feature is selected for aircraft detection. Then, the NSLCM method introduced in Section 5.2.3.3 is adopted to select homogenous ground clutter samples adaptively. Superpixel segmentation is conducted on the two-dimension polarimetric correlation pattern feature. N_c is the compactness coefficient, which ranges from 2 to 10. In this section, $N_c = 5$ is adopted. N_s is the desired number of superpixels and is set according to the size of aircraft to be detected:

$$N_s = \frac{rc}{kA_{\min}} \tag{5.4.1}$$

where r and c represent the size of the PolSAR data in the range and azimuth directions; $A_{\min}$ is the potential pixel number of the smallest aircraft, which can be predefined by expert experience or prior information; and k is the adjustment coefficient, where $k = 2$ is recommended.

After superpixel segmentation, the variance of each superpixel in term of the selected polarimetric feature is calculated. All superpixels can be divided into two classes based on the variance values using the K-means method. The class with higher variances is denoted as the aircraft candidate superpixels, while the rest are clutter superpixels.

Different from the traditional superpixel-level CFAR topology, a new topology is designed to determine non-local homogenous clutter samples. Meanwhile, the logarithmic normal distribution model is used for CFAR detector threshold estimation. Then, with morphological filtering, the final aircraft detection result can be obtained.

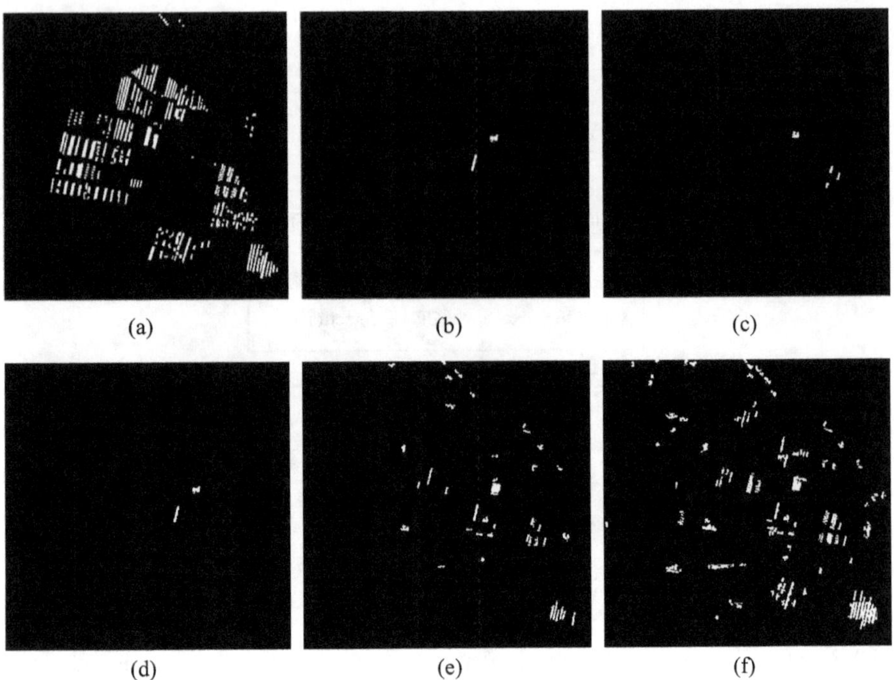

FIGURE 5.4.5 Aircraft detection results with the SO-CFAR detector. (a) Ground-truth image, (b) HH+SO-CFAR method, (c) HV+SO-CFAR method, (d) VV+SO-CFAR method, (e) *SPAN*+SO-CFAR method, (f) $\left|\hat{\gamma}_{\text{(HH-VV)-(HV)}}(\theta)\right|_{\text{contrast}}$ +SO-CFAR method.

In experimental study, the single polarimetric features of HH, HV, and VV, together with the *SPAN* feature, are selected for comparison. These polarimetric features are combined with SO-CFAR and SP-CFAR detectors to be the comparison methods. When the false alarm rate is 10^{-4}, the aircraft detection results with SO-CFAR and SP-CFAR detectors are shown in Figures 5.4.5 and 5.4.6, respectively. The corresponding ground-truth images are also given. For quantitative comparison, the receiver operating characteristic (ROC) curve of each method is shown in Figure 5.4.7. The area under the ROC curve (AUC) can represent the detection performance of different methods. The larger the AUC, the better the detection performance. The AUC values of different methods are summarized in Table 5.4.1.

For the SO-CFAR detector, it can be seen from Figures 5.4.5 (b)–(d) that the aircraft detection methods combined with the single polarimetric features of HH, HV, and VV are usually disturbed by strong points, and a number of missed detections are observed. The corresponding AUC values are 0.8515, 0.8176, and 0.8524, respectively. Missed detections also appear for the *SPAN* feature, as shown in Figure 5.4.5 (e). In comparison, the method combining polarimetric correlation contrast feature

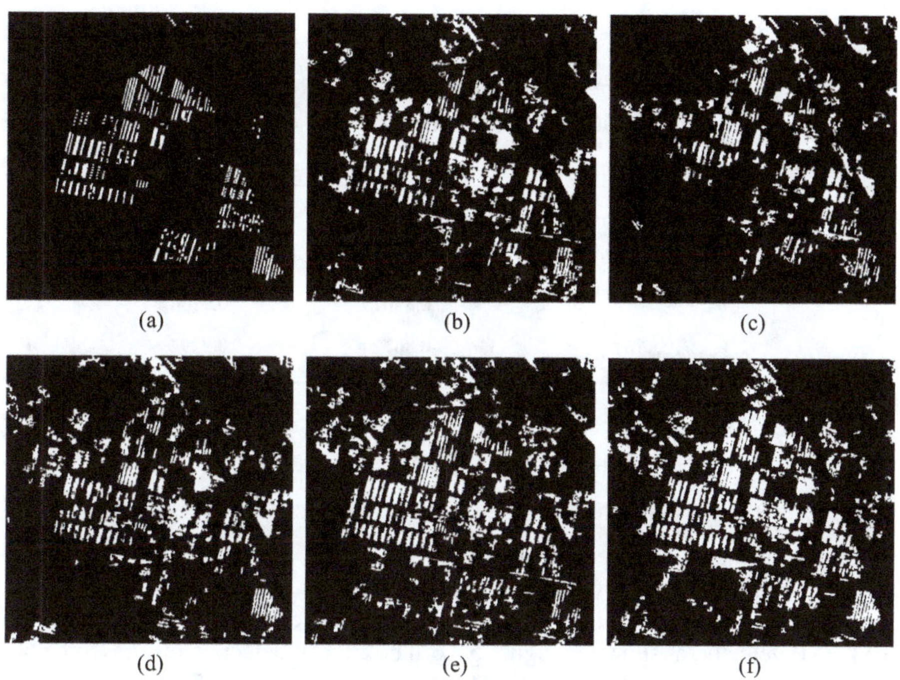

FIGURE 5.4.6 Aircraft detection results with the SP-CFAR detector. (a) Ground-truth image, (b) HH+SP-CFAR method, (c) HV+SP-CFAR method, (d) VV+SP-CFAR method, (e) *SPAN*+SO-CFAR method, (f) $\left|\hat{\gamma}_{(\text{HH-VV})\text{-}(\text{HV})}(\theta)\right|_{\text{contrast}}$ +SO-CFAR method.

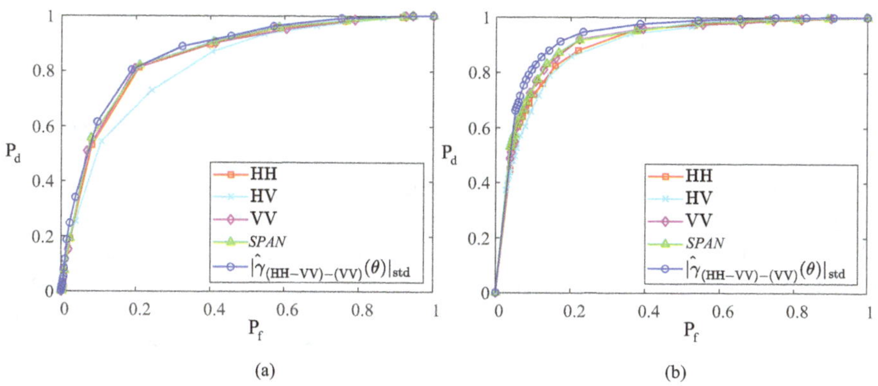

FIGURE 5.4.7 ROC results of different detection methods. (a) SO-CFAR detector, (b) SP-CFAR detector.

TABLE 5.4.1
Quantitative Comparison of Different Detection Methods with the AUC Measure

Feature	SO-CFAR	SP-CFAR
HH	0.8515	0.9028
HV	0.8176	0.8939
VV	0.8524	0.9095
SPAN	0.8567	0.9134
$\left\|\hat{\gamma}_{\text{(HH-VV)-(HV)}}(\theta)\right\|_{\text{contrast}}$	**0.8672**	**0.9297**

$\left|\hat{\gamma}_{\text{(HH-VV)-(HV)}}(\theta)\right|_{\text{contrast}}$ can detect more aircraft pixels, as shown in Figure 5.4.5 (f). Its AUC value is the highest, reaching 0.8672.

Compared with the SO-CFAR detector, the aircraft detection performances are significantly improved for each considered polarimetric feature combined with the SP-CFAR detector, as shown in Figure 5.4.6. The aircraft detection results from polarimetric features HH, HV, VV, and *SPAN* are shown in Figures 5.4.6 (b)–(e). While most of the aircraft pixels are detected, a number of false alarms also occur. Their AUC values are 0.9028, 0.8939, 0.9095, and 0.9134, respectively. The aircraft detection result from the polarimetric correlation contrast feature $\left|\hat{\gamma}_{\text{(HH-VV)-(HV)}}(\theta)\right|_{\text{contrast}}$ is shown in Figure 5.4.6 (f). It can be seen that the aircraft detection result is closer to the ground-truth, and its AUC value achieves the highest value of 0.9297.

In summary, the polarimetric rotation domain aircraft detection method achieves even better detection performance.

5.5 CONCLUSION

This chapter introduces polarimetric rotation domain target detection applications, including ship detection, building detection, and aircraft detection. The core ideas of these detection methods are to disclose the scattering mechanism differences between targets and clutter in the polarimetric rotation domain. Then, suitable polarimetric rotation domain features are selected to enhance candidate targets. Finally, combining the selected features and suitable detectors, target detection results can be obtained. For PolSAR ship detection, two polarimetric rotation domain ship detection methods are introduced based on the two-dimension and three-dimension polarimetric correlation pattern interpretation tools. Comparison experiments are carried out with three Radarsat-2 PolSAR datasets, which validate the efficiency and superiority of the two ship detection methods based on polarimetric rotation domain interpretation tools. For PolSAR building detection, a polarimetric rotation domain building detection method based on the polarimetric coherence pattern interpretation tool is introduced. The contrast between buildings (especially oriented buildings) and natural land covers such as forests can be clearly enhanced using the polarimetric maximum coherence feature, which is adopted for building detection. The performance of this method is demonstrated with PiSAR PolSAR data. For PolSAR aircraft detection, a polarimetric rotation domain aircraft detection method based on the two-dimension polarimetric correlation pattern interpretation tool is introduced, and its efficiency is validated by Radarsat-2 PolSAR data.

REFERENCES

1. X. C. Cui, C. S. Tao, Y. Su, and S. W. Chen, "PolSAR ship detection based on polarimetric correlation pattern," *IEEE Geoscience and Remote Sensing Letters,* vol. 18, no. 3, pp. 471–475, 2021.
2. X. C. Cui, Y. Su, and S. W. Chen, "Polarimetric SAR ship detection based on polarimetric rotation domain features and superpixel technique," *Journal of Radars,* vol. 10, no. 1, pp. 35–48, 2021.
3. L. Zhai, Y. Li, and Y. Su, "Inshore ship detection via saliency and context information in high-resolution SAR images," *IEEE Geoscience and Remote Sensing Letters,* vol. 13, no. 12, pp. 1870–1874, 2016.
4. D. Velotto, F. Nunziata, M. Migliaccio, and S. Lehner, "Dual-polarimetric TerraSAR-X SAR data for target at sea observation," *IEEE Geoscience and Remote Sensing Letters,* vol. 10, no. 5, pp. 1114–1118, 2013.
5. M. D. Li, S. P. Xiao, and S. W. Chen, "Three-dimension polarimetric correlation pattern interpretation tool and its application," *IEEE Transactions on Geoscience and Remote Sensing,* vol. 60, pp. 1–16, 2022.
6. M. D. Li, S. P. Xiao, and S. W. Chen, "Three-dimensional polarimetric orientation domain interpretation tool and ship detection," *Radar Science and Technology,* vol. 20, no. 3, pp. 245–254, 2022.
7. M. D. Li, G. Q. Wu, S. P. Xiao, and S. W. Chen, "A three-dimension polarimetric correlation pattern interpretation tool and its application," in *CIE International Conference on Radar*, Haikou, Hainan, China, 2021, pp. 582–585.
8. M. D. Li, X. C. Cui, and S. W. Chen, "Adaptive superpixel-level CFAR detector for SAR inshore dense ship detection," *IEEE Geoscience and Remote Sensing Letters,* vol. 19, pp. 1–5, 2022.
9. R. Achanata, A. Shaji, and K. Smith, "SLIC superpixels compared to state-of-the-art superpixel methods," *IEEE Transactions on Pattern Analysis and Machine Intelligence,* vol. 34, no. 11, pp. 2274–2282, 2012.
10. V. G. Hansen and J. H. Sawyers, "Detectability loss due to 'greatest of' selection in a cell-averaging CFAR," *IEEE Transactions on Aerospace and Electronic Systems,* vol. 16, no. 1, pp. 115–118, 1980.
11. T. Liu, Z. Y. Yang, T. Zhang, Y. Du, and A. Marino, "A new form of the polarimetric notch filter," *IEEE Geoscience and Remote Sensing Letters,* vol. 19, pp. 1–5, 2022.
12. X. Wang, C. Chen, Z. Pan, and Z. Pan, "Superpixel-based LCM detector for faint ships hidden in strong noise background SAR imagery," *IEEE Geoscience and Remote Sensing Letters,* vol. 16, no. 3, pp. 417–421, 2019.
13. S. P. Xiao, S. W. Chen, Y. L. Chang, Y. Z. Li, and M. Sato, "Polarimetric coherence optimization and its application for manmade target extraction in PolSAR data," *IEICE Transactions on Electronics,* vol. E97C, no. 6, pp. 566–574, 2014.

14. S. W. Chen, Y. Z. Li, and X. S. Wang, "A visualization tool for polarimetric SAR data investigation," in *Proceedings of EUSAR 2016: 11th European Conference on Synthetic Aperture Radar*, Hamburg, Germany, 2016, pp. 579–582.
15. S. W. Chen, "Polarimetric coherence pattern: A visualization and characterization tool for PolSAR data investigation," *IEEE Transactions on Geoscience and Remote Sensing*, vol. 56, no. 1, pp. 286–297, 2018.
16. S. W. Chen, X. S. Wang, S. P. Xiao, and M. Sato, *Target Scattering Mechanism in Polarimetric Synthetic Aperture Radar-Interpretation and Application*. Singapore: Springer, 2018.

6 Polarimetric Rotation Domain Structure Recognition

6.1 INTRODUCTION

Polarimetric radar, which can obtain full-polarization information and provide all-weather and day-night observation for the scene, has become an important sensor within microwave remote sensing [1, 2]. Manmade target recognition is an important application for polarimetric radar. Target structure recognition is the key to distinguishing manmade targets. However, the target scattering diversity effect makes this task challenging.

Generally, polarimetric information provides vital clues to distinguish different manmade structures. At present, there are mainly two kinds of manmade structure discrimination methods. The first is based on scattering center models, such as geometrical theory of diffraction (GTD) model [3], attributed scattering center (ASC) model [4], and canonical scattering feature model [5], and so on. They have been developed to describe the characteristics of complex manmade structures and targets. Scattering center models establish the analytical relationships between the radar backscattering from scattering centers and their physical attributes. The corresponding geometrical features of target structures can be obtained by model parameter inversion. However, most of these models have a number of unknown parameters, and the model inversions are complicated, which restrict their practical applications [6]. Another kind is coherent polarimetric target decompositions, which decompose the polarimetric scattering matrix into a summation of several canonical scattering mechanisms. Representative methods include Pauli decomposition [7], Krogager decomposition [8], and Cameron decomposition [9, 10]. Since their interpretation results have clear physical explanations, coherent polarimetric target decompositions have been widely used in manmade target structure recognition [11–13]. Among them, Cameron decomposition uses a set of six elemental structures to classify the maximum symmetric scattering component of a polarimetric scattering matrix in terms of the trihedral, cylinder, dipole, dihedral, narrow dihedral, and quarter wave device. It should be noted that Cameron decomposition only considers the maximum symmetric scattering matrix, which may have polarimetric information loss [14] and lead to misjudgment of target structures.

In this chapter, two manmade target structure recognition approaches based on polarimetric rotation domain interpretation theory are introduced. The first one aims to extend the polarimetric correlation pattern interpretation tool for manmade structure recognition by developing a novel polarimetric rotation domain feature coding

DOI: 10.1201/9781003461296-6

scheme [15]. The second one presents ideas to explore the null-pol features in the polarimetric rotation domain and establishes a manmade target structure recognition method based on the null-pol response pattern interpretation tool [16].

6.2 TARGET STRUCTURE RECOGNITION WITH POLARIMETRIC ROTATION DOMAIN FEATURE CODING

The polarimetric rotation domain characteristics of a target are closely related to the physical properties, such as the target geometry and orientation. Mining the polarimetric rotation domain characteristics is helpful to improve the target structure recognition performance. In this section, the polarimetric correlation pattern interpretation tool is used to analyze the polarimetric rotation domain characteristics of manmade structures and the influences of polarimetric measurement errors. Then, a target structure recognition method with the polarimetric rotation domain feature coding scheme is introduced.

6.2.1 Target Structure Recognition Methodology

6.2.1.1 Polarimetric Correlation Pattern for Canonical Structures

Polarimetric correlation features between different polarization channels contain abundant target scattering information [17, 18]. According to Chapter 2, combined with the Lexicographic scattering vector, two independent two-dimension polarimetric correlation pattern (denoted as polarimetric correlation pattern for short in this chapter) can be obtained in the polarimetric rotation domain:

$$\left|\hat{\gamma}_{\text{HH-VV}}(\theta)\right| = \left|S_{\text{HH}}(\theta)S_{\text{VV}}^{*}(\theta)\right| \quad \theta \in [-\pi, \pi) \tag{6.2.1}$$

$$\left|\hat{\gamma}_{\text{HH-HV}}(\theta)\right| = \left|S_{\text{HH}}(\theta)S_{\text{HV}}^{*}(\theta)\right| \quad \theta \in [-\pi, \pi) \tag{6.2.2}$$

Considering that the value ranges of polarimetric correlation patterns are different, the maximum value of a polarimetric correlation pattern is used for normalization. Note that such processing is different from the definition of the polarimetric coherence pattern [19]. Canonical structures, including the trihedral, dihedral, dipole, cylinder, narrow dihedral, quarter wave device, and helix, which are consistent with the Cameron decomposition, are also considered. Their ideal polarimetric scattering matrices and the corresponding analytical expressions of normalized polarimetric correlation patterns are shown in Table 6.2.1. The different scattering mechanisms of these structures lead to various patterns of polarimetric correlation features in the polarimetric rotation domain, which provides a theoretical basis to recognize them. The corresponding polarimetric correlation patterns of these canonical structures are shown in Figure 6.2.1. Theoretically, the scattering from a trihedral is polarimetric roll-invariant, while a dihedral exhibits an obvious scattering directivity effect. Taking HH and VV polarization channels, for example, polarimetric correlation pattern of the $\left|\hat{\gamma}_{\text{HH-VV}}(\theta)\right|$ from the trihedral is a circle shape, while that from a dihedral shows a four-lobe shape, shown in Figures 6.2.1 (b1) and (b2). As a result, these

TABLE 6.2.1
Analytical Expressions of Normalized Polarimetric Correlation Patterns for Canonical Structures

Index	Canonical Structure	Polarimetric Scattering Matrix	Polarimetric Correlation Pattern $\lvert\hat{\gamma}_{\text{HH-HV}}(\theta)\rvert$	$\lvert\hat{\gamma}_{\text{HH-VV}}(\theta)\rvert$
1	Trihedral	$\begin{bmatrix}1 & 0\\0 & 1\end{bmatrix}$	0	1
2	Dihedral	$\begin{bmatrix}1 & 0\\0 & -1\end{bmatrix}$	$\lvert\sin(4\theta)\rvert$	$\lvert[1+\cos 4\theta]/2\rvert$
3	Dipole	$\begin{bmatrix}1 & 0\\0 & 0\end{bmatrix}$	$\left\lvert\frac{2\sqrt{2}\sin(2\theta)+\sin(4\theta)}{2\sqrt{2}\sin 64^\circ+\sin 128^\circ}\right\rvert$	$\lvert[1\text{-}\cos 4\theta]/2\rvert$
4	Cylinder	$\begin{bmatrix}1 & 0\\0 & 1/2\end{bmatrix}$	$\left\lvert\frac{6\sin(2\theta)+\sin(4\theta)}{6\sin 74^\circ+\sin 148^\circ}\right\rvert$	$\lvert 17/18-1/18\cos(4\theta)\rvert$
5	Narrow dihedral	$\begin{bmatrix}1 & 0\\0 & -1/2\end{bmatrix}$	$\left\lvert\frac{2\sin(2\theta)+3\sin(4\theta)}{2\sin 52^\circ+3\sin 104^\circ}\right\rvert$	$\lvert 7/16+9/16\cos(4\theta)\rvert$
6	Quarter wave device	$\begin{bmatrix}1 & 0\\0 & \pm j\end{bmatrix}$	$\lvert -1/2\sin(4\theta)+j\sin(2\theta)\rvert$	$\lvert[1+j\cos(2\theta)]^2/2\rvert$
7	Helix	$\begin{bmatrix}1 & \pm j\\\pm j & -1\end{bmatrix}$	1	1

polarimetric correlation patterns are effective to visualize and characterize scattering characteristics and evolution laws of canonical manmade structures in the polarimetric rotation domain. The polarimetric rotation domain characteristic differences of these canonical scattering structures provide important evidence for manmade structures recognition.

6.2.1.2 Generalized Polarimetric Correlation Pattern

Since polarimetric radar is a multi-channel system, it is usually affected by non-ideal factors such as cross-coupling and channel imbalance during practical measurement, which can affect the target structure recognition. With valid polarimetric and radiometric calibrations, the measurement errors of a polarimetric radar system can be significantly mitigated to a relatively low level. Even though, these measurement errors still exist. Therefore, it is necessary to investigate the influences of polarimetric measurement errors on target structure recognition and then establish a more robust target structure recognition method. Considering the isolation and channel imbalance factors, the measured polarimetric scattering matrix can be expressed as [20]

$$\mathbf{M}=\begin{bmatrix}M_{\text{HH}} & M_{\text{HV}}\\M_{\text{VH}} & M_{\text{VV}}\end{bmatrix}=\begin{bmatrix}1 & 0\\0 & a\end{bmatrix}\begin{bmatrix}1 & \delta\\\delta & 1\end{bmatrix}\begin{bmatrix}S_{\text{HH}} & S_{\text{HV}}\\S_{\text{VH}} & S_{\text{VV}}\end{bmatrix}\begin{bmatrix}1 & \delta\\\delta & 1\end{bmatrix}\begin{bmatrix}1 & 0\\0 & a\end{bmatrix}+\mathbf{N} \quad (6.2.3)$$

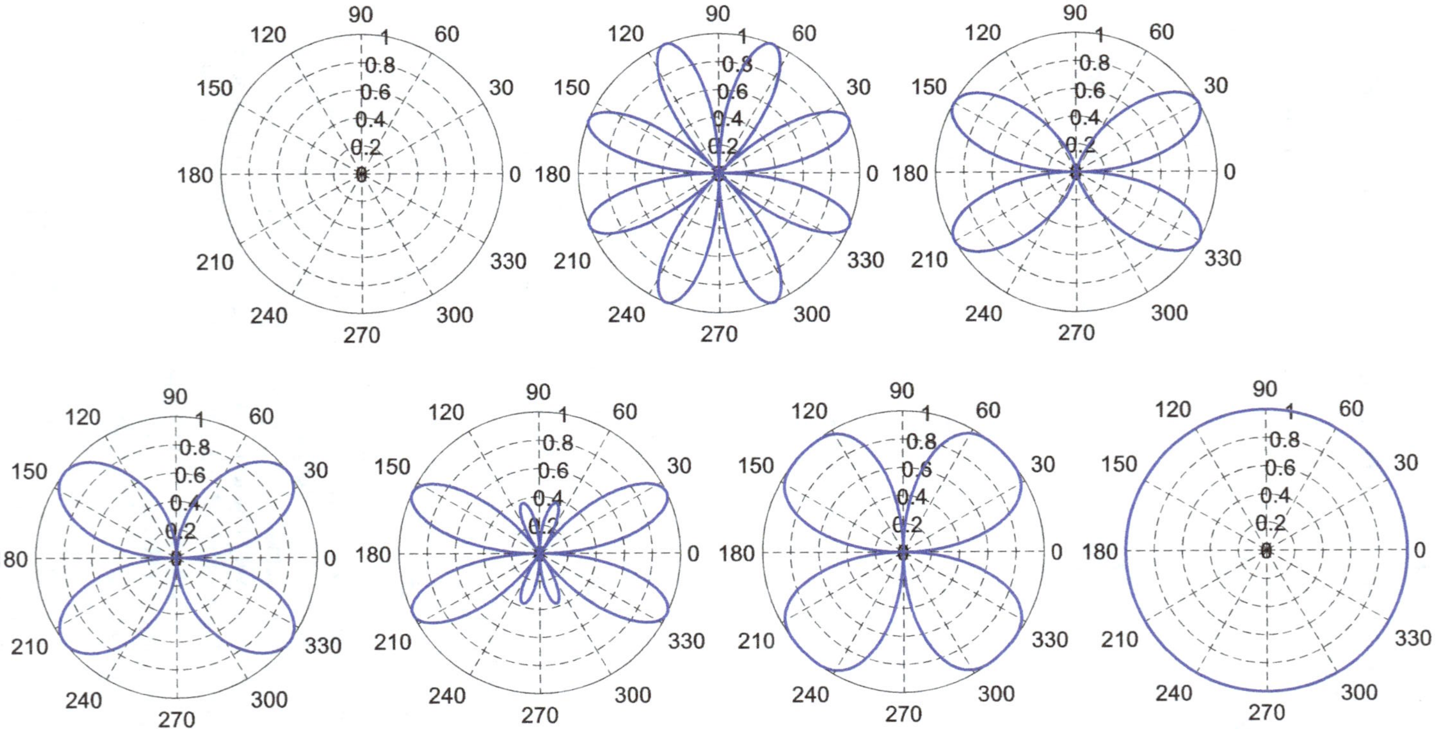

FIGURE 6.2.1 Polarimetric correlation patterns of canonical structures. (a1)–(a7) $\left|\hat{\gamma}_{\text{HH-HV}}(\theta)\right|$, (b1)–(b7) $\left|\hat{\gamma}_{\text{HH-VV}}(\theta)\right|$. The numbers 1–7 indicate the trihedral, dihedral, dipole, cylinder, narrow dihedral, quarter wave device, and helix structures, respectively.

(Continued)

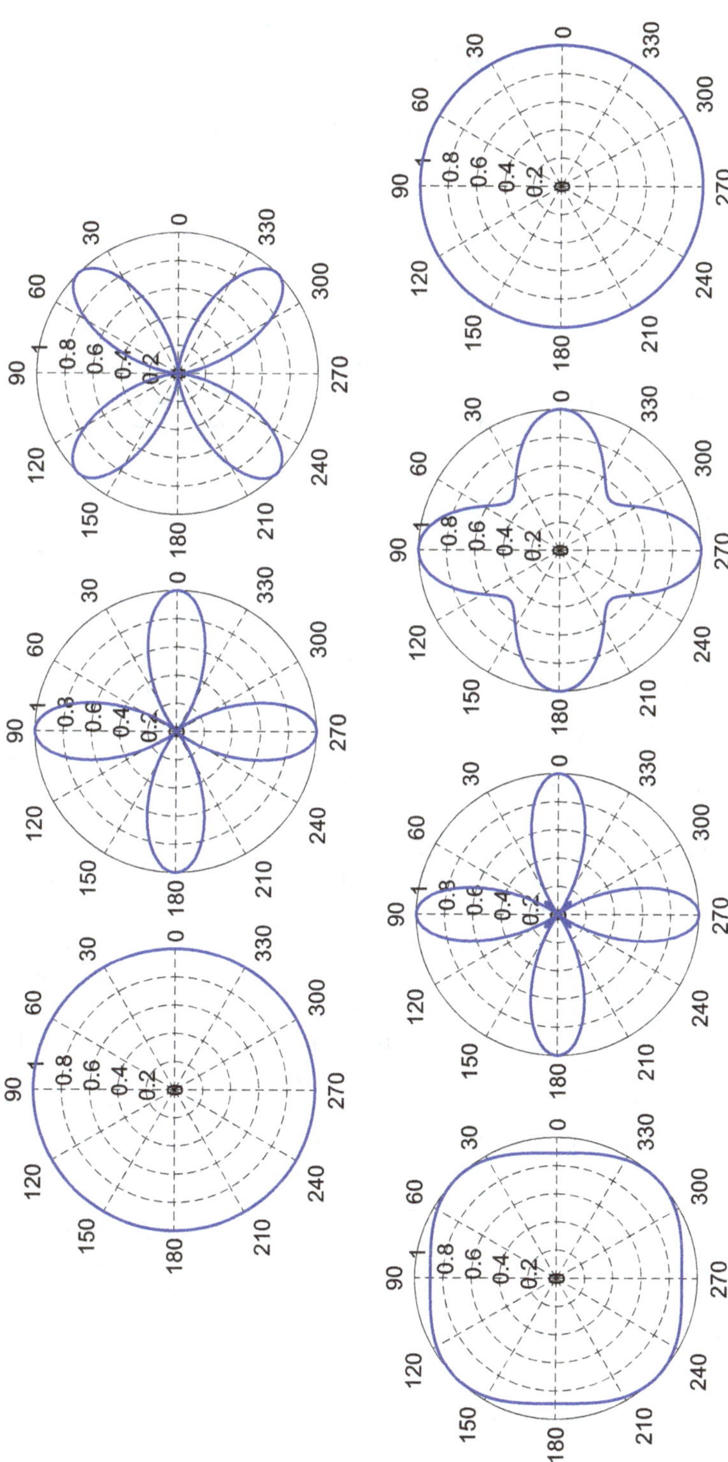

FIGURE 6.2.1 (Continued)

where δ is the polarization isolation; a is a complex value which denotes the polarization channel imbalance in terms of amplitude and phase; and $\mathbf{N}$ represents the system noise term, which is omitted for following analysis.

Therefore, the generalized polarimetric correlation patterns $\left|\tilde{\gamma}_{\text{HH-HV}}(\theta)\right|$ and $\left|\tilde{\gamma}_{\text{HH-VV}}(\theta)\right|$ with polarimetric measurement errors become

$$\left|\tilde{\gamma}_{\text{HH-HV}}(\theta)\right| = \left|M_{\text{HH}}(\theta)M_{\text{HV}}^{*}(\theta)\right| \quad \theta \in [-\pi,\pi) \tag{6.2.4}$$

$$\left|\tilde{\gamma}_{\text{HH-VV}}(\theta)\right| = \left|M_{\text{HH}}(\theta)M_{\text{VV}}^{*}(\theta)\right| \quad \theta \in [-\pi,\pi) \tag{6.2.5}$$

Polarimetric measurement errors can affect the generalized polarimetric correlation patterns of canonical structures. For a certain polarimetric radar system, its polarization isolation and channel imbalance terms can be available after the polarimetric calibration. Therefore, they can be known parameters for (6.2.4) and (6.2.5).

6.2.1.3 Polarimetric Roll-Invariant Feature Coding

From Chapter 2, ten polarimetric features can be derived from each polarimetric correlation pattern [19]. Among them, according to Chapter 3, the correlation degree, the correlation fluctuation, the maximum correlation, the minimum correlation, the correlation contrast, the correlation anisotropy, and the correlation beamwidth are polarimetric roll-invariant features. For two independent polarimetric correlation patterns $\left|\tilde{\gamma}_{\text{HH-HV}}(\theta)\right|$ and $\left|\tilde{\gamma}_{\text{HH-VV}}(\theta)\right|$, there are 14 polarimetric roll-invariant features in total. These polarimetric roll-invariant features are relatively independent of the geometric relationship between the target and radar, and are suitable for target structure recognition.

In order to collect an optimal polarimetric feature set, the class separation distance is adopted as the selection criterion. In detail, the Euclidean distance is utilized to measure the class separation distance. Polarimetric features achieving the largest class separation distance will be chosen, and their selection frequencies will be recorded accordingly. During polarimetric feature selection, seven canonical structures, including the trihedral, dihedral, dipole, cylinder, narrow dihedral, quarter wave device, and helix, are considered. Polarization isolation ranges from –40 dB to –5 dB with an interval of 0.1 dB, amplitude imbalance ranges from 0 dB to 5 dB with an interval of 0.1 dB, and phase imbalance ranges from 0° to 30° with an interval of 0.1° are added into the ideal polarimetric scattering matrices of each canonical structure. Then, a sample dataset in terms of different polarimetric measurement errors can be generated. These seven canonical structures can form 21 structure pairs. The results of polarimetric feature selection are shown in Table 6.2.2. The numbers in parentheses indicate the selection frequency for each polarimetric feature. Finally, five polarimetric roll-invariant features with the top five highest selection frequencies are selected to form the polarimetric feature vector $\mathbf{w}$:

$$\mathbf{w} = \left[\left|\tilde{\gamma}_{\text{HH-HV}}(\theta)\right|_{\text{mean}} \ \left|\tilde{\gamma}_{\text{HH-HV}}(\theta)\right|_{\text{std}} \ \left|\tilde{\gamma}_{\text{HH-HV}}(\theta)\right|_{\text{bw }0.95} \ \left|\tilde{\gamma}_{\text{HH-VV}}(\theta)\right|_{\min} \ \left|\tilde{\gamma}_{\text{HH-VV}}(\theta)\right|_{\text{A}}\right]^{\text{T}} \tag{6.2.6}$$

TABLE 6.2.2
Polarimetric Feature Selection Results

Polarimetric Correlation Pattern	Polarimetric Rotation Domain Feature and Selection Frequency
$\lvert\tilde{\gamma}_{\text{HH-HV}}(\theta)\rvert$	$\lvert\tilde{\gamma}_{\text{HH-HV}}(\theta)\rvert_{\text{bw0.95}}$ (8), $\lvert\tilde{\gamma}_{\text{HH-HV}}(\theta)\rvert_{\text{std}}$ (2), $\lvert\tilde{\gamma}_{\text{HH-HV}}(\theta)\rvert_{\text{mean}}$ (2), $\lvert\tilde{\gamma}_{\text{HH-HV}}(\theta)\rvert_{\text{min}}$ (0), $\lvert\tilde{\gamma}_{\text{HH-HV}}(\theta)\rvert_{\text{max}}$ (0), $\lvert\tilde{\gamma}_{\text{HH-HV}}(\theta)\rvert_{\text{contrast}}$ (0), $\lvert\tilde{\gamma}_{\text{HH-HV}}(\theta)\rvert_{\text{A}}$ (0)
$\lvert\tilde{\gamma}_{\text{HH-VV}}(\theta)\rvert$	$\lvert\tilde{\gamma}_{\text{HH-VV}}(\theta)\rvert_{\text{A}}$ (7), $\lvert\tilde{\gamma}_{\text{HH-VV}}(\theta)\rvert_{\text{min}}$ (2), $\lvert\tilde{\gamma}_{\text{HH-VV}}(\theta)\rvert_{\text{bw0.95}}$ (0), $\lvert\tilde{\gamma}_{\text{HH-VV}}(\theta)\rvert_{\text{mean}}$ (0), $\lvert\tilde{\gamma}_{\text{HH-VV}}(\theta)\rvert_{\text{max}}$ (0), $\lvert\tilde{\gamma}_{\text{HH-VV}}(\theta)\rvert_{\text{contrast}}$ (0), $\lvert\tilde{\gamma}_{\text{HH-VV}}(\theta)\rvert_{\text{std}}$ (0)

These five polarimetric roll-invariant features represent the rich hidden information in the polarimetric rotation domain for given polarization channels. The combination of these features can discriminate different manmade structures. The feature vector corresponding to each canonical structure is defined as the polarimetric roll-invariant feature coding vector (denoted the polarimetric feature coding vector for short). In this vein, the polarimetric feature coding vectors under different polarimetric measurement errors can be available. For example, the polarimetric feature coding vectors with –40 dB polarization isolation are

$$\begin{bmatrix} \mathbf{v}_1 & \mathbf{v}_2 & \mathbf{v}_3 & \mathbf{v}_4 & \mathbf{v}_5 & \mathbf{v}_6 & \mathbf{v}_7 \end{bmatrix} = \begin{bmatrix} 0.64 & 0.64 & 0.49 & 0.61 & 0.47 & 0.72 & 1 \\ 0.31 & 0.31 & 0.36 & 0.31 & 0.32 & 0.30 & 0 \\ 0.64 & 0.32 & 0.46 & 0.58 & 0.37 & 0.89 & 1.57 \\ 1 & 0 & 0 & 0.89 & 0.01 & 0.50 & 1 \\ 0 & 1 & 1 & 0.06 & 0.99 & 0.33 & 0 \end{bmatrix} \tag{6.2.7}$$

where $\mathbf{v}_k$ is the polarimetric feature coding vector of the kth canonical structure. The numbers 1–7 indicate the trihedral, dihedral, dipole, cylinder, narrow dihedral, quarter wave device, and helix structures, respectively.

6.2.1.4 Manmade Structure Recognition Algorithm

With the polarimetric feature coding vector, seven canonical structures can potentially be distinguished. In practical application, the polarimetric features extracted from the measured data are usually different from their theoretical values. In this vein, the similarity distance can be used to measure the closeness among the measured polarimetric feature vector $\mathbf{w}$ and the theoretical polarimetric feature coding vector $\mathbf{v}_k$. The structure type can be determined when the similarity distance reaches the minimum value. Therefore, the structure type determination expression is

$$d = \arg\min_{k} \left\| \mathbf{w} - \mathbf{v}_k \right\|^2 \tag{6.2.8}$$

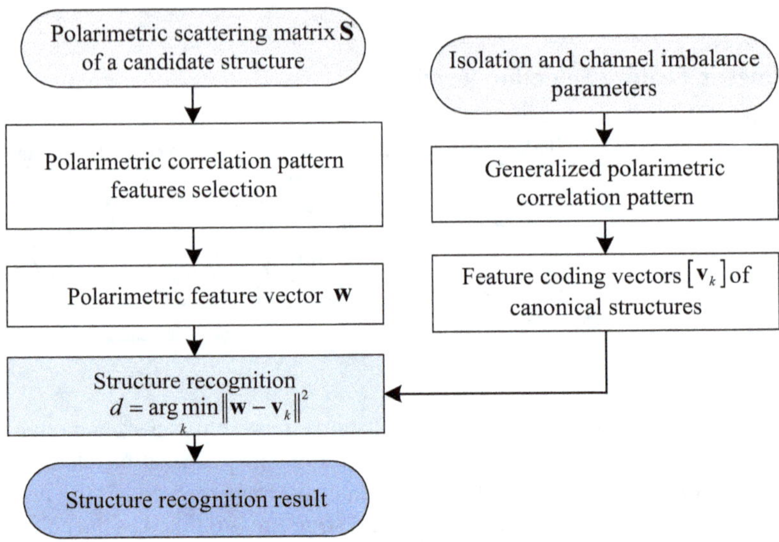

FIGURE 6.2.2 Flowchart of manmade structure recognition with polarimetric rotation domain feature coding.

where $\mathbf{w}$ is the measured polarimetric feature vector, and d indicates the canonical structure corresponding to the theoretical value that minimizes the distance.

Above all, a manmade structure recognition algorithm with polarimetric feature coding vector is developed. First, according to the polarimetric correlation pattern interpretation tool, five optimal polarimetric roll-invariant features $\left|\tilde{\gamma}_{\text{HH-HV}}(\theta)\right|_{\text{mean}}$, $\left|\tilde{\gamma}_{\text{HH-HV}}(\theta)\right|_{\text{std}}$, $\left|\tilde{\gamma}_{\text{HH-HV}}(\theta)\right|_{\text{bw0.95}}$, $\left|\tilde{\gamma}_{\text{HH-VV}}(\theta)\right|_{\text{min}}$ and $\left|\tilde{\gamma}_{\text{HH-VV}}(\theta)\right|_{\text{A}}$ are derived to construct the polarimetric feature vector $\mathbf{w}$. Second, the polarimetric feature coding vectors $\mathbf{v}_k$ of the canonical scattering structures are obtained with the polarimetric measurement error parameters. Finally, according to the structure type determination expression (6.2.8), the target structure type is determined. The flowchart of the manmade structure recognition with polarimetric rotation domain feature coding is shown in Figure 6.2.2.

6.2.2 Experimental Studies

Both the electromagnetic computation data of canonical structures, an unmanned aerial vehicle (UAV) target, and the real PolSAR data are employed for experimental studies. Cameron decomposition [9] is utilized as the comparison method.

6.2.2.1 Canonical Structure Recognition with Electromagnetic Computation Data

The main canonical structures of the plate, trihedral, dihedral, and cylinder are utilized for electromagnetic computation. The simulation models are shown in Figure 6.2.3.

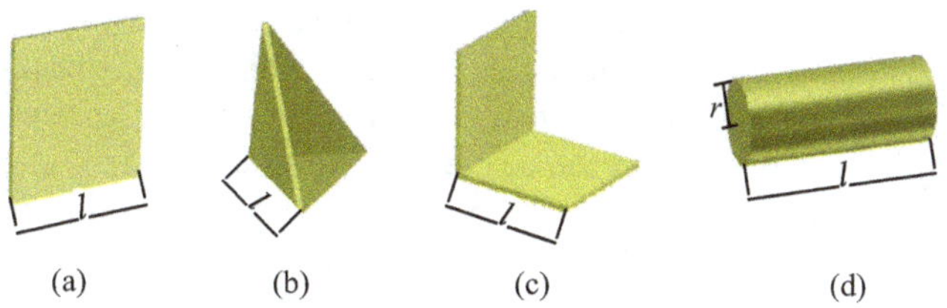

FIGURE 6.2.3 Canonical structure simulation models. (a) Plate, (b) trihedral, (c) dihedral, (d) cylinder.

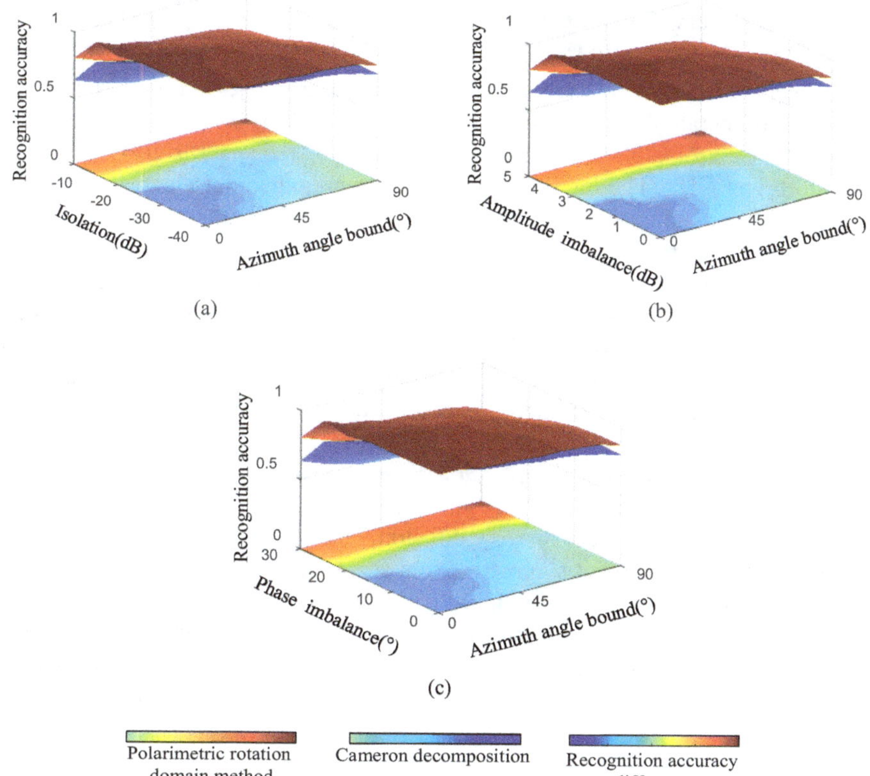

FIGURE 6.2.4 Canonical structure recognition accuracy comparison in terms of azimuth angle bound and polarimetric measurement errors of (a) isolation, (b) amplitude imbalance, and (c) phase imbalance.

For electromagnetic computation, the frequency ranges from 8 GHz to 12 GHz with an interval of 0.1 GHz, the elevation angle ranges from 30° to 60° with an interval of 10°, and the azimuth angle ranges from –45° to 45° with an interval of 1°. The structure size l is set as 1 m, and the radius r is set as 0.25 m.

The recognition results from the Cameron decomposition and the polarimetric rotation domain method are shown in Figure 6.2.4. The projection plane indicates the recognition accuracy differences between the two methods. It is clear that the recognition accuracies of both methods decrease with the increasing of the azimuth angle bound and the polarimetric measurement errors. However, the degree of the recognition accuracy decrease of the polarimetric rotation domain method is less than that of Cameron decomposition. In other word, the polarimetric rotation domain method outperforms Cameron decomposition. Moreover, the polarimetric rotation domain method is less sensitive to polarimetric measurement errors. These comparison studies demonstrate that the polarimetric rotation domain method exhibits superior recognition accuracy and better robustness for oriented manmade structures.

6.2.2.2 UAV Target Structure Recognition with Electromagnetic Computation Data

In this section, the structure recognition experiment is carried out with the electromagnetic computation data of an all-metal model UAV, shown in Figure 6.2.5 (a). During the electromagnetic computation, the center frequency is 10 GHz and the bandwidth is 4 GHz. The Pauli images of the front and squint views are shown in Figures 6.2.5 (b) and (c), respectively. The pixels with strong energy are extracted as the scattering centers.

The recognition results from Cameron decomposition and the polarimetric rotation domain method are shown in Figure 6.2.6. The confusion matrix of the recognition results from both methods is shown in Table 6.2.3. From Table 6.2.3, 84.17% of the recognition results obtained by the polarimetric rotation domain method are the same as those obtained by Cameron decomposition, while 15.83% of the recognition results are inconsistent. For the front view case, the majority of these scattering centers appear as cylinder structures. For the squint view case, the scattering mechanisms become more complicated. Most of these scattering centers are identified as quarter wave device, cylinder, narrow dihedral, and dipole structures. For the front view image, since the UAV is symmetric, four scattering centers at the junction of the wing and fuselage should correspond to the same structure. However, one of these scattering centers is determined as a quarter wave device, while other three scattering centers are identified as narrow dihedral structures by the Cameron decomposition, shown in Figure 6.2.6 (a). In comparison, the polarimetric rotation domain method

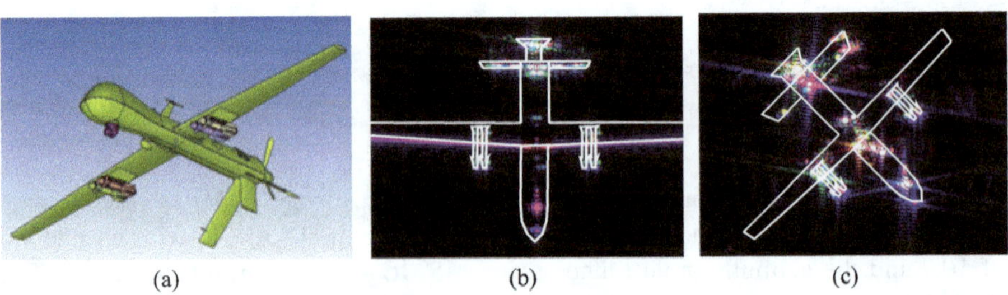

(a) (b) (c)

FIGURE 6.2.5 (a) UAV target simulation model, (b) Pauli image of the UAV's front view, (c) Pauli image of the UAV's squint view.

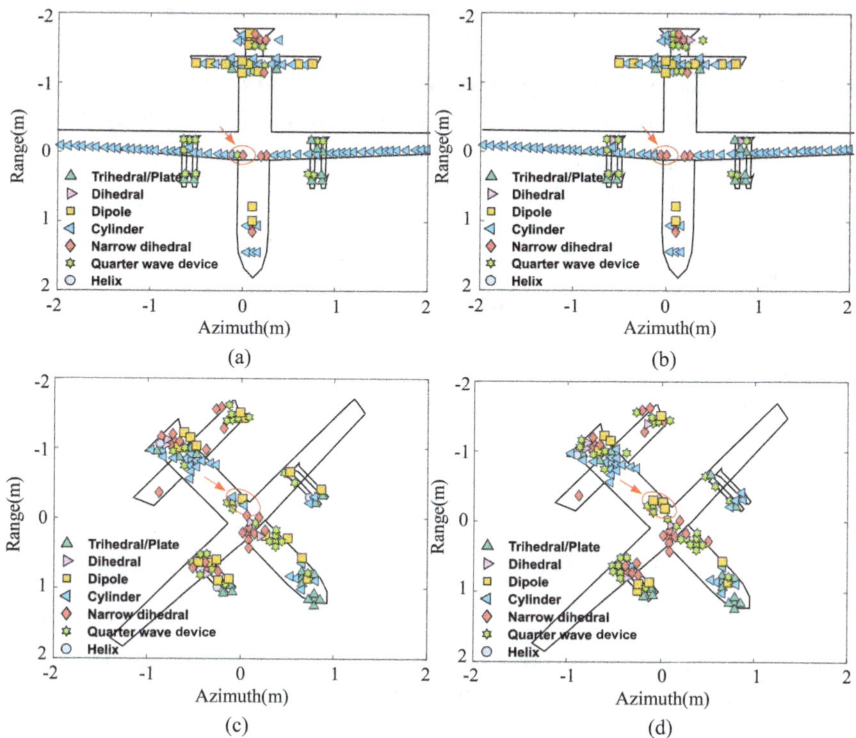

FIGURE 6.2.6 UAV target structure recognition comparison. (a) Cameron decomposition for the front view case, (b) polarimetric rotation domain method for the front view case, (c) Cameron decomposition for the squint view case, (d) polarimetric rotation domain method for the squint view case.

TABLE 6.2.3
Confusion Matrix of UAV Target Structure Recognition Results between the Polarimetric Rotation Domain Method and Cameron Decomposition

View	Cameron Decomposition \ Polarimetric Rotation Domain Method	Trihedral	Dihedral	Dipole	Cylinder	Narrow Dihedral	Quarter Wave	Helix
The front view	Trihedral	10	0	0	0	0	0	0
	Dihedral	0	0	0	0	0	0	0
	Dipole	0	0	9	1	0	4	0
	Cylinder	0	0	1	69	0	1	0
	Narrow dihedral	0	1	0	0	9	0	0
	Quarter wave device	0	2	0	0	1	12	0
	Helix	0	0	0	0	0	0	0

(Continued)

TABLE 6.2.3 ***(Continued)***
Confusion Matrix of UAV Target Structure Recognition Results between the Polarimetric Rotation Domain Method and Cameron Decomposition

View	Cameron Decomposition \ Polarimetric Rotation Domain Method	Trihedral	Dihedral	Dipole	Cylinder	Narrow Dihedral	Quarter Wave	Helix
The squint view	Trihedral	15	0	0	2	0	0	1
	Dihedral	0	4	1	0	0	0	0
	Dipole	1	0	10	3	3	1	0
	Cylinder	1	0	2	21	0	0	0
	Narrow dihedral	0	0	1	0	18	4	0
	Quarter wave device	0	1.	0	0	2	25	0
	Helix	0	0	3	0	0	1	0

successfully identifies them as narrow dihedral structures, shown in Figure 6.2.6 (b), which is more consistent with the UAV's overall symmetry structure. For the squint view case, the area marked with a red ellipse is the UAV's antenna and should be a dipole structure as identified by the polarimetric rotation domain method. However, some of these scattering centers are recognized as cylinders by Cameron decomposition. In addition, most of the differences between the two methods appear at the empennage part. However, due to the very complicated structure of the empennage, the corresponding structure relationships are not clear, and the investigation is omitted here. Generally, the structure recognition results from the polarimetric rotation domain method are more consistent with the actual UAV structure.

6.2.2.3 Ship Target Structure Recognition with Real PolSAR Data

In order to evaluate the polarimetric rotation domain method for more realistic manmade targets, L-band ALOS-2 PolSAR data are utilized. This dataset was acquired on August 21, 2018, over the San Francisco area. The nominal resolutions are 5.1 m and 4.3 m in the range and azimuth directions. A ROI of 50×120 pixels containing a tanker is selected for investigation. The recognition results from Cameron decomposition and the polarimetric rotation domain method are shown in Figure 6.2.7.

The stern and sides of the tanker are mainly composed of a double-bounce scattering mechanism, corresponding to the dihedral and narrow dihedral structure. There are cylinder structures on the deck. Due to the cabin and the corners, the bow has some local structures as quarter wave devices. In the optical image, the areas marked by red ellipses are platform and oil pipelines, which correspond to the plate and cylinder structures. Cameron decomposition misidentifies the plate as a quarter wave device and the oil pipeline as a trihedral and dipole. The polarimetric rotation domain method can recognize the plate structure and the cylinder structure accurately. Therefore, compared with Cameron decomposition, the polarimetric rotation domain method achieves better structure recognition performance for this ship target.

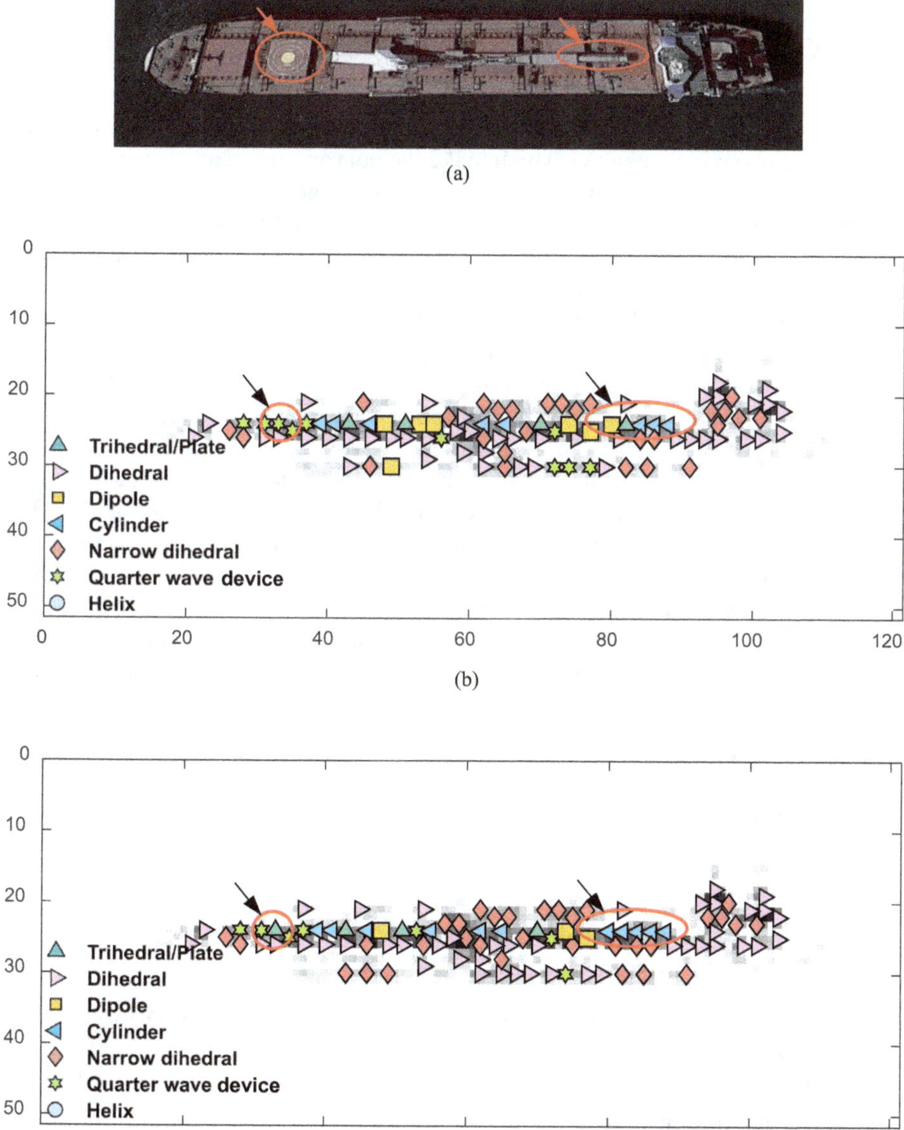

FIGURE 6.2.7 Ship structure recognition comparison. (a) Optical image, (b) Cameron decomposition, (c) polarimetric rotation domain method.

6.3 TARGET STRUCTURE RECOGNITION WITH NULL-POL RESPONSE PATTERN IN THE POLARIMETRIC ROTATION DOMAIN

Full-polarization acquisition is significant for target scattering mechanism interpretation [1, 21]. Target optimal polarization refers to the polarization states of transmitting and receiving antennas, which make the polarimetric radar receiving power reach the maximum or minimum. The concept of target optimal polarization was proposed by Kennaugh in 1952, which mainly considers the monostatic reciprocal condition [22]. Huynen further proposed the radar target phenomenological theory and used the Poincaré sphere and Stokes vector to represent the polarization state of electromagnetic waves. Then, the concept of the polarization fork was proposed [23, 24]. Furthermore, Boerner extended the target optimal polarization to the bistatic, nonreciprocal, and incoherent situations [25–30]. Thereafter, target optimal polarization theory has been developed rapidly [31–33]. Co-polarization null, also known as null-pol, is a special intrinsic polarization, which plays an important role in target optimal polarization theory. Except for the absolute energy, all information of a polarimetric scattering matrix can be recovered by a pair of null-pol points [33, 34]. So far, null-pol research mainly focuses on basic concepts and theoretical derivation [32]. Applications have seldom been reported. Inspired by the polarimetric coherence pattern interpretation tool introduced in Chapter 2, this section presents studies of null-pol features in the polarimetric rotation domain and introduces a target structure recognition method based on the null-pol response pattern features [16].

6.3.1 Null-Pol Response Patterns Interpretation Tool

6.3.1.1 Target Null-Pol Theory

With a given polarimetric scattering matrix, the received polarimetric power is represented as

$$P = \left|\mathbf{h}_{\mathrm{r}}^{\mathrm{T}}\mathbf{S}\mathbf{h}_{\mathrm{t}}\right|^2 \tag{6.3.1}$$

where $\mathbf{h}_{\mathrm{t}}$ and $\mathbf{h}_{\mathrm{r}}$ are the polarization states of transmitting and receiving antennas, respectively.

When $\mathbf{h}_{\mathrm{t}}=\mathbf{h}_{\mathrm{r}}=\mathbf{h}$ is valid, the polarimetric response $P_{\mathrm{co}} = \left|\mathbf{h}^{\mathrm{T}}\mathbf{S}\mathbf{h}\right|^2$ is called the co-polarization response. By adjusting the polarization state of the transmitting and receiving antennas, polarimetric response can achieve the extremum value. Such polarization states of antennas are called the target optimal polarization. Among these optimal polarization states, the null-pols can make the co-polarization response equal to zero, $P_{\mathrm{co}} = 0$.

The null-pol vectors of ten canonical structures are calculated and summarized in Table 6.3.1. These canonical structures are widely used for manmade target structure recognition. As shown in Table 6.3.1, most these polarimetric scattering matrices have two null-pol vectors, except for the horizontal/vertical dipoles and left/right helix. Moreover, the null-pol vectors of left and right helix are equal to those of the trihedral. Therefore, there are 14 unique null-pol vectors for the ten canonical structures.

TABLE 6.3.1
Polarimetric Scattering Matrix and Null-Pol Vectors of Canonical Structures

Structure	Polarimetric Scattering Matrix	Null-Pol Vector
Trihedral	$\begin{bmatrix}1 & 0\\0 & 1\end{bmatrix}$	$\mathbf{h}_1=\frac{1}{\sqrt{2}}\begin{bmatrix}1\\\mathrm{j}\end{bmatrix},\mathbf{h}_2=\frac{1}{\sqrt{2}}\begin{bmatrix}1\\-\mathrm{j}\end{bmatrix}$
Dihedral	$\begin{bmatrix}1 & 0\\0 & -1\end{bmatrix}$	$\mathbf{h}_3=\frac{1}{\sqrt{2}}\begin{bmatrix}1\\1\end{bmatrix},\mathbf{h}_4=\frac{1}{\sqrt{2}}\begin{bmatrix}1\\-1\end{bmatrix}$
Cylinder	$\begin{bmatrix}1 & 0\\0 & 1/2\end{bmatrix}$	$\mathbf{h}_5=\frac{1}{\sqrt{3}}\begin{bmatrix}1\\\sqrt{2}\mathrm{j}\end{bmatrix},\mathbf{h}_6=\frac{1}{\sqrt{3}}\begin{bmatrix}1\\-\sqrt{2}\mathrm{j}\end{bmatrix}$
Narrow dihedral	$\begin{bmatrix}1 & 0\\0 & -1/2\end{bmatrix}$	$\mathbf{h}_7=\frac{1}{\sqrt{3}}\begin{bmatrix}1\\\sqrt{2}\end{bmatrix},\mathbf{h}_8=\frac{1}{\sqrt{3}}\begin{bmatrix}1\\-\sqrt{2}\end{bmatrix}$
Quarter wave device-1	$\begin{bmatrix}1 & 0\\0 & \mathrm{j}\end{bmatrix}$	$\mathbf{h}_9=\frac{1}{2}\begin{bmatrix}\sqrt{2}\\(1+\mathrm{j})\end{bmatrix},\mathbf{h}_{10}=\frac{1}{2}\begin{bmatrix}\sqrt{2}\\-(1+\mathrm{j})\end{bmatrix}$
Quarter wave device-2	$\begin{bmatrix}1 & 0\\0 & -\mathrm{j}\end{bmatrix}$	$\mathbf{h}_{11}=\frac{1}{2}\begin{bmatrix}\sqrt{2}\\(1-\mathrm{j})\end{bmatrix},\mathbf{h}_{12}=\frac{1}{2}\begin{bmatrix}\sqrt{2}\\-(1-\mathrm{j})\end{bmatrix}$
Horizontal dipole	$\begin{bmatrix}1 & 0\\0 & 0\end{bmatrix}$	$\mathbf{h}_{13}=\begin{bmatrix}0\\1\end{bmatrix}$
Vertical dipole	$\begin{bmatrix}0 & 0\\0 & 1\end{bmatrix}$	$\mathbf{h}_{14}=\begin{bmatrix}1\\0\end{bmatrix}$
Left helix	$\begin{bmatrix}1 & \mathrm{j}\\\mathrm{j} & -1\end{bmatrix}$	$\mathbf{h}_1=\frac{1}{\sqrt{2}}\begin{bmatrix}1\\\mathrm{j}\end{bmatrix}$
Right helix	$\begin{bmatrix}1 & -\mathrm{j}\\-\mathrm{j} & -1\end{bmatrix}$	$\mathbf{h}_2=\frac{1}{\sqrt{2}}\begin{bmatrix}1\\-\mathrm{j}\end{bmatrix}$

6.3.1.2 Null-Pol Response Pattern in the Polarimetric Rotation Domain

With a polarimetric scattering matrix $\mathbf{S}(\theta)$ in the polarimetric rotation domain, the received polarimetric power can be extended to the polarimetric rotation domain, and the co-polarization response can be represented as

$$P_{\mathrm{co}}(\theta)=\left|\mathbf{h}^{\mathrm{T}}\mathbf{S}(\theta)\mathbf{h}\right|^2 \tag{6.3.2}$$

When $\mathbf{h}$ is replaced by the null-pol vectors $\mathbf{h}_i$ and $i=1,2,\cdots,14$ shown in Table 6.3.1, $P_{\mathrm{co}}(\theta)$ becomes the null-pol response in the polarimetric rotation domain. In this vein, a new polarimetric rotation domain interpretation tool of the polarimetric rotation domain null-pol response pattern can be further obtained to characterize the null-pol response characteristics in the polarimetric rotation domain. Its definition is

$$\left|\gamma_i(\theta)\right|^{\mathrm{NP}}=\frac{P_{\mathrm{co}}(\theta)}{P_{\mathrm{co}}^{\max}}=\frac{\left|\mathbf{h}_i^{\mathrm{T}}\mathbf{S}(\theta)\mathbf{h}_i\right|^2}{P_{\mathrm{co}}^{\max}},\ \theta\in[-\pi,\pi) \tag{6.3.3}$$

where $P_{\text{co}}^{\max}$ denotes the maximum co-polarization response among all the null-pol vectors and $\gamma_i(\theta) \in [0,1]$.

It can be verified that when the null-pol vector satisfies $\mathbf{h}_m = \mathbf{R}(\theta_j)\mathbf{h}_n$, the null-pol response patterns $|\gamma_m(\theta)|^{\text{NP}}$ and $|\gamma_n(\theta)|^{\text{NP}}$ are equivalent. According to Table 6.3.1, the following equivalent relationships can be obtained

$$\mathbf{h}_3 = \mathbf{R}(\theta_1)\mathbf{h}_7, \qquad \mathbf{h}_4 = \mathbf{R}(\theta_2)\mathbf{h}_8 \tag{6.3.4}$$

$$\mathbf{h}_3 = \mathbf{R}(\theta_3)\mathbf{h}_{13}, \qquad \mathbf{h}_4 = \mathbf{R}(\theta_4)\mathbf{h}_{14} \tag{6.3.5}$$

$$\mathbf{h}_9 = \mathbf{R}(\theta_5)\mathbf{h}_{11}, \qquad \mathbf{h}_{10} = \mathbf{R}(\theta_6)\mathbf{h}_{12} \tag{6.3.6}$$

where $\mathbf{R}(\theta_j)$ is the polarimetric rotation matrix, while $\theta_1 \approx 9.7°$, $\theta_2 \approx -9.7°$, θ_3=45°, θ_4=−45°, $\theta_5 \approx 70.5°$, and $\theta_6 \approx 70.5°$.

In this vein, $|\gamma_3(\theta)|^{\text{NP}}$, $|\gamma_4(\theta)|^{\text{NP}}$, $|\gamma_7(\theta)|^{\text{NP}}$, $|\gamma_8(\theta)|^{\text{NP}}$, $|\gamma_{13}(\theta)|^{\text{NP}}$ and $|\gamma_{14}(\theta)|^{\text{NP}}$, and $|\gamma_9(\theta)|^{\text{NP}}$, $|\gamma_{10}(\theta)|^{\text{NP}}$, $|\gamma_{11}(\theta)|^{\text{NP}}$ and $|\gamma_{12}(\theta)|^{\text{NP}}$ are equivalent, respectively. Therefore, among the 14 null-pol vectors of canonical structures in Table 6.3.1, only $\mathbf{h}_1$, $\mathbf{h}_2$, $\mathbf{h}_3$, $\mathbf{h}_4$, $\mathbf{h}_5$, $\mathbf{h}_6$, $\mathbf{h}_9$, and $\mathbf{h}_{10}$ can be used to derive eight unique null-pol response patterns $|\gamma_i(\theta)|^{\text{NP}}, i = 1,2,\cdots,8$.

6.3.1.3 Visualization and Characterization

The null-pol response pattern provides a visualization tool to view the characteristics of null-pol response in the polarimetric rotation domain. For each target, there are eight independent null-pol response patterns in the polarimetric rotation domain. For example, the eight independent null-pol response patterns of a narrow dihedral in the polarimetric rotation domain are shown in Figure 6.3.1. It can be observed that different null-pol response patterns exhibit quite different characteristics in the polarimetric rotation domain.

In order to quantitatively characterize the null-pol response pattern in the polarimetric rotation domain, similar to the polarimetric coherence pattern, ten polarimetric features can be defined as follows:

1) Original Null-Pol Response $|\gamma_i(\theta)|_{\text{org}}^{\text{NP}}$: it is the null-pol response without any rotation processing, as

$$|\gamma_i(\theta)|_{\text{org}}^{\text{NP}} = |\gamma_i(0)|^{\text{NP}} \tag{6.3.7}$$

2) Null-Pol Response Degree $|\gamma_i(\theta)|_{\text{mean}}^{\text{NP}}$: it is the mean value of the null-pol response in the polarimetric rotation domain, as

$$|\gamma_i(\theta)|_{\text{mean}}^{\text{NP}} = \frac{1}{2\pi}\int_{-\pi}^{\pi} |\gamma_i(\theta)|^{\text{NP}}\, d\theta \tag{6.3.8}$$

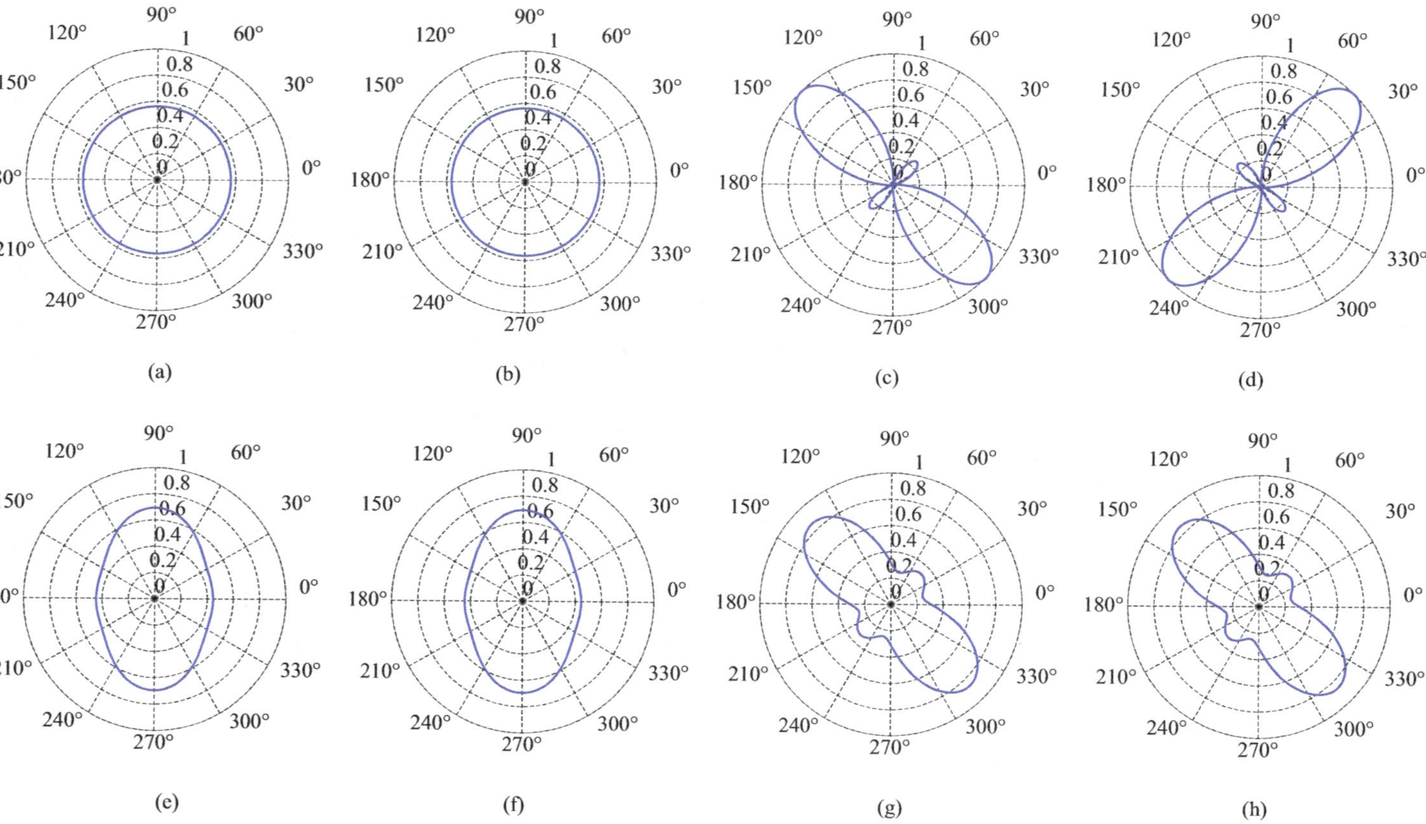

FIGURE 6.3.1 The null-pol response patterns of a narrow dihedral in the polarimetric rotation domain in terms of null-pol vectors (a) $\mathbf{h}_1$, (b) $\mathbf{h}_2$, (c) $\mathbf{h}_3$, (d) $\mathbf{h}_4$, (e) $\mathbf{h}_5$, (f) $\mathbf{h}_6$, (g) $\mathbf{h}_9$, and (h) $\mathbf{h}_{10}$, respectively.

3) Null-Pol Response Fluctuation $\left|\gamma_i(\theta)\right|_{\text{std}}^{\text{NP}}$: it is the standard deviation of the null-pol response in the polarimetric rotation domain, as

$$\left|\gamma_i(\theta)\right|_{\text{std}}^{\text{NP}} = \sqrt{\frac{1}{2\pi}\int_{-\pi}^{\pi}\left(\left|\gamma_i(\theta)\right|^{\text{NP}} - \left|\gamma_i(\theta)\right|_{\text{mean}}^{\text{NP}}\right)^2 \mathrm{d}\theta} \tag{6.3.9}$$

4) Maximum Null-Pol Response $\left|\gamma_i(\theta)\right|_{\max}^{\text{NP}}$: it is the maximum null-pol response in the polarimetric rotation domain, as

$$\left|\gamma_i(\theta)\right|_{\max}^{\text{NP}} = \max\left(\left|\gamma_i(\theta)\right|^{\text{NP}}\right) \tag{6.3.10}$$

5) Minimum Null-Pol Response $\left|\gamma_i(\theta)\right|_{\min}^{\text{NP}}$: it is the minimum null-pol response in the polarimetric rotation domain, as

$$\left|\gamma_i(\theta)\right|_{\min}^{\text{NP}} = \min\left(\left|\gamma_i(\theta)\right|^{\text{NP}}\right) \tag{6.3.11}$$

6) Null-Pol Response Contrast $\left|\gamma_i(\theta)\right|_{\text{contrast}}^{\text{NP}}$: it is defined as

$$\left|\gamma_i(\theta)\right|_{\text{contrast}}^{\text{NP}} = \left|\gamma_i(\theta)\right|_{\max}^{\text{NP}} - \left|\gamma_i(\theta)\right|_{\min}^{\text{NP}} \tag{6.3.12}$$

7) Null-Pol Response Anisotropy $\left|\gamma_i(\theta)\right|_{\text{A}}^{\text{NP}}$: it is defined as

$$\left|\gamma_i(\theta)\right|_{\text{A}}^{\text{NP}} = \frac{\left|\gamma_i(\theta)\right|_{\max}^{\text{NP}} - \left|\gamma_i(\theta)\right|_{\min}^{\text{NP}}}{\left|\gamma_i(\theta)\right|_{\max}^{\text{NP}} + \left|\gamma_i(\theta)\right|_{\min}^{\text{NP}}} \tag{6.3.13}$$

8) Null-Pol Response Beamwidth $\left|\gamma_i(\theta)\right|_{\text{bw0.95}}^{\text{NP}}$: it is defined as the rotation angle range within which the null-pol response values are no less than $0.95\times\left|\gamma_i(\theta)\right|_{\max}^{\text{NP}}$, which is

$$\left|\gamma_i(\theta)\right|_{\text{bw0.95}}^{\text{NP}} = \left\{\theta \,\middle|\, \left|\gamma_i(\theta)\right|_{\max}^{\text{NP}} \geq \left|\gamma_i(\theta)\right| \geq 0.95\times\left|\gamma_i(\theta)\right|_{\max}^{\text{NP}}\right\} \tag{6.3.14}$$

9) Maximum Rotation Angle $\theta_{\left|\gamma_i(\theta)\right|_{\max}^{\text{NP}}}$: it is defined as the rotation angle within the main range which produces the maximum null-pol response $\left|\gamma_i(\theta)\right|_{\max}^{\text{NP}}$, which is

$$\theta_{\left|\gamma_i(\theta)\right|_{\max}^{\text{NP}}} = \left\{\theta \,\middle|\, \left|\gamma_i(\theta)\right|^{\text{NP}} = \left|\gamma_i(\theta)\right|_{\max}^{\text{NP}}\right\} \tag{6.3.15}$$

10) Minimum Rotation Angle $\theta_{\left|\gamma_i(\theta)\right|_{\min}^{\text{NP}}}$: it is defined as the rotation angle within the main range which produces the minimum null-pol response $\left|\gamma_i(\theta)\right|_{\min}^{\text{NP}}$, which is

$$\theta_{|\gamma_i(\theta)|^{NP}_{min}} = \left\{ \theta \middle| |\gamma_i(\theta)|^{NP} = |\gamma_i(\theta)|^{NP}_{min} \right\} \quad (6.3.16)$$

6.3.2 Target Structure Recognition Methodology

6.3.2.1 Polarimetric Feature Coding of Target Null-Pol Response Pattern

Among null-pol response pattern features, the maximum rotation angle, the minimum rotation angle, and the null-pol response beamwidth are angle features. The other seven null-pol response pattern features are amplitude features, which are mainly considered in this section. Moreover, the null-pol response contrast and null-pol response anisotropy can be derived from the maximum and minimum null-pol response features. Thus, only five features, the original null-pol response $|\gamma_i(\theta)|^{NP}_{org}$, null-pol response degree $|\gamma_i(\theta)|^{NP}_{mean}$, null-pol response fluctuation $|\gamma_i(\theta)|^{NP}_{std}$, maximum null-pol response $|\gamma_i(\theta)|^{NP}_{max}$, and minimum null-pol response $|\gamma_i(\theta)|^{NP}_{min}$, are adopted to construct the polarimetric feature coding vector. In addition, with the independent null-pol vectors of canonical structures, there are eight null-pol response patterns. In this vein, the null-pol response pattern feature coding vector $\mathbf{F}$ is defined as

$$\mathbf{F} = \left\{ |\gamma_i(\theta)|^{NP}_{org}, |\gamma_i(\theta)|^{NP}_{mean}, |\gamma_i(\theta)|^{NP}_{std}, |\gamma_i(\theta)|^{NP}_{max}, |\gamma_i(\theta)|^{NP}_{min} \right\}, \quad i = 1, 2, \cdots, 8 \quad (6.3.17)$$

It is clear that $\mathbf{F}$ is a matrix of five rows and eight columns. An example of the cylinder's null-pol response pattern feature coding vector is shown in Table 6.3.2.

6.3.2.2 Manmade Target Structure Recognition Scheme

This section introduces the target structure recognition method based on the polarimetric feature coding vector of the null-pol response pattern (denoted as polarimetric

TABLE 6.3.2
The Polarimetric Feature Coding Vector of Null-Pol Response Patterns in the Polarimetric Rotation Domain from a Cylinder Structure

Polarimetric Feature Coding Vector	Null-Pol Response Patterns in the Polarimetric Rotation Domain							
	$\lvert\gamma_1(\theta)\rvert^{NP}$	$\lvert\gamma_2(\theta)\rvert^{NP}$	$\lvert\gamma_3(\theta)\rvert^{NP}$	$\lvert\gamma_4(\theta)\rvert^{NP}$	$\lvert\gamma_4(\theta)\rvert^{NP}$	$\lvert\gamma_6(\theta)\rvert^{NP}$	$\lvert\gamma_7(\theta)\rvert^{NP}$	$\lvert\gamma_8(\theta)\rvert^{NP}$
$\lvert\gamma_i(\theta)\rvert^{NP}_{org}$	0.06	0.06	0.56	0.56	0	0	0.31	0.31
$\lvert\gamma_i(\theta)\rvert^{NP}_{mean}$	0.06	0.06	0.59	0.59	0.12	0.12	0.33	0.33
$\lvert\gamma_i(\theta)\rvert^{NP}_{std}$	0	0	0.27	0.27	0.09	0.09	0.19	0.19
$\lvert\gamma_i(\theta)\rvert^{NP}_{max}$	0.06	0.06	1	1	0.25	0.25	0.61	0.61
$\lvert\gamma_i(\theta)\rvert^{NP}_{min}$	0.06	0.06	0.25	0.25	0	0	0.08	0.08

rotation domain null-pol recognition method). Its core idea is to recognize the scattering structure type based on the distance among different scattering structures' null-pol response pattern features [16]. The flowchart of polarimetric rotation domain null-pol recognition method is shown in Figure 6.3.2.

For the input polarimetric scattering matrix of a candidate target, the null-pol response pattern can be constructed according to (6.3.3). Then, the null-pol response pattern feature coding vector is $\mathbf{F}$. In order to compare the difference of scattering characteristics between the target and canonical scattering structures, the Frobenius norm is adopted as the distance measure.

In this vein, the similarity distance can be used to measure the closeness among the measured null-pol response pattern feature coding vector $\mathbf{F}$ and the theoretical null-pol response pattern feature coding vector $\mathbf{F}_m^{\mathrm{c}}$ of a canonical scattering structure. The structure type can be determined when the similarity distance reaches the minimum value. Therefore, the structure type determination expression is

$$d = \arg\min_{m} \left\{ \left\| \mathbf{F} - \mathbf{F}_m^{\mathrm{c}} \right\|_{\mathrm{F}} \right\} \tag{6.3.18}$$

where $\|\cdot\|_{\mathrm{F}}$ is the Frobenius norm of a matrix; $\mathbf{F}_m^{\mathrm{c}}$ is the theoretical null-pol response pattern feature coding vector of the mth canonical scattering structure and $m = 1,2,\ldots 10$; and d indicates the canonical structure corresponding to the theoretical null-pol response pattern feature coding vector that minimizes the distance.

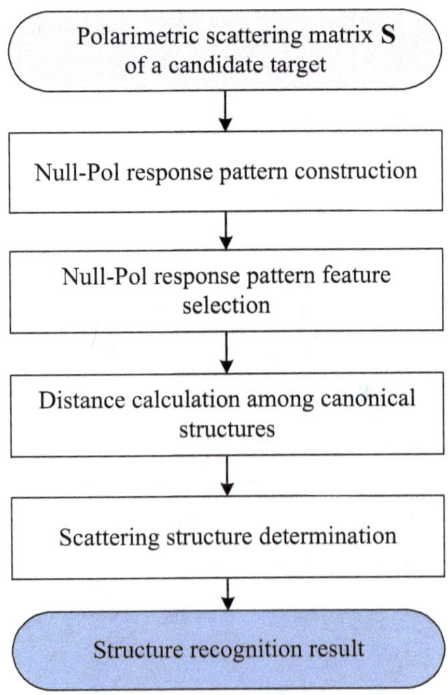

FIGURE 6.3.2 Flowchart of polarimetric rotation domain null-pol recognition method.

6.3.3 Experimental Studies

The simulation data of canonical structures, the electromagnetic computation data, and the anechoic chamber measured data of UAV targets are adopted for experimental studies. Cameron decomposition is also utilized for comparison.

6.3.3.1 Canonical Structures Recognition with Simulation Data

The simulation data contains ten canonical structures, the trihedral, cylinder, horizontal dipole, vertical dipole, dihedral, narrow dihedral, quarter wave device, left helix, and right helix. The Pauli image and the ground-truth of these structures are shown in Figure 6.3.3. The structure recognition results from the Cameron decomposition and the polarimetric rotation domain null-pol recognition method are shown in Figure 6.3.4. It is clear that the Cameron decomposition cannot distinguish horizontal and vertical dipoles, while the left and right helix structures are not correctly recognized. In comparison, the polarimetric rotation domain null-pol recognition method can recognize all these structures correctly.

In order to verify the recognition performance in terms of different polarimetric measurement errors, the polarimetric scattering matrix data with different polarization isolation, amplitude, and phase imbalances are simulated for quantitative analysis. For each polarimetric measurement error and canonical structure, 1000 samples in terms of the signal-to-noise ratio (SNR) of 15~30 dB are generated. The recognition accuracy values of the polarimetric rotation domain null-pol recognition method and Cameron decomposition are shown in Figure 6.3.5. From Figure 6.3.5 (a), when the polarization isolation is lower than –15dB, the recognition accuracies of both methods are relatively stable. Since the Cameron decomposition cannot discriminate vertical dipole and helix, its recognition accuracy is 0.7, and that of the polarimetric rotation domain null-pol recognition method is 1. When the polarization isolation is over –15 dB, the recognition performance of both methods deteriorates. Anyway, the recognition accuracies of the polarimetric rotation domain null-pol recognition method are always higher than

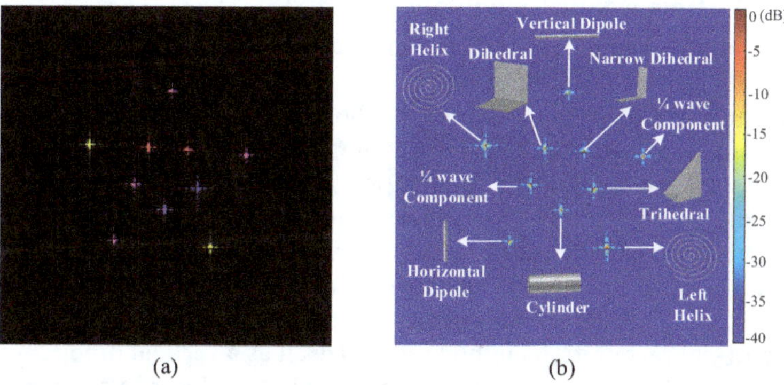

FIGURE 6.3.3 The simulation data of canonical structures. (a) Pauli image, (b) ground-truth of these structures.

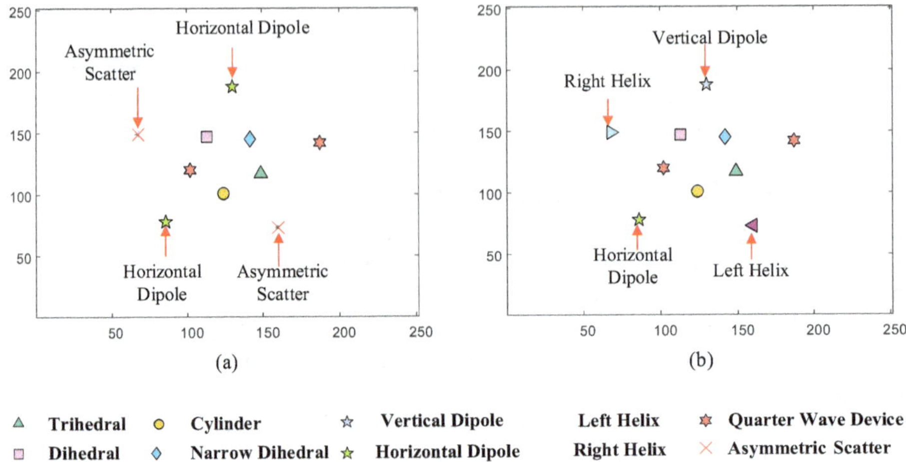

FIGURE 6.3.4 Structure recognition comparison with simulation data. (a) Cameron decomposition, (b) polarimetric rotation domain null-pol recognition method.

those of Cameron decomposition. From Figure 6.3.5 (b), both methods are highly influenced by the amplitude imbalance. Both methods can only maintain the best performance within a small range where the amplitude imbalance is lower than 0.5 dB. When the amplitude imbalance reaches 2 dB, the recognition accuracies of both methods are lower than 0.5, even though the recognition performance of the polarimetric rotation domain null-pol recognition method is still better than Cameron decomposition. As shown in Figure 6.3.5 (c), both methods are insensitive to phase imbalance. As a result, compared with Cameron decomposition, the polarimetric rotation domain null-pol recognition method achieves better and more stable recognition performance.

6.3.3.2 UAV Target Structure Recognition with Electromagnetic Computation Data

The UAV electromagnetic computation data described in Section 6.2.2.2 are used for comparison. The structure recognition results from the polarimetric rotation domain null-pol recognition method and the Cameron decomposition are shown in Figure 6.3.6. Compared with Cameron decomposition, the polarimetric rotation domain null-pol recognition method exhibits better performance for symmetric components of the UAV target (indicated by the red arrow), since it correctly recognizes these symmetric components as the same scattering structure. For the scattering center indicated by the dashed arrow, Cameron decomposition identifies it as a horizontal dipole, while the polarimetric rotation domain null-pol recognition method recognizes it as a vertical dipole. According to its normalized polarimetric scattering matrix, it is closer to a vertical dipole. Generally, majority of the UAV structure recognition results from both methods are consistent. The recognition results from the polarimetric rotation domain

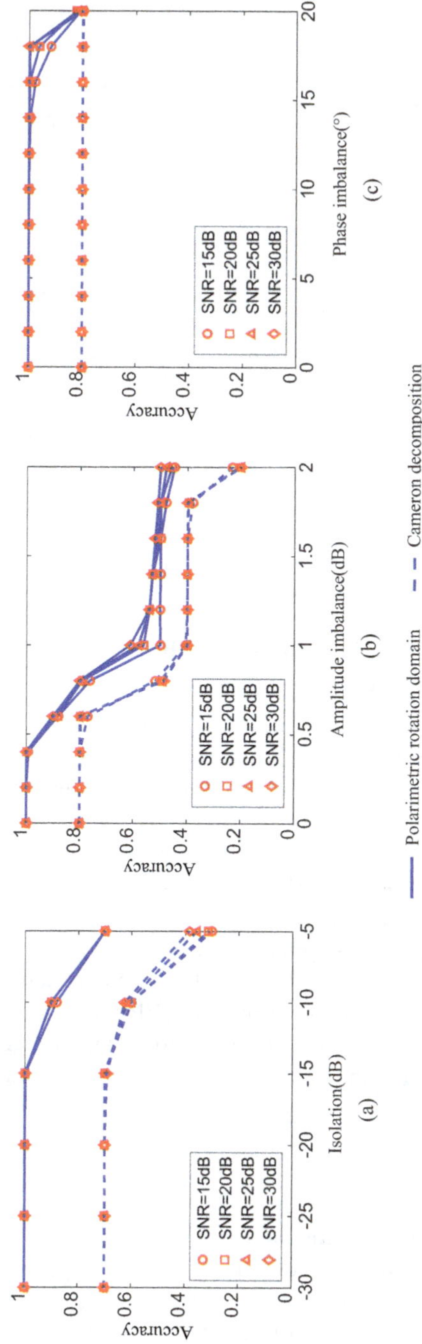

FIGURE 6.3.5 The structure recognition accuracies in terms of polarimetric measurement errors of (a) polarization isolation, (b) amplitude imbalance, (c) phase imbalance.

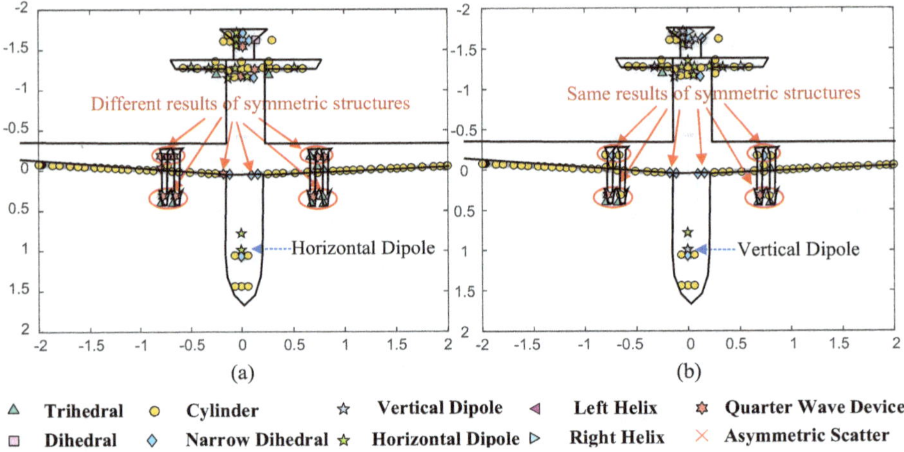

FIGURE 6.3.6 UAV target structure recognition comparison. (a) Cameron decomposition, (b) polarimetric rotation domain null-pol recognition method.

null-pol recognition method are closer to the actual structures, especially for the symmetric components of the UAV target.

6.3.3.3 UAV Target Structure Recognition with Anechoic Chamber Measured Data

The anechoic chamber measured data of a fixed wing UAV target are adopted for comparison. The length of the UAV is 2.3 m, and its wingspan is 2.9 m. During the measurement, the center frequency is 10 GHz, while the bandwidth is 4 GHz. The measurement scene and the UAV model are shown in Figure 6.3.7. The recognition results are shown in Figure 6.3.8. From the recognition results, the polarimetric rotation domain null-pol recognition method can correctly recognize the two connecting rods between the head and the empennage as cylinder structures. In comparison, Cameron decomposition misjudges some of them as trihedral structures. For the head and wings of the UAV target, compared with the polarimetric rotation domain null-pol recognition method, the recognition results from Cameron decomposition are more chaotic. In this vein, the polarimetric rotation domain null-pol recognition method exhibits more accurate and robust recognition performance over the same structures of the UAV target.

6.4 SUMMARY

In this chapter, two polarimetric rotation domain manmade target structure recognition methods are introduced. The first method is based on the polarimetric correlation pattern and realizes manmade target structure recognition by polarimetric feature coding vectors. The second method realizes manmade target structure recognition by designing null-pol response pattern feature coding vectors. The core idea of these two methods is to mine and utilize polarimetric features with high discrimination

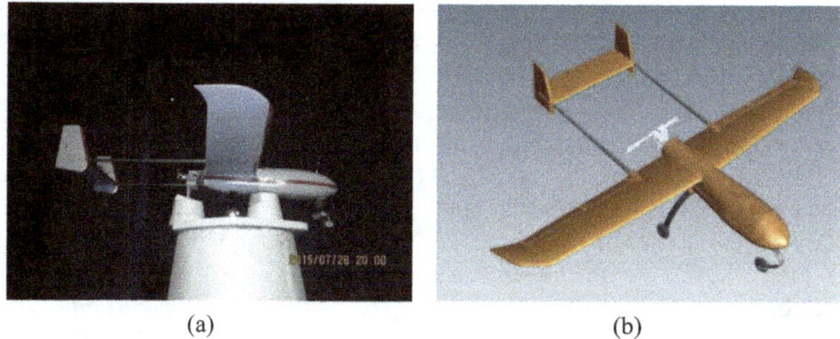

(a) (b)

FIGURE 6.3.7 The measurement scene and the UAV model. (a) Measurement scene in the anechoic chamber, (b) UAV model.

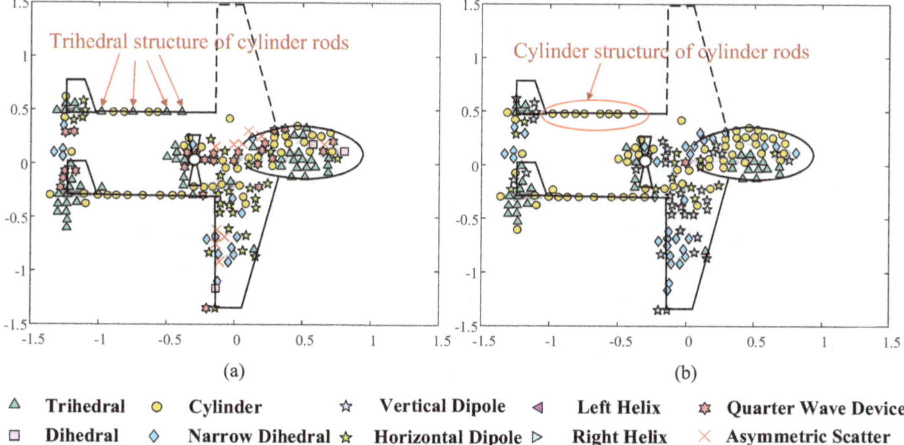

FIGURE 6.3.8 UAV target structure recognition comparison. (a) Cameron decomposition, (b) polarimetric rotation domain null-pol recognition method.

properties obtained in the polarimetric rotation domain. Then, with polarimetric feature coding vectors and the similarity distance measures, the target structures can be recognized. Experimental studies are carried out with electromagnetic computation data, anechoic chamber measured data, and real PolSAR data. Comparison studies demonstrate the effectiveness and the superiority of both polarimetric rotation domain manmade structure recognition methods.

REFERENCES

1. J. S. Lee and E. Pottier, *Polarimetric Radar Imaging: From Basics to Applications*. Boca Raton: CRC Press, 2009.
2. Y. Yamaguchi, *Polarimetric SAR Imaging: Theory and Applications*. Boca Raton: CRC Press, 2020.

3. L. C. Potter, C. Da Ming, R. Carriere, and M. J. Gerry, "A GTD-based parametric model for radar scattering," *IEEE Transactions on Antennas and Propagation,* vol. 43, no. 10, pp. 1058–1067, 1995.
4. L. C. Potter and R. L. Moses, "Attributed scattering centers for SAR ATR," *IEEE Transactions on Image Processing,* vol. 6, no. 1, pp. 79–91, 1997.
5. J. A. Jackson, B. D. Rigling, and R. L. Moses, "Canonical scattering feature models for 3D and bistatic SAR," *IEEE Transactions on Aerospace and Electronic Systems,* vol. 46, no. 2, pp. 525–541, 2010.
6. M. J. Gerry, L. C. Potter, I. J. Gupta, and A. V. D. Merwe, "A parametric model for synthetic aperture radar measurements," *IEEE Transactions on Antennas and Propagation,* vol. 47, no. 7, pp. 1179–1188, 1999.
7. S. R. Cloude and E. Pottier, "A review of target decomposition theorems in radar polarimetry," *IEEE Transactions on Geoscience and Remote Sensing,* vol. 34, no. 2, pp. 498–518, 1996.
8. E. Krogager, "New decomposition of the radar target scattering matrix," *Electronics Letters,* vol. 26, no. 18, pp. 1525–1527, 1990.
9. W. L. Cameron, N. N. Youssef, and L. K. Leung, "Simulated polarimetric signatures of primitive geometrical shapes," *IEEE Transactions on Geoscience and Remote Sensing,* vol. 34, no. 3, pp. 793–803, 1996.
10. W. L. Cameron and H. Rais, "Conservative polarimetric scatterers and their role in incorrect extensions of the Cameron decomposition," *IEEE Transactions on Geoscience and Remote Sensing,* vol. 44, no. 12, pp. 3506–3516, 2006.
11. M. Xing, R. Guo, C. W. Qiu, L. Liu, and Z. Bao, "Experimental research of unsupervised Cameron/maximum-likelihood classification method for fully polarimetric synthetic aperture radar data," *IET Radar Sonar and Navigation,* vol. 4, no. 1, pp. 85–95, 2010.
12. M. Martorella, E. Giusti, A. Capria, F. Berizzi, and B. Bates, "Automatic target recognition by means of polarimetric ISAR images and neural networks," *IEEE Transactions on Geoscience and Remote Sensing,* vol. 47, no. 11, pp. 3786–3794, 2009.
13. E. Giusti, M. Martorella, and A. Capria, "Polarimetrically-persistent-scatterer-based automatic target recognition," *IEEE Transactions on Geoscience and Remote Sensing,* vol. 49, no. 11, pp. 4588–4599, 2011.
14. R. Paladini, L. F. Famil, E. Pottier, M. Martorella, F. Berizzi, and E. Dalle Mese, "Point target classification via fast lossless and sufficient Ω–Ψ–Φ decomposition of high-resolution and fully polarimetric SAR/ISAR data," *Proceedings of the IEEE,* vol. 101, no. 3, pp. 798–830, 2013.
15. H. L. Li, M. D. Li, X. C. Cui, and S. W. Chen, "Man-made target structure recognition with polarimetric correlation pattern and roll-invariant feature coding," *IEEE Geoscience and Remote Sensing Letters,* vol. 19, pp. 1–5, 2022.
16. G. Q. Wu, S. W. Chen, Y. Z. Li, and X. S. Wang, "Null-pol response pattern in polarimetric rotation domain: Characterization and application," *IEEE Geoscience and Remote Sensing Letters,* vol. 19, pp. 1–5, 2022.
17. X. C. Cui, C. S. Tao, Y. Su, and S. W. Chen, "PolSAR ship detection based on polarimetric correlation pattern," *IEEE Geoscience and Remote Sensing Letters,* vol. 18, no. 3, pp. 471–475, 2021.
18. M. D. Li, S. P. Xiao, and S. W. Chen, "Three-dimension polarimetric correlation pattern interpretation tool and its application," *IEEE Transactions on Geoscience and Remote Sensing,* vol. 60, pp. 1–16, 2022.
19. S. W. Chen, "Polarimetric coherence pattern: A visualization and characterization tool for PolSAR data investigation," *IEEE Transactions on Geoscience and Remote Sensing,* vol. 56, no. 1, pp. 286–297, 2018.

20. J. J. Van. Zyl, "Calibration of polarimetric radar images using only image parameters and trihedral corner reflector responses," *IEEE Transactions on Geoscience and Remote Sensing,* vol. 28, no. 3, pp. 337–348, 1990.
21. S. W. Chen, X. S. Wang, S. P. Xiao, and M. Sato, *Target Scattering Mechanism in Polarimetric Synthetic Aperture Radar-Interpretation and Application.* Singapore: Springer, 2018.
22. E. M. Kennaugh, "Polarization properties of radar reflections," Ph.D. dissertation, The Ohio State University, Columbus, 1952.
23. J. R. Huynen, "The Stokes matrix parameters and their interpretation in terms of physical target properties," in *Proceedings of the SPIE,* York, 1991, 15–18 April, pp. 195–207.
24. E. Pottier, "Dr. JR Huynen's main contributions in the development of polarimetric radar techniques and how the 'radar targets phenomenological concept' becomes a theory," *Proceedings of the SPIE,* 1993, vol. 1748, pp. 72–85.
25. W. M. Boerner, "Use of polarization in electromagnetic inverse scattering," *Radio Science,* vol. 16, no. 6, pp. 1037–1045, 1981.
26. M. Davidovitz and W. M. Boerner, "Extension of Kennaugh's optimal polarization concept to the asymmetric scattering matrix case," *IEEE Transactions on Antennas and Propagation,* vol. 34, no. 4, pp. 569–574, 1986.
27. A. P. Agrawal and W. M. Boerner, "Redevelopment of Kennaugh's target characteristic polarization state theory using the polarization transformation ration formalism for the coherent case," *IEEE Transactions on Geoscience and Remote Sensing,* vol. 27, no. 1, pp. 3–14, 1989.
28. W. M. Boerner and A. Q. Xi, "The characteristic radar target polarization state theory for the coherent monostatic and reciprocal case using the generalized polarization transformation ratio formulation," *Aeu-Archiv Fur Elektronik Und Ubertragungstechnik–International Journal of Electronics and Communications,* vol. 44, no. 4, pp. 273–281, 1990.
29. W. M. Boerner, W. L. Yan, A. Q. Xi, and Y. Yamaguchi, "On the basic principles of radar polarimetry-the target characteristic polarization state theory of Kennaugh, Huynen's polarization fork concept, and its extension to the partially polarized case," *Proceedings of the IEEE,* vol. 79, no. 10, pp. 1538–1550, 1991.
30. Y. Yamaguchi, W. M. Boerner, H. J. Eom, M. Sengoku, S. Motooka, and T. Abe, "On characteristic polarization states in the cross-polarized radar channel," *IEEE Transactions on Geoscience and Remote Sensing,* vol. 30, no. 5, pp. 1078–1080, 1992.
31. J. Yang, Y. N. Peng, Y. Yamaguchi, and W. M. Boerner, "Optimal polarisation problem for the multistatic radar case," *Electronics Letters,* vol. 36, no. 19, pp. 1647–1649, 2000.
32. J. Yang, Y. Yamaguchi, H. Yamada, W. M. Boerner, H. Mott, and Y. N. Peng, "Development of target null theory," *IEEE Transactions on Geoscience and Remote Sensing,* vol. 39, no. 2, pp. 330–338, 2001.
33. J. Yang, Y. Yamaguchi, H. Yamada, Z. H. Czyz, W. M. Boerner, H. Mott, et al., "The characteristic polarization states and the equi-power curves," *IEEE Transactions on Geoscience and Remote Sensing,* vol. 40, no. 2, pp. 305–313, 2002.
34. Z. W. Zhuang, S. P. Xiao, and X. S. Wang, *Radar Polarization Information Processing and Application.* Beijing: National Defense Industry Press, 1999.

7 Polarimetric Rotation Domain Urban Damage Mapping

7.1 INTRODUCTION

PolSAR, which can acquire full-polarization information of an imaging scene, is becoming mainstream in microwave remote sensing [1, 2]. With the advantages of working day and night, cloud penetration, and wide observation swath, PolSAR is very useful for natural disaster monitoring and evaluation [3–5].

Large-scale earthquakes and tsunamis usually lead to huge damage to human beings. When such disasters happen, the urban area, which directly relates to human life and economic loss, becomes one of the focuses for damage evaluation and rescue planning. Damage ranges and damage levels are vitally important information that can greatly assist the rescue operation and the following reconstruction project. With proper scattering mechanism modeling and interpretation techniques, target (e.g. building) changes induced by disaster damages can be sensed and identified using PolSAR data. Eigenvalue-eigenvector-based and model-based decompositions [6–10] have been applied for urban damage investigation [4, 5, 11–14]. As expected, the derived polarimetric features have good potential for damaged area detection and evaluation. Comparison studies in terms of a set of these classical polarimetric features over damaged urban areas have been carried out [15]. Case studies for urban damage investigation using post-event PolSAR data have also been reported [16]. With these advances, damage ranges of urban areas can be effectively detected and extracted.

Currently, obtaining damage level information for urban areas is still challenging. For relatively low-resolution PolSAR data which are suitable for huge area damage investigation, the main changes of damaged urban patches are the decrease of ground-wall structures. The destruction of ground-wall structures leads to the reduction of the dominant double-bounce scattering mechanism and the homogeneity destruction of polarization orientation angles [17]. Two polarimetric damage indexes have been established with a deep investigation of target scattering mechanisms [17]. Both of them can successfully discriminate urban patches with various damage levels, which are defined as the percentage of collapsed buildings within a local urban patch. As a further advance, a rapid urban damage level mapping technique that can automatically detect damaged urban areas and provide detailed damage level information has been reported recently [18].

In addition to polarimetric features derived from scattering mechanism modeling, cross-channel polarimetric correlation is an important source and has the potential

DOI: 10.1201/9781003461296-7

to reveal the physical properties of the imaging scene [19]. The magnitude of the polarimetric correlation coefficient, also called polarimetric coherence, has been used for PolSAR data investigation. Polarimetric coherence strongly depends on the polarization combination, local scatterer type, and target orientation relative to the PolSAR illumination direction. Although the target scattering diversity phenomenon may induce ambiguities in scattering mechanism modeling and interpretation [20], it also provides valuable insights for physical parameter retrieval [21–24]. In order to fully represent, understand, and utilize the target scattering diversity, polarimetric rotation domain interpretation theory is introduced in Chapter 2. An interpretation tool of the two-dimension polarimetric coherence pattern has been reported for polarimetric coherence characterization in the polarimetric rotation domain [25–27]. Valuable features have been derived and explored to represent the signatures of polarimetric coherence in the polarimetric rotation domain. Some of them have been successfully used for crop type discrimination [28], land cover classification [29, 30], manmade target detection [31, 32], manmade target structure recognition [33, 34], and so on, which are summarized in Chapters 4–6.

This chapter aims at urban damage mapping using polarimetric rotation domain techniques. Buildings before and after collapse damage are the main focus of this study. The key difference between an intact building and a collapsed building is the change of the ground-wall structure. The collapsed building areas may change into relatively rough surfaces [17]. The different geometries of ground-wall structures and rough surfaces result in quite different directivity effects in the polarimetric rotation domain. The directivity effects of ground-wall structures are much more obvious than those of rough surfaces. In this vein, the corresponding two-dimension polarimetric coherence patterns from buildings before and after collapse damage are differentiable. In addition, ground-wall structure and rough surfaces mainly exhibit double-bounce and odd-bounce scattering mechanisms, which directly relate to the combination of co-polarization components from Pauli decomposition viewpoint. Therefore, this chapter introduces the progresses of exploring the co-polarization coherence pattern for urban damage investigation. New polarimetric damage index and damage level inversion relationships are found [35]. Then, an urban damage level mapping approach is developed. Compared with scattering mechanism modeling and interpretation technique, mainly in terms of polarimetric target decomposition [6, 17, 18], this chapter provides an alternative solution for urban damage investigation. The main novelty lies in that it is not affected by the target orientation effect. Instead, it utilizes target scattering diversity and firstly explores the rich hidden polarimetric coherence information in the polarimetric rotation domain for urban damage evaluation.

7.2 CO-POLARIZATION COHERENCE PATTERN INVESTIGATION

7.2.1 Co-Polarization Coherence Pattern

The combinations of co-polarization components in terms of $S_{HH} + S_{VV}$ and $S_{HH} - S_{VV}$ respectively relate to canonical surface scattering and double-bounce scattering from the Pauli decomposition viewpoint. In this vein, co-polarization components

$S_{\rm HH}$ and $S_{\rm VV}$ are the emphasis of the following investigations, and their expressions in the polarimetric rotation domain are

$$S_{\rm HH}(\theta) = S_{\rm HH}\cos^2\theta + S_{\rm HV}\cos\theta\sin\theta + S_{\rm VH}\cos\theta\sin\theta + S_{\rm VV}\sin^2\theta \quad (7.2.1)$$

$$S_{\rm VV}(\theta) = S_{\rm HH}\sin^2\theta - S_{\rm HV}\cos\theta\sin\theta - S_{\rm VH}\cos\theta\sin\theta + S_{\rm VV}\cos^2\theta \quad (7.2.2)$$

In practice, sample averaging of sufficient samples with similar properties is used to estimate polarimetric coherence [19, 36]. Co-polarization coherence without rotation is

$$|\gamma_{\text{HH-VV}}| = \frac{\left|\left\langle S_{\rm HH} S_{\rm VV}^* \right\rangle\right|}{\sqrt{\left\langle |S_{\rm HH}|^2 \right\rangle \left\langle |S_{\rm VV}|^2 \right\rangle}} \quad (7.2.3)$$

where $S_{\rm VV}^*$ is the conjugate of $S_{\rm VV}$ and $\langle\cdot\rangle$ indicates sample averaging. The value of $|\gamma_{\text{HH-VV}}|$ is within $[0,1]$.

The two-dimension polarimetric coherence pattern interpretation tool, which was first proposed in [25–27], is effective to visualize and characterize polarimetric coherence characteristics in the polarimetric rotation domain. The core idea is to extend polarimetric coherence at a certain geometry to the whole polarimetric rotation domain. In this vein, the definition of co-polarization coherence pattern $|\gamma_{\text{HH-VV}}(\theta)|$ is

$$|\gamma_{\text{HH-VV}}(\theta)| = \frac{\left|\left\langle S_{\rm HH}(\theta) S_{\rm VV}^*(\theta) \right\rangle\right|}{\sqrt{\left\langle |S_{\rm HH}(\theta)|^2 \right\rangle \left\langle |S_{\rm VV}(\theta)|^2 \right\rangle}}, \quad \theta \in [-\pi, \pi) \quad (7.2.4)$$

Note that the rotation angle θ covers the full range of the polarimetric rotation domain other than a specific angle. When the speckle effect is well smoothed during the estimation procedure, the rotation signature of the co-polarization coherence is fully determined by the elements $S_{\rm HH}(\theta)$ and $S_{\rm VV}(\theta)$. In addition, it can be proved that the period of $|\gamma_{\text{HH-VV}}(\theta)|$ in the polarimetric rotation domain is $\frac{\pi}{2}$.

7.2.2 Demonstration and Investigation

Pi-SAR X-band PolSAR data are utilized for co-polarization coherence pattern demonstration and investigation. The study area contains various types of land covers. The corresponding optical image from Google Earth is shown in Figure 7.2.1 (a), while the PolSAR RGB composite image with Pauli basis is shown in Figure 7.2.1 (b). The SimiTest filter [36] has been applied to smooth the speckle effect.

Buildings before and after collapse damage are the main focus of this work. The main difference between an intact building and a collapsed building is the change of the ground-wall structure. For a completely collapsed building, the ground-wall structure is totally destroyed. Especially for coarse-resolution PolSAR data, which are suitable for large-area investigation, the collapsed building areas may change

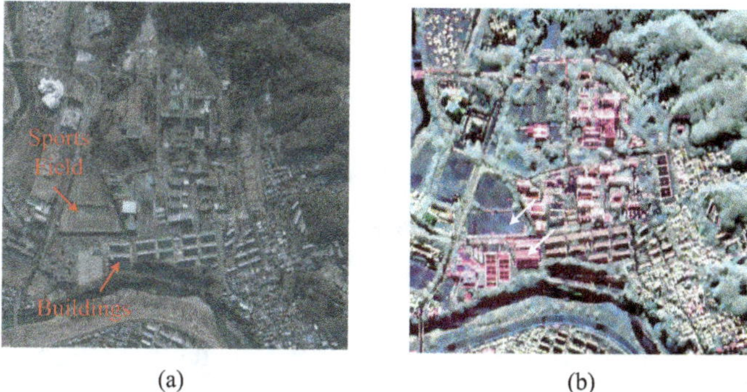

(a) (b)

FIGURE 7.2.1 Study area. (a) Optical image, (b) Pi-SAR PolSAR Pauli image.

into relatively rough surfaces. In this vein, in order to investigate the performances of co-polarization coherence patterns over buildings before and after damage, two samples are randomly selected from a rough surface of a sports field and a building area, indicated by arrows in Figure 7.2.1. Indeed, the geometric structures of a rough surface and a dihedral corner reflector, which is normally used to model building's ground-wall structure, are quite different. As a result, their polarimetric coherence signatures in the polarimetric rotation domain can be differentiable. Theoretically, relatively flat surfaces could be roll-invariant, and the co-polarization coherence fluctuation can be very minor in the polarimetric rotation domain. In comparison, dihedral structures have an obvious directivity effect and so are co-polarization coherence patterns. The corresponding co-polarization coherence patterns are shown in Figure 7.2.2. For the sample from a rough surface, the shape of the co-polarization coherence pattern is close to a circle with very minor fluctuations. Meanwhile, the co-polarization coherence pattern of the sample from a building mainly shows a four-lobe shape with some weak side-lobe effects. It is clear that the co-polarization coherence fluctuations between them are obviously different. According to Chapter 2, the definition of co-polarization coherence fluctuation $\left|\gamma_{\text{HH-VV}}(\theta)\right|_{\text{std}}$ is given as

$$\left|\gamma_{\text{HH-VV}}(\theta)\right|_{\text{std}} = \sqrt{\frac{1}{2\pi}\int_{-\pi}^{\pi}\left(\left|\gamma_{\text{HH-VV}}(\theta)\right| - \left|\gamma_{\text{HH-VV}}(\theta)\right|_{\text{mean}}\right)^2 \mathrm{d}\theta} \tag{7.2.5}$$

The co-polarization coherence fluctuations $\left|\gamma_{\text{HH-VV}}(\theta)\right|_{\text{std}}$ are calculated for the full-scene and shown in Figure 7.2.3. It is observed that the values of $\left|\gamma_{\text{HH-VV}}(\theta)\right|_{\text{std}}$ from buildings are generally much higher than those from sports fields with rough surfaces. These demonstrations are consistent with the previous analysis.

The interpretation tool of co-polarization coherence pattern $\left|\gamma_{\text{HH-VV}}(\theta)\right|$ indeed contains rich information for target characteristics in the polarimetric rotation domain. The derived feature of the co-polarization coherence fluctuations $\left|\gamma_{\text{HH-VV}}(\theta)\right|_{\text{std}}$ has the potential to differentiate intact buildings and collapsed buildings. Note that this

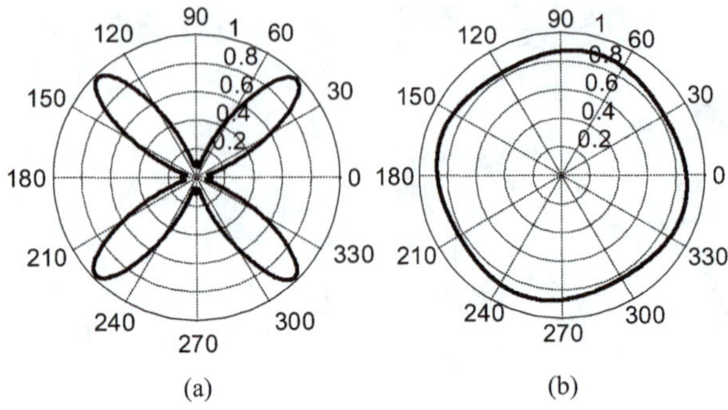

FIGURE 7.2.2 Co-polarization coherence pattern $\left|\gamma_{\text{HH-VV}}(\theta)\right|$ of samples from (a) building area, (b) rough surface area.

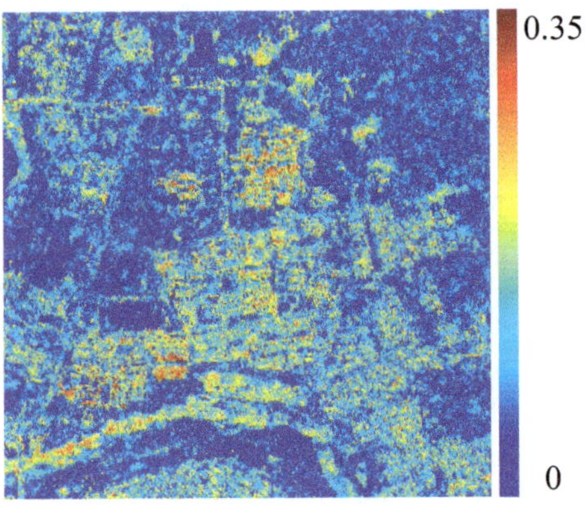

FIGURE 7.2.3 Co-polarization coherence fluctuation $\left|\gamma_{\text{HH-VV}}(\theta)\right|_{\text{std}}$.

demonstration is more or less a kind of schematic illustration to show the efficiency of co-polarization coherence pattern technique. The following investigations are carried out for real damaged urban patches using the co-polarization coherence pattern technique and the derived features.

7.3 URBAN DAMAGE INVESTIGATION

7.3.1 Study Area and Data Description

The great 3.11 East Japan earthquake and tsunami induced huge damage at seashore areas [5]. More than 128,000 buildings were completely destroyed. The Miyagi prefecture is one of the main affected areas and is adopted for investigation. Multi-temporal ALOS/PALSAR PolSAR datasets are co-registered and used for investigation. Radiometric and polarimetric calibrations have been accomplished for these datasets, and the calibration details can be found in [37]. The pre-event data acquired on November 21, 2010, are shown in Figure 7.3.1 (a), while the post-event data acquired on April 8, 2011, are shown in Figure 7.3.1 (b). The very seriously

(a) (b)

FIGURE 7.3.1 ALOS/PALSAR PolSAR Pauli images. (a) Pre-event (Nov. 21, 2010), (b) post-event (Apr. 8, 2011).

damaged Ishinomaki city, Onagawa-cho, and Minamisanriku-cho are highlighted by the rectangular boxes. The resolutions of the azimuth and range directions are 4.45 and 23.14 m, respectively [37], for full-polarization mode. The datasets have been 8-look processed in the azimuth direction to make the pixel size comparable for both the azimuth and range directions. In this way, it is easier to compare the corresponding PolSAR and optical images. Then, the SimiTest filter [36] is implemented to smooth the speckle effect and is utilized for co-polarization coherence estimation. Generally, sample averaging of more pixels with similar properties can lead to unbiased coherence estimation with lower standard deviations [19]. In order to select more samples, a larger searching window is preferred. The challenge lies in how to reject pixels in a large search window with different characteristics from the pixel which is to be filtered. SimiTest filter is one such method which is dedicated to this issue. A similarity test for polarimetric matrices has been found to select similar pixels. Meanwhile, pixels with different characters are rejected. In this way, it is suitable to be applied to a large search area to accept a number of homogeneous pixels.

7.3.2 Investigation for Damaged Urban Patches

For this relatively coarse-resolution PolSAR data, instead of single buildings, urban patches are the investigation unit. The percentage of flushed or collapsed buildings in a local urban patch is defined as the damage level, which is also consistent with the investigations in [17, 18]. The seriously affected Ishinomaki city containing urban patches with various damage levels is adopted for damage index development. In order to illustrate the local damage conditions, pan-sharpened true-color images with finer resolution of 2.5 m for both the range and azimuth directions are generated from the datasets acquired by ALOS optical sensors PRISM and AVNIR-2 [38]. The observation dates are respectively August 23, 2010, and April 10, 2011, for pre- and post-event. Co-registered to the corresponding PolSAR data, these enlarged images clearly show the destruction situations, displayed in Figures 7.3.2 (a) and 7.3.2 (b). Furthermore, the ground-truth of building damage map is shown in Figure 7.3.2 (c).

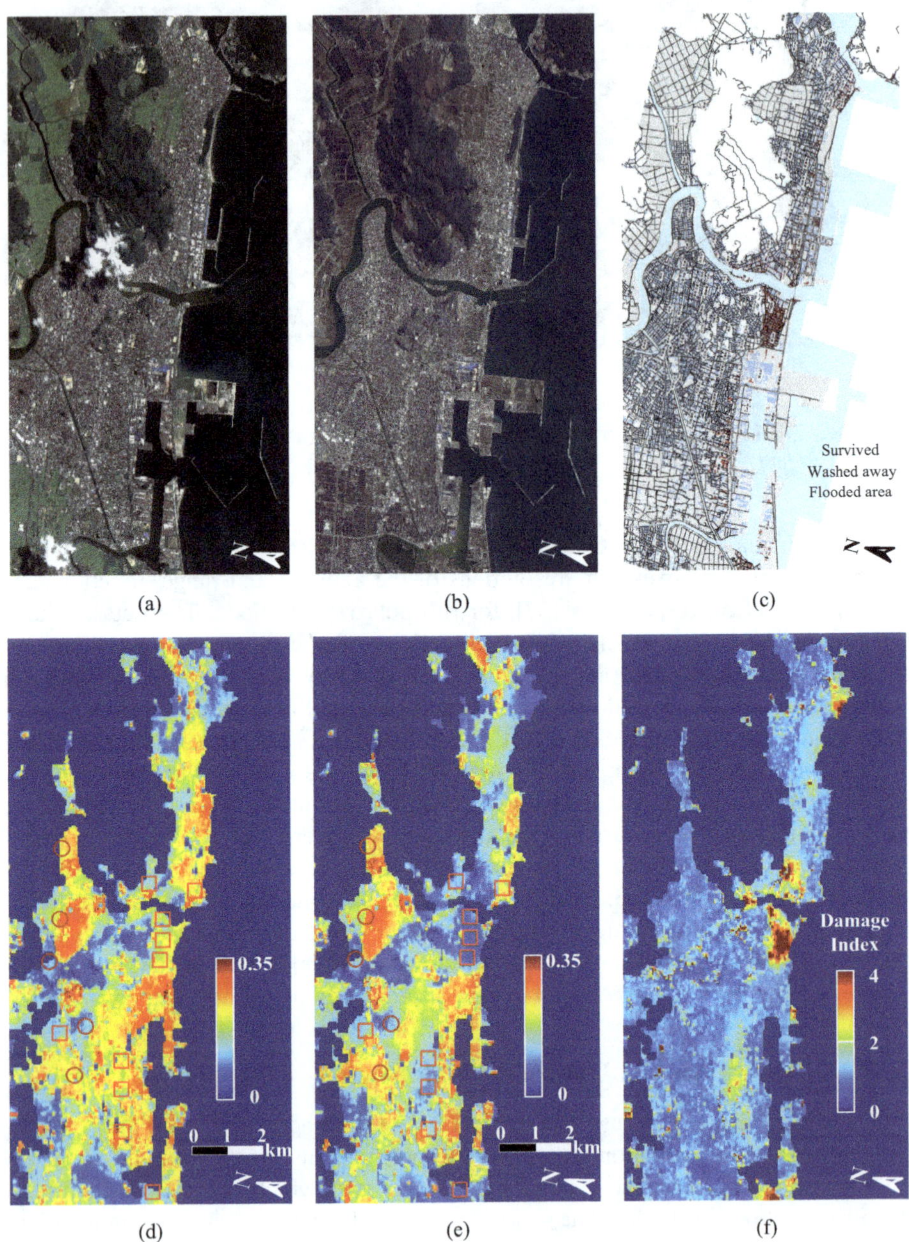

FIGURE 7.3.2 Investigation of the seriously damaged Ishinomaki city. (a) and (b) Pan-sharpened true-color images for pre-event (Aug. 23, 2010) and post-event (Apr. 10, 2011), (c) ground-truth of building damage map, (d) and (e) derived coherence fluctuation features $\left|\gamma_{\text{HH-VV}}(\theta)\right|_{\text{std}}^{\text{pre}}$ and $\left|\gamma_{\text{HH-VV}}(\theta)\right|_{\text{std}}^{\text{post}}$ for pre-event (Nov. 21, 2010) and post-event (Apr. 8, 2011), (f) estimated damage level index $\text{Ratio}_{\text{std}}$.

The red and blue colors indicate the flushed or collapsed buildings and surviving buildings, respectively, while the gray color identify the flooded range.

For quantitative evaluation, ten small urban patches with damage levels ranging from 5%–95% are selected from Figure 7.3.2, indicated by rectangular boxes. Meanwhile, five urban patches that were only flooded, indicated by circles in Figure 7.3.2, are selected for comparison. The ground-truth damage maps for these patches are shown in Figures 7.3.3 and 7.3.4, respectively. The co-polarization coherence pattern $|\gamma_{\text{HH-VV}}(\theta)|$ for these only flooded and damaged urban patches is shown and compared in Figures 7.3.5 and 7.3.6, accordingly. For only flooded urban patches with almost no ground-wall structure destruction, it is clear that co-polarization coherence patterns $|\gamma_{\text{HH-VV}}(\theta)|$ are mainly the same before and after the event. Also, these co-polarization coherence patterns generally show four-lobe shapes with high coherence fluctuations, illustrated in Figure 7.3.5. It can be interpreted from the narrow directivity of a dihedral corner reflector which can model the ground-wall structure. In comparison, after the earthquake and tsunami, the destroyed ground-wall

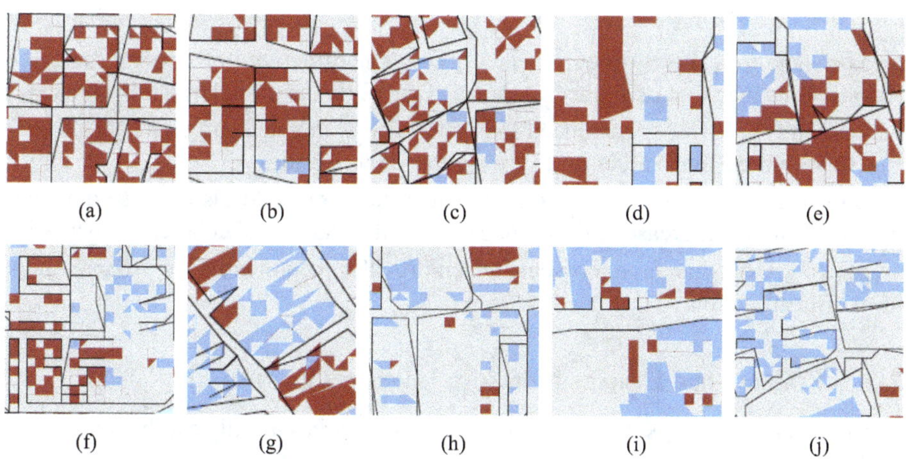

FIGURE 7.3.3 Ground-truth of urban patches with different damage levels. (a) 95%, (b) 90%, (c) 80%, (d) 75%, (e) 70%, (f) 60%, (g) 40%, (h) 25%, (i) 20%, (j) 5%.

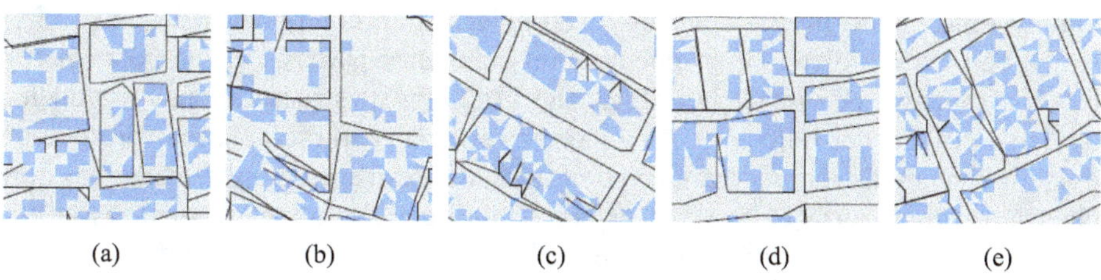

FIGURE 7.3.4 Ground-truth of flooded urban patches. (a)–(e) patches 1–5, respectively.

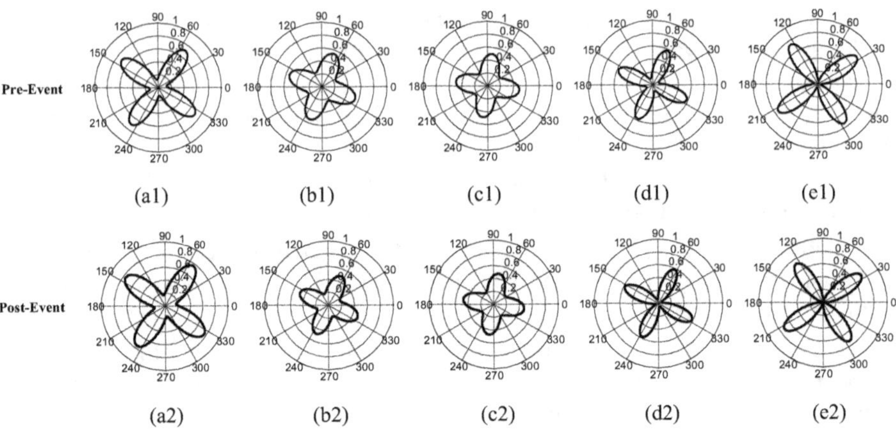

FIGURE 7.3.5 Co-polarization coherence patterns of $\left|\gamma_{\text{HH-VV}}(\theta)\right|$ for flooded urban patches pre- and post-event. (a) Patch 1, (b) patch 2, (c) patch 3, (d) patch 4, (e) patch 5. Numbers 1–2 indicate pre- and post-event, respectively.

areas change to relatively rough surface areas. The directivity of a rough surface is less sensitive than that of a dihedral model. Therefore, the co-polarization coherence fluctuation $\left|\gamma_{\text{HH-VV}}(\theta)\right|_{\text{std}}$ is reduced after the event for damaged urban patches, which can be observed and validated from Figure 7.3.6. Besides, the changes of coherence fluctuation $\left|\gamma_{\text{HH-VV}}(\theta)\right|_{\text{std}}$ can also reflect the damage levels of these damaged patches. With the increase of the damage levels, the changes of coherence fluctuation $\left|\gamma_{\text{HH-VV}}(\theta)\right|_{\text{std}}$ become more obvious, which provides the potential for urban damage level discrimination.

7.3.3 Polarimetric Damage Index and Inversion Relationship

For relatively coarse-resolution PolSAR data, urban areas usually exhibit a dominant double-bounce scattering mechanism due to the ground-wall structures. Ground-wall structures are reduced among damaged urban areas during an earthquake or tsunami natural disaster. Then, the dominant scattering mechanism changes from double-bounce scattering to surface scattering over these damaged areas after a destructive disaster. The changes of these two dominant scattering mechanisms are mainly represented by the changes of co-polarization components S_{HH} and S_{VV}, which can lead to different co-polarization coherence patterns in the polarimetric rotation domain. For intact buildings, the ground-wall structure can be modeled by a dihedral corner reflector with a narrow directivity in the polarimetric rotation domain. In comparison, for damaged buildings, the destroyed ground-wall areas may change to rough surfaces to some extent. The directivity of rough surfaces is less sensitive than that of a dihedral corner reflector. In this vein, as also demonstrated in Figure 7.3.6, the fluctuation of co-polarization coherence $\left|\gamma_{\text{HH-VV}}(\theta)\right|$ is reduced after the damage for the damaged urban patches and has the potential to disclose the changes of urban patches before and after the damage.

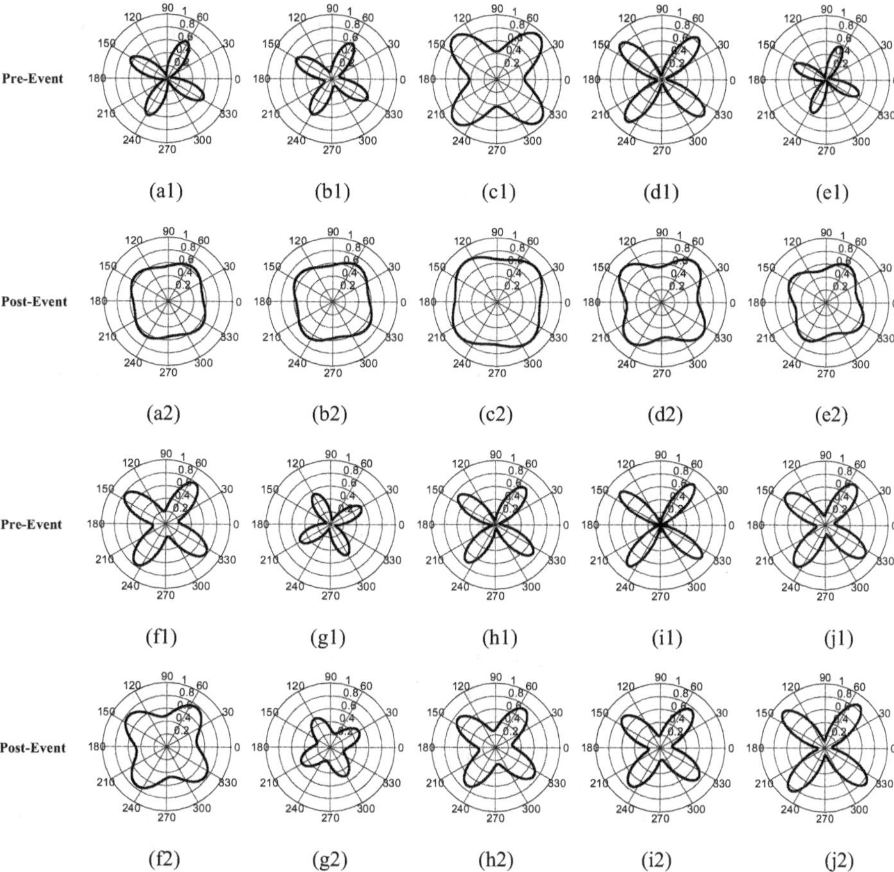

FIGURE 7.3.6 Co-polarization coherence patterns $\left|\gamma_{\text{HH-VV}}(\theta)\right|$ for damaged urban patches pre- and post-event. The damage levels are (a) 95%, (b) 90%, (c) 80%, (d) 75%, (e) 70%, (f) 60%, (g) 40%, (h) 25%, (i) 20%, (j) 5%. Numbers 1–2 indicate the pre- and post-event, respectively.

The co-polarization coherence fluctuation $\left|\gamma_{\text{HH-VV}}(\theta)\right|_{\text{std}}$ has the capability to characterize the coherence fluctuation signature in the polarimetric rotation domain. Using the method proposed in [18, 31] (introduced in Section 5.3), buildings in PolSAR images can be extracted, and a building mask can be generated. The co-polarization coherence fluctuation $\left|\gamma_{\text{HH-VV}}(\theta)\right|_{\text{std}}$ of the considered full-scene Ishinomaki city is calculated and shown in Figures 7.3.2 (d) and 7.3.2 (e) for pre- and post-event, respectively. Compared with pre-event data, the values of the co-polarization coherence fluctuation $\left|\gamma_{\text{HH-VV}}(\theta)\right|_{\text{std}}$ over damaged regions obviously decrease for the post-event case. Meanwhile, as expected, the reduction degree of $\left|\gamma_{\text{HH-VV}}(\theta)\right|_{\text{std}}$ shows a closely corresponding relationship to the damage levels of local urban patches. In order to highlight this relationship, the ratio of the co-polarization coherence fluctuation

$\left|\gamma_{\text{HH-VV}}(\theta)\right|_{\text{std}}^{\text{pre}}$ and $\left|\gamma_{\text{HH-VV}}(\theta)\right|_{\text{std}}^{\text{post}}$ before and after the damage is adopted as the damage index for urban damage level indication [35]:

$$\text{Ratio}_{\text{std}} = \frac{\left|\gamma_{\text{HH-VV}}(\theta)\right|_{\text{std}}^{\text{pre}}}{\left|\gamma_{\text{HH-VV}}(\theta)\right|_{\text{std}}^{\text{post}}} \tag{7.3.1}$$

where $\left|\gamma_{\text{HH-VV}}(\theta)\right|_{\text{std}}^{\text{pre}}$ and $\left|\gamma_{\text{HH-VV}}(\theta)\right|_{\text{std}}^{\text{post}}$ correspond to pre- and post-event cases, respectively. For undamaged or very slightly damaged urban patches, the damage index $\text{Ratio}_{\text{std}}$ is close to 1, while it is over 1 for damaged urban patches.

The damage index $\text{Ratio}_{\text{std}}$ is calculated for the damaged Ishinomaki city, shown in Figure 7.3.2 (f). Compared with the ground-truth of the building damage map shown in Figure 7.3.2 (c), the damage index $\text{Ratio}_{\text{std}}$ clearly identifies damaged urban patches. Furthermore, a well-corresponding relationship between the values of $\text{Ratio}_{\text{std}}$ and the truth damage levels can be observed. Quantitative comparisons are carried out for the selected ten damaged urban patches and five only flooded urban patches. The mean and standard deviation of the damage index are obtained within a 3×3 local window. The error bar plots to demonstrate the relationship between the damage index $\text{Ratio}_{\text{std}}$ and the truth damage level are given in Figure 7.3.7. For damaged urban patches, it is clear that the damage index $\text{Ratio}_{\text{std}}$ increases with the increase of the damage levels. Meanwhile, for only flooded urban patches, the values of the damage index $\text{Ratio}_{\text{std}}$ are all around 1, indicating an undamaged condition, which agrees with the ground-truth. In detail, for the urban patch with a damage level of 5%, the co-polarization coherence patterns in Figures 7.3.6 (j1) and 7.3.6 (j2) are almost the same, and the corresponding damage index $\text{Ratio}_{\text{std}}$ is around 1, illustrated in Figure 7.3.7 (a). On the other hand, for the most seriously damaged urban patch with a damage level of 95%, the damage index $\text{Ratio}_{\text{std}}$ is around 5.3, which

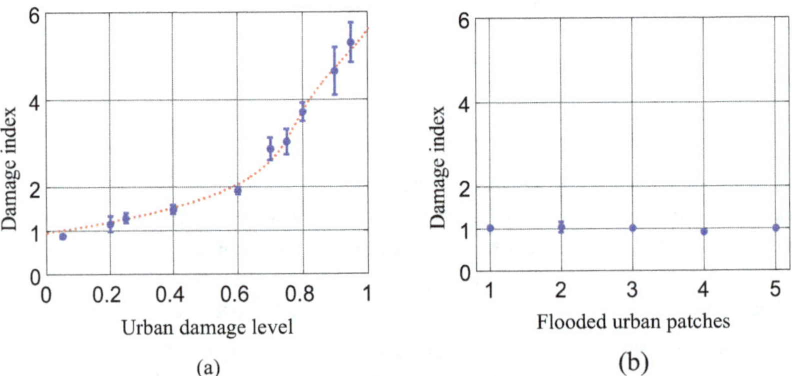

FIGURE 7.3.7 Relationship demonstration between the damage index $\text{Ratio}_{\text{std}}$ and the truth damage level. (a) Urban patches with different damage levels, (b) flooded urban patches.

clearly reflects the changes of its co-polarization coherence patterns before and after the damage shown in Figures 7.3.6 (a1) and 7.3.6 (a2). From Figure 7.3.7, the damage index $\text{Ratio}_{\text{std}}$ is very effective to discriminate these urban patches with damage levels over 20%. For urban patches with damage levels below 20%, the damage index $\text{Ratio}_{\text{std}}$ may not be sensitive enough to identify them. These results further validate the efficiency of the damage index $\text{Ratio}_{\text{std}}$.

In order to obtain the analytical inversion relationship, a fourth order polynomial model is used for fitting:

$$DL = \begin{cases} p_1 x^4 + p_2 x^3 + p_3 x^2 + p_4 x + p_5 & \text{Ratio}_{\text{std}} > 1 \\ 0 & \text{Ratio}_{\text{std}} \leq 1 \end{cases} \tag{7.3.2}$$

where DL is the urban damage level, and $x = \text{Ratio}_{\text{std}}$ and p_i, $i = 1, 2, \cdots, 5$ are fitting parameters.

Using the selected urban patches with known damage levels, as shown in Figure 7.3.3, the fitting parameters can be estimated as $p_1 = -0.0076$, $p_2 = 0.1254$, $p_3 = -0.7532$, $p_4 = -2.0360$, and $p_5 = -1.3560$. The fitting curve with dashed line is shown in Figure 7.3.7 (a). It is observed that the fitting performance is satisfactory. Since there is no clear evidence to prefer one fitting model to others, this work adopts the polynomial fitting for simplicity. Indeed, other fitting methods can also be effective.

7.4 URBAN DAMAGE LEVEL MAPPING

7.4.1 Urban Damage Level Mapping Methodology

With the damage index $\text{Ratio}_{\text{std}}$ and damage level inversion relationship (7.3.2), an urban damage level mapping approach is established [35]. The processing flowchart is shown in Figure 7.4.1. The inputs are co-registered pre- and post-event PolSAR data. Then, multi-looking is an optional step to make the range and azimuth resolutions comparable. Sample averaging is necessary to provide an unbiased and accurate estimation of polarimetric coherence. In this work, the SimiTest speckle filter [36], which is especially suitable for polarimetric coherence estimation, is adopted to smooth the speckle effect. Alternatively, other advanced speckle filters can also be utilized. An urban mask is generated from the speckle filtered pre-event data. Morphological post-processing, mainly including the closing and opening operations, is used to fill the holes and rule out minor isolated areas in the binary image. The urban mask generation method is adopted from [18, 31]. Then, the co-polarization coherence pattern $|\gamma_{\text{HH-VV}}(\theta)|$ and co-polarization coherence fluctuations $|\gamma_{\text{HH-VV}}(\theta)|_{\text{std}}^{\text{pre}}$ and $|\gamma_{\text{HH-VV}}(\theta)|_{\text{std}}^{\text{post}}$ are calculated over urban candidates for pre- and post-event datasets, respectively. Finally, the damage index $\text{Ratio}_{\text{std}}$ can be obtained and the final urban damage level map can be generated with the damage level inversion relationship (7.3.2).

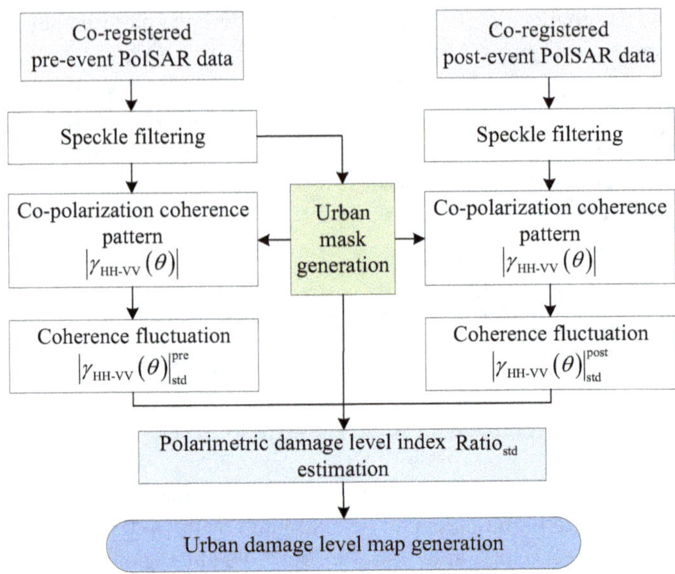

FIGURE 7.4.1 Flowchart of urban damage level mapping.

7.4.2 Full-Scene Urban Damage Level Mapping Investigation and Validation

Using the established approach, full-scene urban damage level mapping result of the study area is automatically generated and shown in Figure 7.4.2. The mapping result is superimposed on the image of total backscattering power $SPAN$. As shown in Figure 7.3.7, the damage index $\text{Ratio}_{\text{std}}$ may not effectively discriminate only flooded urban patches and damaged urban patches with damage levels below 20%. In order to avoid interpretation ambiguity, the estimated damage levels are limited to within 20%–100%. Visually, this damage level mapping approach can successfully identify the locations of damaged urban areas, which are mainly along the seashore. Meanwhile, their detailed damage levels are clearly labeled. Mainly due to the temporal changes during the two acquisitions, some false alarms are also observed. Furthermore, the damage level mapping results of the seriously damaged Ishinomaki city, Onagawa-cho, and Minamisanriku-cho, indicated by rectangular boxes, are enlarged and also shown in Figure 7.4.2, respectively. Compared with their ground-truth damage maps, the estimated damage levels show high accuracy. In addition, the seriously damaged Ohkawa elementary school, which was almost completely destroyed and can be seen from the photo in Figure 7.4.2, is also successfully identified by the methodology, indicated by the circle in Figure 7.4.2.

For further quantitative validation, the pre- and post-event optical images, ground-truth damage maps, and co-polarization coherence fluctuations together with the estimated damage indexes $\text{Ratio}_{\text{std}}$ for Onagawa-cho and Minamisanriku-cho are enlarged and shown in Figures 7.4.3 and 7.4.4, respectively. Three small urban patches with damage levels ranging from 35%–80% are selected from the damaged

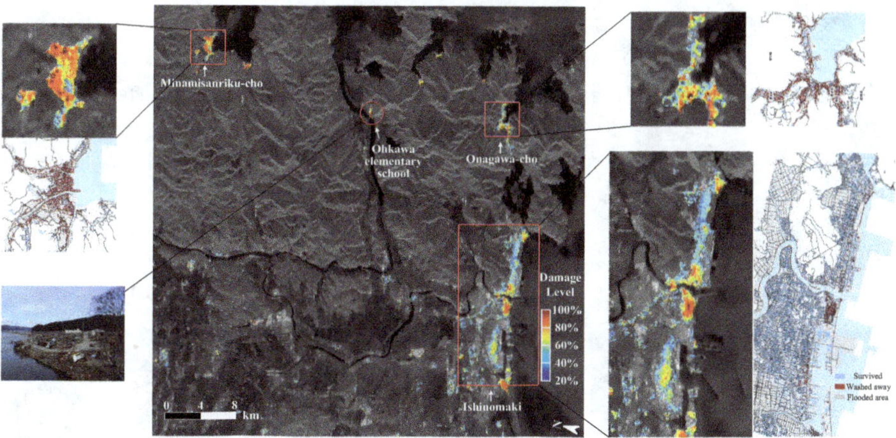

FIGURE 7.4.2 Generated full-scene urban damage level map.

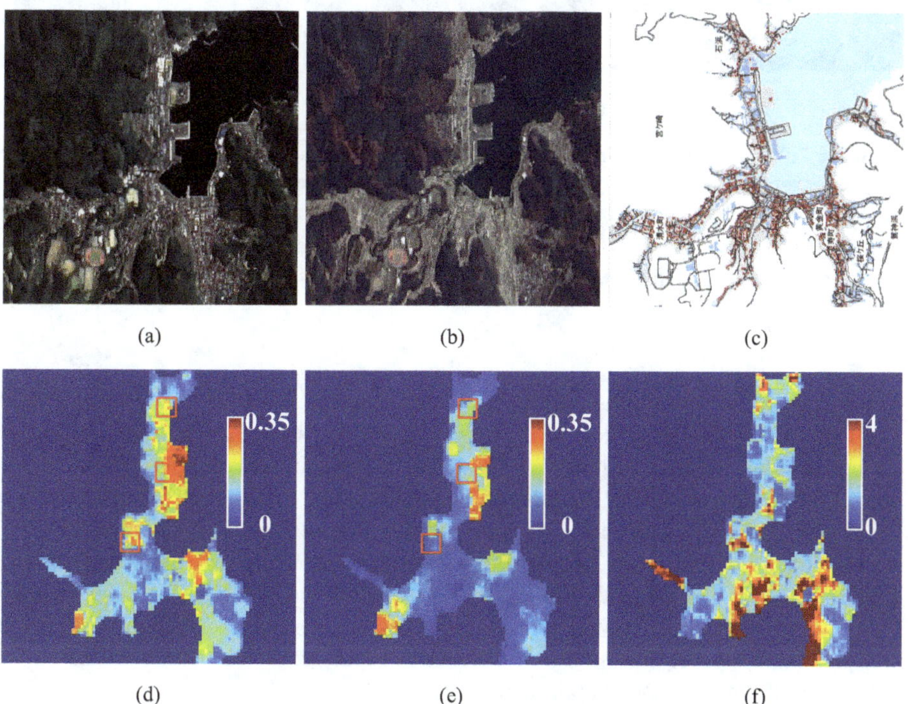

FIGURE 7.4.3 Further validation with the damaged Onagawa-cho. (a) Pan-sharpened true-color images for pre-event (Aug. 23, 2010), (b) pan-sharpened true-color images for post-event (Apr. 10, 2011), (c) ground-truth of building damage map, (d) coherence fluctuation $\left|\gamma_{\text{HH-VV}}(\theta)\right|_{\text{std}}^{\text{pre}}$ for pre-event (Nov. 21, 2010), (e) coherence fluctuation $\left|\gamma_{\text{HH-VV}}(\theta)\right|_{\text{std}}^{\text{post}}$ for post-event (Apr. 8, 2011), (f) estimated damage level index $\text{Ratio}_{\text{std}}$.

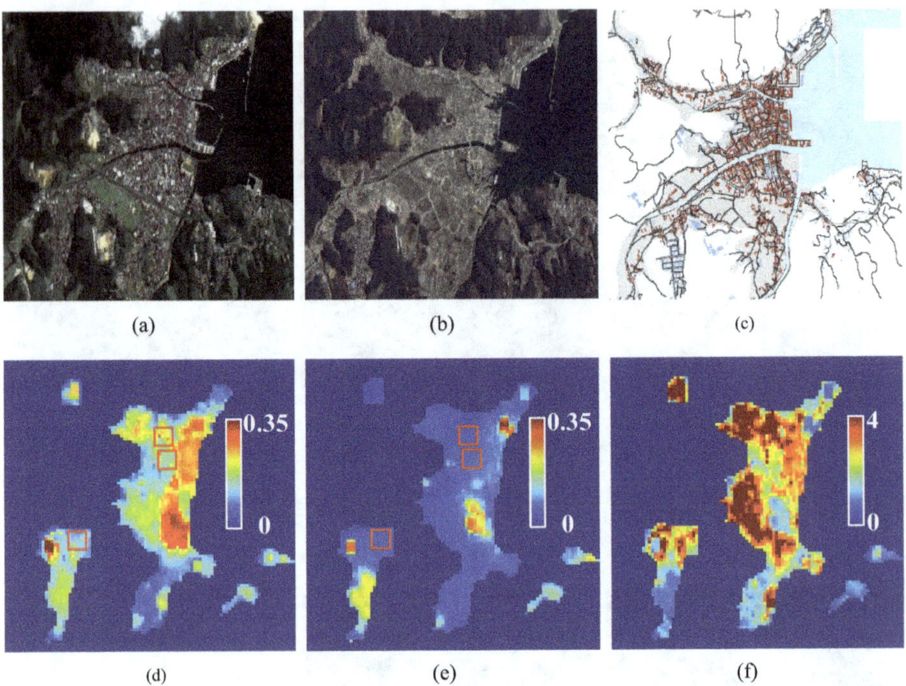

FIGURE 7.4.4 Further validation with the damaged Minamisanriku-cho. (a) Pan-sharpened true-color images for pre-event (Aug. 23, 2010), (b) pan-sharpened true-color images for post-event (Apr. 10, 2011), (c) ground-truth of building damage map, (d) coherence fluctuation $\left|\gamma_{\text{HH-VV}}(\theta)\right|_{\text{std}}^{\text{pre}}$ for pre-event (Nov. 21, 2010), (e) coherence fluctuation $\left|\gamma_{\text{HH-VV}}(\theta)\right|_{\text{std}}^{\text{post}}$ for post-event (Apr. 8, 2011), (f) the estimated damage level index $\text{Ratio}_{\text{std}}$.

Onagawa-cho, while another three small urban patches with damage levels ranging from 65%–90% are selected from the damaged Minamisanriku-cho, indicated by rectangular boxes in Figures 7.4.3 and 7.4.4. The ground-truth damage maps for these patches are shown in Figures 7.4.5 and 7.4.6. For these six urban patches, the mean and standard deviation of damage indexes $\text{Ratio}_{\text{std}}$ are also calculated within a 3×3 local window. Error bar plots showing the relationship between the estimated damage indexes and the truth damage levels are presented in Figure 7.4.7. The dashed line is the urban damage level inversion relationship expression (7.3.2). It is observed that the damage level inversion performance is accurate, which further validates the efficiency of the introduced approach.

The existence of target scattering diversity usually makes scattering mechanism modeling and interpretation difficult [6]. The incorporation of orientation compensation can enhance the performance of polarimetric target decompositions [9, 39]. However, as pointed out in [20], even with orientation compensation processing, scattering mechanism ambiguity still occurs for buildings with large orientation angles. On the other hand, target scattering diversity also contains rich information which

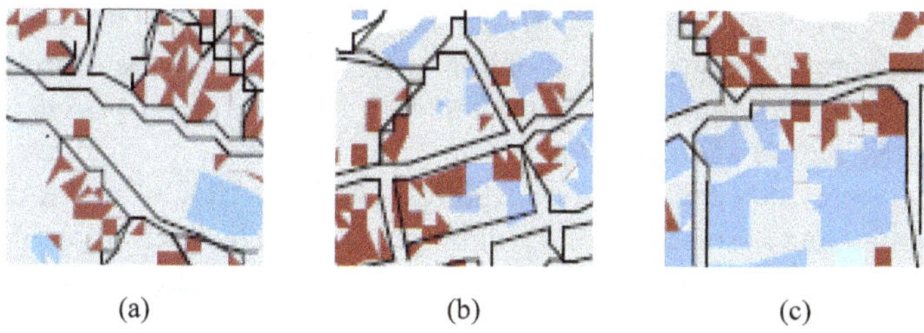

FIGURE 7.4.5 Ground-truth of urban patches with different damage levels from the damaged Onagawa-cho. (a) 80%, (b) 55%, (c) 35%.

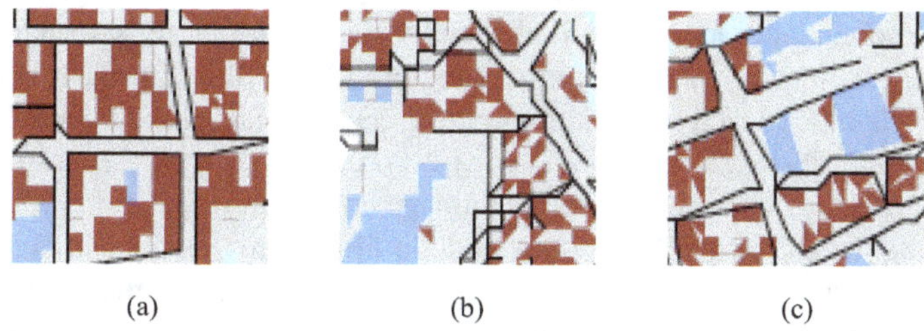

FIGURE 7.4.6 Ground-truth of urban patches with different damage levels from the damaged Minamisanriku-cho. (a) 90%, (b) 75%, (c) 65%.

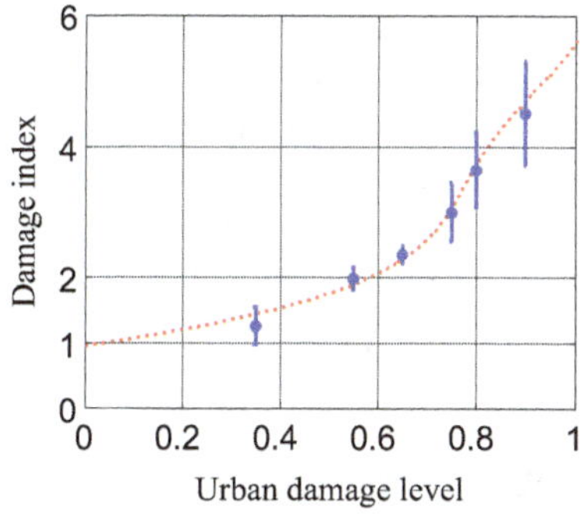

FIGURE 7.4.7 Relationship demonstration between the estimated damage level index and the truth damage level.

can potentially reflect targets' intrinsic properties with suitable interpretation tools, such as the uniform polarimetric matrix rotation theory and polarimetric coherence pattern technique, which are introduced in Chapter 2. As a complement to polarimetric target decomposition techniques, this work investigates and demonstrates the efficiency of the co-polarization coherence pattern tool for urban damage level mapping.

In this study, the temporal and spatial baselines pre- and post-event are respectively 138 days and 1747 meters for the used ALOS/PALSAR data. Such baseline configurations produce serious decorrelation effects, which lead to very poor interferometric coherence. However, it should be pointed out that when InSAR imaging mode is available with suitable baseline configurations, the complex coherence of the interferometric pair is another valuable source to understand the damage conditions [40]. Furthermore, as a combination of PolSAR and InSAR, PolInSAR can provide even more complete observation of the damaged areas. The integration of PolInSAR coherence with polarimetric target decomposition [41] would be a promising method for urban damage evaluation.

7.5 SUMMARY

This chapter introduces a polarimetric rotation domain urban damage level estimation and mapping approach, which is especially suitable for large-scale natural disasters such as earthquakes and tsunamis. The core idea of this approach is to explore and utilize the hidden rich information within the co-polarization coherence pattern in the polarimetric rotation domain. An interpretation tool of the co-polarization coherence pattern is explored for urban damage condition investigation.

Scattering mechanisms from damaged buildings in the polarimetric rotation domain are disclosed. The physical principle that the co-polarization coherence pattern is sensitive to urban damage situations is clarified. In this vein, a damage index defined as the ratio of co-polarization coherence fluctuations pre- and post-event is introduced for urban damage level discrimination. The damage level inversion relationship is established by investigating measured earthquake data and ground-truth data. Then, an urban damage level map for the full-scene is generated. Both the damaged ranges and detailed damage levels can be simultaneously available. Experimental studies demonstrate the efficiency and accuracy of the introduced method. The implementation of the established mapping method is simple, and the underlying principle is clear. In addition, according to this study, the rich information contained in target scattering diversity and the validity of the polarimetric rotation domain interpretation theory are further confirmed.

REFERENCES

1. J. S. Lee and E. Pottier, *Polarimetric Radar Imaging: From Basics to Applications*. Boca Raton, US: CRC Press, 2009.
2. S. W. Chen, X. S. Wang, S. P. Xiao, and M. Sato, *Target Scattering Mechanism in Polarimetric Synthetic Aperture Radar-Interpretation and Application*. Singapore: Springer, 2018.
3. F. Dell'Acqua and P. Gamba, "Remote sensing and earthquake damage assessment: Experiences, limits, and perspectives," *Proceedings of the IEEE,* vol. 100, no. 10, pp. 2876–2890, 2012.
4. Y. Yamaguchi, "Disaster monitoring by fully polarimetric SAR data acquired with ALOS-PALSAR," *Proceedings of the IEEE,* vol. 100, no. 10, pp. 2851–2860, 2012.
5. M. Sato, S. W. Chen, and M. Satake, "Polarimetric SAR analysis of tsunami damage following the March 11, 2011 East Japan earthquake," *Proceedings of The IEEE,* vol. 100, no. 10, pp. 2861–2875, 2012.

6. S. W. Chen, Y. Z. Li, X. S. Wang, S. P. Xiao, and M. Sato, "Modeling and interpretation of scattering mechanisms in polarimetric synthetic aperture radar: Advances and perspectives," *IEEE Signal Processing Magazine,* vol. 31, no. 4, pp. 79–89, 2014.
7. S. R. Cloude and E. Pottier, "An entropy based classification scheme for land applications of polarimetric SAR," *IEEE Transactions on Geoscience and Remote Sensing,* vol. 35, no. 1, pp. 68–78, 1997.
8. A. Freeman and S. L. Durden, "A three-component scattering model for polarimetric SAR data," *IEEE Transactions on Geoscience and Remote Sensing,* vol. 36, no. 3, pp. 963–973, 1998.
9. Y. Yamaguchi, A. Sato, W. M. Boerner, R. Sato, and H. Yamada, "Four-component scattering power decomposition with rotation of coherency matrix," *IEEE Transactions on Geoscience and Remote Sensing,* vol. 49, no. 6, pp. 2251–2258, 2011.
10. S. W. Chen, X. S. Wang, S. P. Xiao, and M. Sato, "General polarimetric model-based decomposition for coherency matrix," *IEEE Transactions on Geoscience and Remote Sensing,* vol. 52, no. 3, pp. 1843–1855, 2014.
11. H. D. Guo, X. Y. Wang, X. W. Li, G. Liu, L. Zhang, and S. Y. Yan, "Yushu earthquake synergic analysis using multimodal SAR datasets," *Chinese Science Bulletin,* vol. 55, no. 31, pp. 3499–3503, 2010.
12. M. Watanabe, T. Motohka, Y. Miyagi, C. Yonezawa, and M. Shimada, "Analysis of urban areas affected by the 2011 off the pacific coast of Tohoku earthquake and Tsunami with L-band SAR full-polarimetric mode," *IEEE Geoscience and Remote Sensing Letters,* vol. 9, no. 3, pp. 472–476, 2012.
13. G. Singh, Y. Yamaguchi, W. M. Boerner, and S. E. Park, "Monitoring of the March 11, 2011, Off-Tohoku 9.0 earthquake with super-tsunami disaster by implementing fully polarimetric high-resolution POLSAR techniques," *Proceedings of the IEEE,* vol. 101, no. 3, pp. 831–846, 2013.
14. S. E. Park, Y. Yamaguchi, and D. J. Kim, "Polarimetric SAR remote sensing of the 2011 Tohoku earthquake using ALOS/PALSAR," *Remote Sensing of Environment,* vol. 132, pp. 212–220, 2013.
15. L. Shi, W. Sun, J. Yang, P. Li, and L. Lu, "Building collapse assessment by the use of post earthquake Chinese VHR airborne SAR," *IEEE Geoscience and Remote Sensing Letters,* vol. 12, no. 10, pp. 2021–2025, 2015.
16. W. Zhai and C. Huang, "Fast building damage mapping using a single post-earthquake PolSAR image: A case study of the 2010 Yushu earthquake," *Earth Planets and Space,* vol. 68, no.1, pp. 1–12, 2016.
17. S. W. Chen and M. Sato, "Tsunami damage investigation of built-up areas using multitemporal spaceborne full polarimetric SAR images," *IEEE Transactions on Geoscience and Remote Sensing,* vol. 51, no. 4, pp. 1985–1997, 2013.
18. S. W. Chen, X. S. Wang, and M. Sato, "Urban damage level mapping based on scattering mechanism investigation using fully polarimetric SAR data for the 3.11 East Japan earthquake," *IEEE Transactions on Geoscience and Remote Sensing,* vol. 54, no. 12, pp. 6919–6929, 2016.
19. R. Touzi, A. Lopes, J. Bruniquel, and P. W. Vachon, "Coherence estimation for SAR imagery," *IEEE Transactions on Geoscience and Remote Sensing,* vol. 37, no. 1, pp. 135–149, 1999.
20. S. W. Chen, M. Ohki, M. Shimada, and M. Sato, "Deorientation effect investigation for model-based decomposition over oriented built-up areas," *IEEE Geoscience and Remote Sensing Letters,* vol. 10, no. 2, pp. 273–277, 2013.
21. D. L. Schuler, J. S. Lee, D. Kasilingam, and E. Pottier, "Measurement of ocean surface slopes and wave spectra using polarimetric SAR image data," *Remote Sensing of Environment,* vol. 91, no. 2, pp. 198–211, 2004.

22. J. S. Lee, D. L. Schuler, T. L. Ainsworth, E. Krogager, D. Kasilingam, and W. M. Boerner, "On the estimation of radar polarization orientation shifts induced by terrain slopes," *IEEE Transactions on Geoscience and Remote Sensing,* vol. 40, no. 1, pp. 30–41, 2002.
23. H. Kimura, "Radar polarization orientation shifts in built-up areas," *IEEE Geoscience and Remote Sensing Letters,* vol. 5, no. 2, pp. 217–221, 2008.
24. K. Iribe and M. Sato, "Analysis of polarization orientation angle shifts by artificial structures," *IEEE Transactions on Geoscience and Remote Sensing,* vol. 45, no. 11, pp. 3417–3425, 2007.
25. S. W. Chen, Y. Z. Li, and X. S. Wang, "A visualization tool for polarimetric SAR data investigation," in *The 11th European Synthetic Aperture Radar Conference*, Hamburg, Germany, 2016, pp. 579–582.
26. S. W. Chen and X. S. Wang, "Polarimetric coherence pattern: A visualization tool for PolSAR data investigation," in *IEEE International Geoscience and Remote Sensing Symposium*, Beijing, China, 2016, pp. 7509–7512.
27. S. W. Chen, "Polarimetric coherence pattern: A visualization and characterization tool for PolSAR data investigation," *IEEE Transactions on Geoscience and Remote Sensing,* vol. 56, no. 1, pp. 286–297, 2018.
28. S. W. Chen, Y. Z. Li, and X. S. Wang, "Crop discrimination based on polarimetric correlation coefficients optimization for PolSAR data," *International Journal of Remote Sensing,* vol. 36, no. 16, pp. 4233–4249, 2015.
29. C. S. Tao, S. W. Chen, Y. Z. Li, and S. P. Xiao, "PolSAR land cover classification based on roll-invariant and selected hidden polarimetric features in the rotation domain," *Remote Sensing,* vol. 9, no. 7, p. 660, 2017.
30. S. W. Chen and C. S. Tao, "PolSAR image classification using polarimetric-feature–driven deep convolutional neural network," *IEEE Geoscience Remote Sensing Letters,* vol. 15, no. 4, pp. 627–631, 2018.
31. S. P. Xiao, S. W. Chen, Y. L. Chang, Y. Z. Li, and M. Sato, "Polarimetric coherence optimization and its application for manmade target extraction in PolSAR data," *IEICE Transactions on Electronics,* vol. E97C, no. 6, pp. 566–574, 2014.
32. X. C. Cui, C. S. Tao, Y. Su, and S. W. Chen, "PolSAR ship detection based on polarimetric correlation pattern," *IEEE Geoscience and Remote Sensing Letters,* vol. 18, no. 3, pp. 471–475, 2021.
33. H. L. Li, M. D. Li, X. C. Cui, and S. W. Chen, "Man-made target structure recognition with polarimetric correlation pattern and roll-invariant feature coding," *IEEE Geoscience and Remote Sensing Letters,* vol. 19, pp. 1–5, 2022.
34. G. Q. Wu, S. W. Chen, Y. Z. Li, and X. S. Wang, "Null-pol response pattern in polarimetric rotation domain: Characterization and application," *IEEE Geoscience and Remote Sensing Letters,* vol. 19, pp. 1–5, 2022.
35. S. W. Chen, X. S. Wang, and S. P. Xiao, "Urban damage level mapping based on co-polarization coherence pattern using multitemporal polarimetric SAR data," *IEEE Journal of Selected Topics in Applied Earth Observations and Remote Sensing,* vol. 11, no. 8, pp. 2657–2667, 2018.
36. S. W. Chen, X. S. Wang, and M. Sato, "PolInSAR complex coherence estimation based on covariance matrix similarity test," *IEEE Transactions on Geoscience and Remote Sensing,* vol. 50, no. 11, pp. 4699–4710, 2012.
37. M. Shimada, O. Isoguchi, T. Tadono, and K. Isono, "PALSAR radiometric and geometric calibration," *IEEE Transactions on Geoscience and Remote Sensing,* vol. 47, no. 12, pp. 3915–3932, 2009.
38. T. Tadono, M. Shimada, H. Murakami, and J. Takaku, "Calibration of PRISM and AVNIR-2 onboard ALOS 'Daichi'," *IEEE Transactions on Geoscience and Remote Sensing,* vol. 47, no. 12, pp. 4042–4050, 2009.

39. W. An, C. Xie, X. Yuan, Y. Cui, and J. Yang, "Four-component decomposition of polarimetric SAR images with deorientation," *IEEE Geoscience and Remote Sensing Letters,* vol. 8, no. 6, pp. 1090–1094, 2011.
40. G. A. Arciniegas, W. Bijker, N. Kerle, and V. A. Tolpekin, "Coherence- and amplitude-based analysis of seismogenic damage in Bam, Iran, using ENVISAT ASAR data," *IEEE Transactions on Geoscience and Remote Sensing,* vol. 45, no. 6, pp. 1571–1581, 2007.
41. S. W. Chen, X. S. Wang, Y. Z. Li, and M. Sato, "Adaptive model-based polarimetric decomposition using PolInSAR coherence," *IEEE Transactions on Geoscience and Remote Sensing,* vol. 52, no. 3, pp. 1705–1718, 2014.

Index

Note: Page numbers in *italics* indicate a figure on the corresponding page.

R

S

T

U

For Product Safety Concerns and Information please contact our EU representative GPSR@taylorandfrancis.com
Taylor & Francis Verlag GmbH, Kaufingerstraße 24, 80331 München, Germany

www.ingramcontent.com/pod-product-compliance
Lightning Source LLC
LaVergne TN
LVHW010550110826
845149LV00003B/614

* 9 7 8 1 0 3 2 6 0 9 5 9 1 *